Society 5.0

The book will help the readers in the field of the Internet of Things, and especially its convergence with artificial intelligence, which has given rise to the new paradigm of artificial intelligence of things (AIoT). It covers important concepts such as intelligent space and human-centred robotics and its effect on human well-being and human-centred aviation automation.

This book:

- Supports the advancement in artificial intelligence and the Internet of Things used in societal applications.
- Discusses the role of modelling human factors in designing smart systems as highlighted in Industry 4.0.
- Covers big data scheduling and the global standard method applied to smart maintenance.
- Presents human-centred aviation automation, human-centred processes, and decision support systems.
- Highlights the importance of data privacy and secure communication in society 5.0.

The text is primarily written for senior undergraduate, graduate students, and academic researchers in diverse fields, including electrical engineering, electronics and communications engineering, computer science and engineering, and information technology.

Intelligent Data-Driven Systems and Artificial Intelligence
Series Editor: Harish Garg

Cognitive Machine Intelligence
Applications, Challenges, and Related Technologies
Inam Ullah Khan, Salma El Hajjami, Mariya Ouaissa, Salwa Belqziz and Tarandeep Kaur Bhatia

Artificial Intelligence and Internet of Things based Augmented Trends for Data Driven Systems
Anshu Singla, Sarvesh Tanwar, Pao-Ann Hsiung

Modelling of Virtual Worlds Using the Internet of Things
Edited by Simar Preet Singh and Arun Solanki

Data-Driven Technologies and Artificial Intelligence in Supply Chain
Tools and Techniques
Mahesh Chand, Vineet Jain and Puneeta Ajmera

Society 5.0
A Transformation towards Human-Centred Artificial Intelligence
Edited by Biswaranjan Acharya, Madhu Shukla, Chandreyee Chowdhury and Harish Garg

For more information about this series, please visit: www.routledge.com/Intelligent-Data-Driven-Systems-and-Artificial-Intelligence/book-series/CRCIDDSAAI

Society 5.0
A Transformation towards Human-Centred Artificial Intelligence

Edited by
Biswaranjan Acharya
Madhu Shukla
Chandreyee Chowdhury
Harish Garg

CRC Press
Taylor & Francis Group
Boca Raton London New York

CRC Press is an imprint of the
Taylor & Francis Group, an **informa** business

Designed cover image: Shutterstock

First edition published 2025
by CRC Press
2385 NW Executive Center Drive, Suite 320, Boca Raton FL 33431

and by CRC Press
4 Park Square, Milton Park, Abingdon, Oxon, OX14 4RN

CRC Press is an imprint of Taylor & Francis Group, LLC

ISBN: 978-1-032-66759-1 (hbk)
ISBN: 978-1-041-07987-3 (pbk)
ISBN: 978-1-003-64316-6 (ebk)

DOI: 10.1201/9781003643166

Typeset in Sabon
by SPi Technologies India Pvt Ltd (Straive)

Contents

Preface

The book centres concept of "Intelligent Data-Driven Systems and Artificial Intelligence". Bringing together how AI and Intelligent systems helps society in achieving new state of well-being and security aligning with the fundamentals of Society 5.0. Different chapter uncovers the different ways in which technology today can be a driving factor for the people in different day to day sectors of life. Because there is a limit to what people can do, the task of finding the necessary information from overflowing information and analysing it was a burden, and the labour and scope of action were restricted due to age and varying degrees of ability. Also, due to various restrictions on issues such as a decreasing birthrate and ageing population and local depopulation, it was difficult to respond adequately.

Social reform (innovation) in Society 5.0 will achieve a forward-looking society that breaks down the existing sense of stagnation, a society whose members have mutual respect, transcending the generations, and a society in which each person can lead an active and enjoyable life through technological advancement. This book aims to bring more innovative ideas for reality.

While the gross theme of the book is technologies and their adoption for Society 5.0, the main features include a thorough analysis of human-centred artificial intelligence and human-centred robotics, and the technological potential of solving social problems of not only developed societies but also of developing countries like India. The major application domains of aviation, e-commerce and health will also be reviewed w.r.t Society 5.0.

This volume consists of 12 chapters that revolve around the various aspects of Society 5.0 while AI and ML take the forefront. The relation of various Industry standards is also dealt with. Healthcare domain has been explored by many of the authors that usher in a brighter future through inclusive growth and better quality of life.

Chapter 1 explores Society 5.0's impact on areas like healthcare, security, and social development, highlighting historical context, benefits, challenges, and ways to balance quality of life with technological and economic growth. Chapter 2 examines the challenges and reforms in India's healthcare system, focusing on resource access, infrastructure improvements, government initiatives like Ayushman Bharat, and the role of healthcare managers in addressing workforce shortages and quality concerns. Despite progress, achieving equitable, high-quality healthcare requires sustained collaboration among all stakeholders. Chapter 3 explains how early intervention and dropout prevention (EIDP) strategies use data-driven approaches and machine learning to identify at-risk students, enabling timely interventions like mentoring and tutoring to enhance retention, foster equity, and reduce dropout rates. Chapter 4 describes Resume Analyzer, which simplifies the process of tailoring resumes to job descriptions by using Google's Gemini-Pro LLM to extract and align key details. This tool saves time, ensures accuracy, and offers a user-friendly solution for job seekers.

Chapter 5 depicts how Artificial Intelligence (AI) and Machine Learning (ML) are transforming drug discovery, personalised treatment, and diagnostics by integrating experimental and computational techniques. These technologies, including methods like Artificial Neural Networks and CRISPR, enhance accuracy in drug repurposing, disease diagnosis, and clinical decision-making. Despite challenges, AI's role in precision medicine is expanding, offering potential solutions for improving treatment outcomes and predictions. Chapter 6, Social Network Analysis (SNA), explores trust in online interactions, which is crucial for building relationships and collaboration. Machine learning models and fuzzy logic are used to predict and evaluate trust based on factors such as influence, followers, and engagement. Chapter 7 highlights the importance of social responsibility, sustainability, and overcoming challenges such as data privacy and ethical concerns in implementing a human-centric, innovative Society 5.0 framework. Chapter 8 critically overviews Medicine 4.0, which integrates advanced technologies like AI, telemedicine, and wearable devices to revolutionise patient care and healthcare processes. While offering personalised, efficient care and early disease detection, it also raises ethical and privacy concerns, requiring collaboration to address regulatory and security challenges.

Chapter 9 explores the role of topology in digital image processing, focusing on different topologies constructed on Z^2. It presents a comparative analysis of these topologies and their impact on pattern recognition and image analysis. Industry 5.0 is reshaping employment by combining human skills with advanced technologies, requiring new skill development. Chapter 10 examines its impact on job roles, product quality, and the need for effective policies and training to ensure inclusive growth. Society 5.0 envisions a future where technology enhances human life, with Human-Centred AI at its core. Chapter 11 explores the role of Neurosymbolic AI in integrating neural networks with symbolic reasoning, its applications, ethical concerns, and its potential to revolutionise sectors like healthcare, education, and smart cities. Chapter 12 provides a comprehensive evaluation of autism screening tools for toddlers, emphasising the importance of early detection of autism spectrum disorder (ASD) for timely intervention. It critically analyses tools like M-CHAT-R/F, STAT, Q-CHAT-10, and ASQ, assessing their performance based on sensitivity, specificity, and predictive values while highlighting their strengths, limitations, and implications for early ASD detection and intervention.

Dr. Biswaranjan Acharya
Dr. Madhu Shukla
Dr. Chandreyee Chowdhury
Dr. Harish Garg

About the editors

Biswaranjan Acharya (Senior Member, IEEE) received his M.Tech. degree in Computer Science and Engineering from Biju Patnaik University of Technology (BPUT), Rourkela, Odisha, India, in 2012, and a PhD in Computer Science from Veer Surendra Sai University of Technology (VSSUT), Burla, Odisha, India, in 2024. He is currently an Assistant Professor in the Department of Computer Engineering-AI and BDA at Marwadi University, Gujarat, India. He also received a research fellowship at INTI International University from December 15, 2023, to December 31, 2025. He has more than ten years of academic experience at reputed institutions such as Ravenshaw University and has also worked in the software development industry. He has co-authored more than 85 research articles in internationally reputed journals, 10 edited books and serves as a reviewer for several peer-reviewed journals. Additionally, he holds more than 50 patents. His research interests include multiprocessor scheduling, data analytics, computer vision, machine learning, and the Internet of Things (IoT). He is currently serving as a secondary IEEE Computer Society representative to the IEEE Nanotechnology Council (NTC) Administrative Committee and as an observer of the IEEE P2851 Standard for Functional Safety Data Format. He is also associated with various educational and research societies, including IACSIT, CSI, IAENG, and ISC.

Madhu Shukla received her PhD in Computer Engineering from RK University, Rajkot, India, in 2019. She is currently the Head of the Department of Computer Engineering—AI and Data Science Programs (UG and PG) at Marwadi University, Rajkot, India. With over 15 years of academic and administrative experience, she has made significant contributions to research, teaching, and curriculum development. Her research interests include machine learning, artificial intelligence, data stream clustering, and privacy-preserving algorithms in cloud environments. She has published extensively in Scopus and SCI-indexed journals, covering topics such as AI in healthcare, explainable deep learning, and clustering algorithms. She has also chaired sessions at prestigious conferences and served on the technical program committees of several international events. Dr Shukla is actively involved in mentoring PhD scholars and has guided over 150 undergraduate projects. Her accolades include being recognised as a Certified Trainer by Oracle and receiving awards for excellence in teaching and research contributions.

Chandreyee Chowdhury is a professor in the department of Computer Science and Engineering at Jadavpur University, India. She received her PhD degree in 2013 from the university. Her research interests include machine learning, indoor localisation, and human activity recognition. She was awarded Post Doctoral Fellowship by Erasmus Mundus in 2014 to carry out research work at Northumbria University, UK. She has published more than 130 research papers that are in more than 40 SCI-indexed journals, including Q1 journals with above 10 impact factors. She was granted 2 software copyrights and 2 patents. She has successfully guided 5 PhD scholars. She has undertaken DST-sponsored research project on Indoor Localisation and is currently working on a localisation project funded by ISRO. She is a member of IEEE and IEEE Computer Society.

Harish Garg is working as an Associate Professor at Thapar Institute of Engineering & Technology, Patiala, Punjab, India. He is ranked in the World's Top 2% Scientist List and India Rank #1 & World Rank #163 published by Stanford University in the consecutive five years 2020–2024. He received the Most Outstanding Researcher award in the field of Mathematics from Carrer 360 Academy. He is also the recipient of the International Obada-Prize 2022 – Young Distinguished Researchers. He is also the recipient of the Top-Cited paper by an India-based author (2015–2019) from Elsevier Publisher. He also serves as an advisory board member of the Universal Scientific Education and Research Network (USERN). He is a Research Fellow of INTI International University, Malaysia.

Dr Garg's research interests include Computational Intelligence, Multi-criteria decision making, Evolutionary algorithms, Reliability analysis, Expert systems and decision support systems, Computing with words and Soft Computing. He has authored more than 530 papers (over 510 are SCI) published in refereed International Journals. His Google citations are over 27600 with H-index- 92. He is one of the leading researchers in the world related to the MCDM and soft computing approaches. Dr. Garg also serves on editorial boards of several leading international journals, including the Founding Editor-in-Chief of the Journal of Computational and Cognitive Engineering. He is also the Associate Editor of Alexandria Engineering Journal, Journal of Industrial & Management Optimization, Machine Intelligence Research, CAAI Transactions on Intelligence Technology, etc.

Contributors

Biswaranjan Acharya
Department of Computer Engineering – AI & BDA
Marwadi University
Rajkot, Gujarat, India

Aniruddha Banerjee
Kalinga Institute of Industrial Technology
Bhubaneswar, Odisha, India

Anuraag Banerjee
Kalinga Institute of Industrial Technology
Bhubaneswar, Odisha, India

Debabrata Barik
School of Computer Science
Swami Vivekananda University
Barrackpore, West Bengal, India

Dulari Bhatt
Edunet Foundation
Bengaluru, India

Olga Carvalho
Faculty of Medicine
University of Coimbra
Coimbra, Portugal

Nayanika Chatterjee
Kalinga Institute of Industrial Technology
Bhubaneswar, Odisha, India

Abhranil Chattopadhyay
Kalinga Institute of Industrial Technology
Bhubaneswar, Odisha, India

Bhumika Choksi
Mathematics Division
School of Advanced Sciences and Languages
VIT Bhopal University
Sehore, Madhya Pradesh, India

Madhuri Chopade
Gandhinagar University
Gujarat, India

Kalindi Delawala
Department of Mathematics
Marwadi University
Rajkot, Gujarat, India

Prerna Reddy Ganga
Stanley College of Engineering and Technology for Women
Hyderabad, Telangana, India

Inês Maria Gouveia
Egas Moniz School of Health and Science
Caparica, Portugal

Somsubhra Gupta
School of Computer Science
Swami Vivekananda University
Barrackpore, West Bengal, India

Tapesh Chandra Gupta
Dept. of Commerce
Govt. J Yoganandam C.G College
Pt. Ravishankar University
Raipur, India

Purohit Harsh
Department of Microbiology
School of Science
RK University
Rajkot, Gujarat, India

Minakshi Biswas Hathiwala
Department of Mathematics
Marwadi University
Rajkot, Gujarat, India

Ashishkumar Kyada
Department of Pharmaceutical Sciences
Faculty of Health Sciences
Marwadi University
Rajkot, Gujarat, India

Kari J Lippert
University of South Alabama
Mobile, Alabama

Rinku Manvani
Department of Microbiology
Marwadi University
Rajkot, Gujarat, India

David Miranda
Faculty of Medicine
University of Coimbra
Coimbra, Portugal

Biswaroop Nath
Kalinga Institute of Industrial Technology
Bhubaneswar, Odisha, India

Adithya P
Department of Computer Science Engineering – AI
Amrita School of Engineering
Bangalore, India

Chintan Patel
IIM Ahmedabad
Ahmedabad, Gujarat, India

Ramesh Parmar
Department of Pharmaceutical Sciences
Faculty of Health Sciences
Marwadi University
Rajkot, Gujarat. India

Chittaranjan Pradhan
Kalinga Institute of Industrial Technology
Bhubaneswar, Odisha, India

Kishor Kumar Reddy C
Stanley College of Engineering and Technology for Women
Hyderabad, Telangana, India

Bernardo Flôr-Rodrigues
Faculty of Medicine
University of Coimbra
Coimbra, Portugal

Minirani S
Department of BSH
MPSTME
NMIMS Deemed to be University
Mumbai, India

Anindita Saha
Techno Main Saltlake
Kolkata, India

Chita Ranjan Sahoo
ICMR-Regional Medical Research Centre
Department of Health Research
Ministry of Health and Family Welfare
Government of India
Bhubaneswar, India

Minal Shukla
MGL Group
Rajkot, Gujarat, India

Sejal Shah
Department of Bioinformatics
Marwadi University
Rajkot, Gujarat, India

Subhranil Som
Bhairab Ganguly College
Kolkata, West Bengal, India

Manoj Kumar Srivastav
School of Computer Science
Swami Vivekananda University
Barrackpore, West Bengal, India

Vijaykumar Sutariya
Department of Pharmaceutical Sciences
USF Health Taneja College of Pharmacy
University of South Florida
Tampa, Florida

Sushma Tiwari
Pt. Ravishankar University
Raipur, India

Nasir Vadia
Department of Pharmaceutical Sciences
Faculty of Health Sciences
Marwadi University
Rajkot, Gujarat, India

Turyomurugyendo Vicent
Department of Mathematics
Marwadi University
Rajkot, Gujarat, India

Dhanashree Yevle
Gujarat Technological University
Ahmedabad, Gujarat, India

Chapter 1

Society 5.0

Research, well-being/healthcare, human security, and social development

Bernardo Flôr-Rodrigues, David Miranda, Inês Maria Gouveia, and Olga Carvalho

1.1 INTRODUCTION

The world we experience nowadays has changed dramatically, and it is almost impossible to imagine life without the many technological devices and gadgets developed over the last few decades. Nowadays, we live in a globalised, multicultural, and plural world, so it is possible to communicate with and see people from the other side of the globe in real-time, instead of using letters or photographs and waiting a long time for a reply. However, beyond simply having access to numerous technological tools, it is both relevant and essential to understand how to use technology appropriately — and, above all, to recognise the importance of employing it ethically, avoiding its misuse of harm or destabilisation.

What is Society 5.0, and why is it such a significant topic of discussion? These are just some of the questions that may arise, and it is hoped that this book — and this chapter in particular — will provide meaningful answers. Our discussion will focus on the areas of research, well-being/healthcare, human security, and social development.

Indeed, we can already observe multiple technological advances transforming different sectors, such as research, healthcare services, business, etc. Moreover, if technology affects the domains that guide society, it also impacts social development. Nevertheless, it is crucial to consider whether technological development offers only benefits. Upon reflection, we can understand that while technology brings considerable advantages, it also generates certain negative consequences. These include natural disasters and the depletion of resources, often resulting from poor management practices. Therefore, our primary objective should be to seek a sustainable balance between human desires and the needs of our planet.

1.2 SOCIAL AND HISTORICAL CONTEXT

Understanding Society 5.0 requires more than simply knowing how this concept works; it is also deeply rooted in historical, social, and economic

DOI: 10.1201/9781003643166-1

evolution. To grasp its significance fully, it is essential to firstly examine its development over time, before exploring each concept separately.

After all, how could our ancestors — living in prehistoric societies — have imagined a world like ours today, surrounded by advanced technologies rather than relying on rudimentary tools or rock art?

Historically, human development has been categorised into several stages, each corresponding to a specific type of society. The earliest was Society 1.0, in which human beings coexisted closely with nature, living as nomads and surviving through hunting and gathering. This was followed by Society 2.0, characterised by agrarian practices, where humans began to settle and form more structured communities based on agriculture.

Subsequently, Society 3.0 — the Industrial Society — emerged and was marked by the rise of large-scale manufacturing and industrialisation, transforming social and economic structures.

Currently, we are living in the era of Society 4.0 — the Information Society. This period is distinguished by the exponential growth of knowledge, the expansion of digital life, global connectivity, and increasing levels of industrialisation. However, this technological and industrial progress has also given rise to numerous global challenges, such as air and water pollution, the endangerment of species, the depletion of natural resources, and rising global temperatures (Fukuyama, 2018).

In this context, could Society 5.0 be the key to helping us solve this global challenge and also a way to expand human expertise/skills?

1.3 IS SOCIETY 4.0 TURNING INTO SOCIETY 5.0?

On 22 January 2016, Japan organised the *5th Basic Plan for Science and Technology*, an event started in 1956 to improve Technology and Science. This meeting involved the work of different Ministries, namely, the Ministry of Education (responsible for the basic study); the Ministry of Industrial Trade, and the Ministry of Agriculture and Forestry (accountable for the agricultural technology research). It is also worth noting that the Science and Technology Agency was established to promote the further development of atomic energy (Hayashi, 2019).

The Japanese government suggested that the Super Intelligent Society is the fifth stage in the development of human society, following hunter-gatherer, pastoral/agrarian, industrial and informational societies, as previously mentioned. It consists of a technologically sustainable civilisation where all requirements are satisfied (Shiroishi *et al.*, 2018).

In Society 5.0, humans and machines collaborate to enhance industrial output. The manufacturing sector is becoming more productive through the combined efforts of human labour and universal robotics. Executive teams within manufacturing companies are expected to design production lines, monitor key performance indicators (KPIs), and ensure that all processes are executed smoothly

The improvement of cognitive computing and artificial intelligence technologies is accelerating the pace of manufacturing and enhancing organisational efficiency. This could bring more advances in sustainability because they intend to create a system that is reliant on renewable energy. Importantly, in Society 5.0, human beings are placed at the centre of a society that integrates the cyberspace and the physical space, encouraging sharing data and establishing interpersonal, technological, and corporate relationships (Adel, 2022; Deguchi *et al.*, 2020; Dinana, 2020; Fukuyama, 2018).

The pretension is to merge the virtual space—cyberspace—and the physical space.

In the Information Society (Society 4.0), members seek, obtain, and evaluate information or data from cloud services/databases in cyberspace, using the Internet. In contrast, in Society 5.0, vast amounts of data collected through sensors in the physical environment are stored online, analysed by artificial intelligence in cyberspace, and subsequently communicated to humans in the physical world in multiple formats.

In summary, this plan, developed in Japan, primarily seeks to create a society in which individuals can fully enjoy life, combining economic growth with technological advancement. In parallel, it aims to address global social challenges and contribute to overall prosperity. Society 5.0 places particular emphasis on collecting, analysing, and applying real-world data.

1.4 SOCIETY 5.0: SHORT ANALYSIS OF PROS VERSUS CONS

The concept underpinning Society 5.0 is to create an industry that is resilient, human-centred, and sustainable, so that advancement may contribute to the improvement of society as a whole. Beyond its techno-economic component, it aims to promote and supplement the paradigm of Society 4.0 (Bigerna *et al.*, 2023).

The principal focus of Society 5.0 is the worker, therefore, the technologies implemented within organisations must serve to advance this objective without infringing upon workers' fundamental rights. Industrial development is intended to become more resilient to external shocks, while the circular economy forms the foundation of sustainable development. An analysis based on top-cited papers emphasises the benefits and challenges of Industry 5.0 in terms of resilience and sustainability. The potential advantages for the sector are considerable, ranging from enhanced talent attraction and energy savings to greater resilience against future disruptions. This suggests that industry may eventually achieve a balance between sustainability, profitability, and productivity.

Society 5.0 presents numerous advantages when compared to contemporary society, largely driven by technological innovation. By embracing cutting-edge technologies, it becomes possible to develop innovative solutions capable of addressing a wide range of challenges. This next-generation society aims to integrate technology seamlessly into daily life, potentially improving quality of life through the automation of tasks, enhanced security, and smarter infrastructure. In addition; it aspires to foster sustainability, enabling the prevention and mitigation of climate change and resource depletion. Smart cities, for instance, can lead to more efficient and environmentally friendly urban environments. This new generation may drive big economic growth and job creation by fostering innovation and entrepreneurship among companies. A truly inclusive society, in which technological advancements benefit all members, may prove revolutionary in sectors such as healthcare, facilitating personalised care, and, consequently, improving health outcomes. Moreover, disaster preparedness and response are expected to play a crucial role in this next-generation society.

On the other hand, there are some cons, namely the complexity of some procedures and the demanding requirements of prevention against cyberattacks or the use of collaborative robots, which reduces the number of people needed for each task. IoT technologies are intimately tied to the digital revolution described in Society 5.0. Consequently, there is an exponential increase in the amount of crucial corporate and personal data that can be compromised. In addition, hardware gets outdated as technology advances, contributing to the gradual emergence of environmental contamination (Sá *et al.*, 2022; Narvaez-Rojas *et al.*, 2021; Palacin *et al.*, 2021).

1.5 SOCIETY 5.0: RESEARCH

There are numerous technological developments that contributed to the research field, and the microscope is probably one of the most remarkable among them. If we remember the History of Science, we easily understand how much it increased over the centuries.

The first microscopes were created because human beings were curious and they felt the need to discover and learn more about structures invisible to the eyes, for example, the composition of our skin or identifying microorganisms. It is no coincidence that when we want to see something better, we first move closer to get a closer look (Bradbury, 2014).

From the simple manual lenses of the early Egyptians to the Dutch designs of the 16th century, to the magnificent microscope of 18th and 19th century Europe, to the latest microprocessor models available today, the microscope's journey has been extraordinary. It all started when the Egyptians cut and polished stones, making it an art around 2600 BC. Until the end of the Roman Empire (31 BC), the Greeks and Romans continued to use the same lenses (Lavanya, 2012).

Microscope works by forming an intermediate image through the optical objective lens, which magnifies the original real image. Each objective was established to observe at a certain distance from the specimen slide and this is called the working distance (Sluder & Nordberg, 2013).

The most common is the optical microscope, which uses lenses to refract the visible light that passes through a thinly sectioned sample to produce an observable image. The initial microscope images were distorted due to the poor quality of their lenses, but progress created the good quality of the actual apparatus. Sometime after, the electron microscope appeared, giving the possibility to study more details, such as cell components.

But not much has been done after the improvement of the microscope, being that the progress has slowed down in recent years, since the optical principles are well understood, and, to a certain extent, the limits have been reached.

However, once again, technological improvement promoted results, giving new tools. As a personal example, we can investigate the research being done in the field of histology because, unlike in the past, when we could only view our samples under a microscope, we now have access to other technologies like *Axioscan®*.

This system allows us to have high-resolution visualisation, accompanied by high-speed scanning, along with multiple image modes through a fully automated and easy-to-operate system. Consequently, it allows us to work more efficiently and access details that would easily go unnoticed, even though they have immense clinical significance.

1.6 SOCIETY 5.0: WELL-BEING/HEALTHCARE

1.6.1 Surgery

In times of primitive surgery, long before man could read or write, as early as 10000 BC, surgeons performed the operation of trephination or trepanning, drilling or cutting rings or squares out of skull bones. Circumcision was practised in Ancient Egypt by the priests' assistants on the clergyman and members of the royal families, and may well be considered the oldest "elective" operation (Ellis, 2019).

The development of new instruments, anaesthesia and aseptic methods has never stopped changing the way surgeries are performed, and the modern technological advances continue to revolutionise surgical practice in different fields. The use of minimally invasive surgery techniques, which speed up recovery and reduce morbidity is one of the main milestones of contemporary surgery. Surgery is facilitated by tiny incisions with high-resolution magnified vision. However, this technique has inherent limitations related to finer and more precise manipulation, which can make it quite difficult to perform operations on complex steps or difficult anatomical sites (Ellis, 2019).

Many fundamental changes can be expected over the next few years. Average life expectancy has been increasing in developed countries, which means that, in the near future, we can predict more surgeries due to more elderly people with chronic comorbidities. There will be an increase in cases of cancer, heart disease, degenerative diseases, etc. Nevertheless, these increases can be offset by disease prevention that depends mainly on the use of new technologies, namely artificial intelligence, which can help with prevention and calculating the risk of each disease, thus improving public health (Alderson, 2019; Barsom *et al.*, 2016).

With the general technologies evolution available today, many conditions and surgical treatments will present more positive outcomes for patients.

The use of augmented reality in the operating room will enhance new imaging and robotic technologies. Three-dimensional planning and printing will probably be used to guide surgery and create unique implants. Another area where significant improvements are expected is transplantation. Organ bioengineering or xenotransplantation is a developing field, but organ availability and patient survival could be impacted by improved perfusion methods, ex vivo immunological manipulation of organs, and other advances in immunology. In summary, it seems that the use of minimally invasive techniques will increase (Martelli *et al.*, 2016; Jessop *et al.*, 2017).

Another potential area to explore and develop is micro-robotics and nanotechnologies, whose clinical uses in surgery are interesting, given that we already use nanosensors at the diagnostic level. Furthermore, it is necessary to emphasise all the ethical and legal issues that the implementation of these methodologies involves, always aiming for safety combined with effectiveness, without neglecting the importance of prevention. One of the main warnings to be made is for all sensors that are developed for individual and daily use of the patient, giving them greater autonomy, but they imply a proportional effort in technological communication skills (fundamental in a world increasingly adept of technology).

An important example of technological advancement in clinical practice is the use of the Da Vince surgical robot. This is very useful, for example, in terms of prostate cancer, a very common male pathology, and in partial nephrectomy in kidney cancer, allowing a greater precision, less invasiveness, requiring only small incisions, less postoperative pain and faster recovery, whether for discharge or resumption of professional activity.

1.6.2 Mental health

Mental health is critical today and is, therefore, of most relevance to be treated accordingly in Society 5.0. There are plenty of challenges in a connected Artificial Intelligence (AI) based society, like the role played by chatbots in people's actual relationships. We should be frightened that one day people can establish relationships with some AI model like a chatbot, and therefore, it is important to be prepared for good and bad things that can come from the propagation of these systems across society. If we look at what we could face ahead, we realise the crucial aspect of the emotional connection that people may develop with these AI models. Humans are social beings and the prospect of developing emotional bonds with chatbots raises either ethical or psychological considerations or challenges.

Firstly, AI-powered conversational agents have the ability to simulate human-like emotions by feeling empathy, which gives humans a sense of support and companionship and therefore, people can feel the need to confide their feelings and thoughts to a chatbot. This is a two-sided situation. On one hand, it can alleviate loneliness, mainly in the most vulnerable and isolated individuals. But, on the other hand, it can reduce people's need to get out and be with real people, establishing real-life conversations and relationships and, therefore, contributing to a bigger loneliness and isolation, even though these individuals will not realise it. That is because people can satisfy their expectations of social needs by talking to AI models, sharing their feelings and emotions with feedback (Klslev, 2023).

Secondly, the ethical implications that come up when talking about emotional bonds between humans and AI-powered applications are not to be dismissed. Chatbots should not have manipulative practices, inducing people into decisions and contributing to a bigger engagement. They must prioritise only the user's well-being and, for this reason, clear regulations providing the guarantee that possible emotional bonds can appear and are based on informed consent, healthy boundaries and mutual understanding of the risks. Again, it is also of extreme importance to sum up the security issues when facing AI-systems by establishing relationships with them. Users' privacy and data must be secure at all times, which can be challenging because chatbots rely on analysing big amounts of data collected from individuals in order to provide personalised interactions. It is, therefore, crucial to establish security protocols that guarantee the safeguarding of individuals' data and the transparency of its usage, keeping the AI model from sharing personal data between users (Klslev, 2023).

It is also important to establish big programs of education and community awareness, explaining the actual advantages, disadvantages, issues and risks of AI-powered systems, in order to be sure that all users are in possession of the knowledge that can keep them safe, either physically or psychologically, when using a chatbot. Information is knowledge, and knowledge is power to make people able to use all the capabilities that can come from an AI-based society and stay safe at the same time.

1.7 SOCIETY 5.0: HUMAN SECURITY

A blockchain-based Society 5.0 brings many challenges, but also many solutions to daily problems. Human security in this next-generation society can take a major benefit from using a blockchain system. Privacy shortness is one of the biggest problems we face nowadays on the Internet. Personal data is required for many things and almost every website will take some of our data from the IP address, the email and even credit card information. A big part of the blockchain concept refers to the anonymity of the individual in the network. A smart contract-based society provides people the guarantee of privacy they need and, at the same time, gives them access to the information they are looking for (Tyagi *et al.*, 2023).

Nevertheless, a blockchain system has downsides and can be hackedlike other network systems. Today, we find ourselves confronted with news of cyberattacks on a daily basis, which tend to become more prevalent at the same rate, as we use more connected devices. Smart homes and smart cities have a lot of connected devices that are susceptible to a cyberattack. It is therefore, becoming easier than ever to commit remote crimes by hackers and cyberterrorists. Society 5.0 will have a major challenge in order to keep its vulnerability at safe levels once cyber risks transcend traditional concerns of reputational and financial impact by becoming life-threatening or environmental risks (Ivezic, 2020).

This society is necessarily bonded to new technologies as well as old ones applied differently to solve either today's daily problems or long-term challenges. Being it financial, environmental, social or health crises, there are many challenges current society faces. Society 5.0 aims to solve these using technology in a people-centric way in contrast with Society 4.0 by merging cyberspace with physical space. Next generation society is therefore intimately correlated to the blockchain concept and all its applications in a modern society.

1.7.1 Cyberattacks and threats to critical infrastructure

The growing dependence on smart technologies and connected systems that is leading to Society 5.0 has ushered in an era of unprecedented convenience and efficiency. However, this increasing digitisation of critical infrastructure brings with it a heightened vulnerability to cyberattacks, posing significant challenges. Looking at some essential infrastructure, we see, for example:

- Power Grids: one of the top concerns in the realm of critical infrastructure. Society 5.0 heavily depends on a stable and continuous supply of electricity to function, and, cyberattacks targeting power grids could lead to widespread blackouts, disrupting daily life, commerce, and national security (Bader *et al.*, 2023).
- Water and utilities systems: clean water supply and other utilities are essential for community health and well-being, so it is critical to prevent cyberattacks on these systems in order to avoid disruptions on water treatment plants, sewage systems and waste management and, therefore, health and environmental hazards.
- National Security: critical infrastructure is often intertwined with national security. Cyberattacks on key infrastructure components can have far-reaching implications for a nation's defence and sovereignty.

- Transportation: means of mobility are becoming increasingly dependent on interconnected and blockchain-based technology. From autonomous vehicles, traffic management systems to modern logistics networks, all rely on digital communication and automation. Hackers' attacks targeting transportation systems could lead to accidents, disruptions in supply chains and compromised safety.
- Financial Infrastructure: the financial sector is another vital component of essential infrastructure. Society 5.0 relies heavily on blockchain-based cryptocurrency and digital financial transactions, so cyberattacks on banks, payment systems, or cryptocurrency exchanges can have severe economic and societal consequences (Srilaksmi *et al.*, 2023).
- Healthcare: health systems play a critical role in Society 5.0, particularly in the context of telemedicine and patient data management. A hacker attack on a healthcare network could lead to compromised patient data, disrupted healthcare services, and endanger lives (Cyberattacks on Healthcare: A Global Threat That Can't Be Ignored, 2024).
- Communication Networks: communication networks underpin all aspects of Society 5.0, connecting companies, individuals, devices, and essential systems. A cyberattack on these networks would disrupt communication between parts, leading to data breaches and compromising global security.
- Emergency Response Systems: in times of crisis like natural disasters, terrorist attacks or infrastructure damage, emergency response systems are critical for public safety, and so it is crucial to keep these systems secure from attacks, meaning they can hamper the ability to respond effectively to disasters or emergencies (Channa & Ahmed, 2010).

1.7.2 AI and machine learning vulnerabilities

The integration of artificial intelligence (AI) and machine learning (ML) into the fabric of Society 5.0 will represent a game-changing in technology's role in our lives. These two technologies promise to enhance efficiency, decision-making and automation across various sectors, from healthcare and finance to transportation and education. Nevertheless, as AI and ML systems become increasingly prevalent, these systems have vulnerabilities that can be exploited by hackers. Cybercriminals can manipulate AI algorithms through adversarial attacks, subtly altering inputs to cause systems to make incorrect decisions (RoX, 2025).

For instance, an attacker could modify an image or voice command to deceive the system, posing significant security risks. Being trained on biased data, AI systems can perpetuate existing biases, leading to discriminatory outcomes. These biases can have ethical and legal implications, requiring continuous scrutiny and auditing of AI systems to ensure fairness and equity. They often rely on vast datasets, raising concerns about the privacy of personal information used for training. Unauthorised access or breaches of datasets can result in privacy violations. Attackers can inject malicious data into training datasets, compromising the integrity of AI models. As it plays an increasingly crucial role in decision-making, such attacks can have far-reaching consequences. Many AI and ML algorithms are considered "black boxes" because they lack transparency in how they arrive at decisions. This opacity poses challenges in holding AI accountable for its actions, especially in critical domains like healthcare and autonomous vehicles. To address these vulnerabilities, continuous research and development of AI security measures are essential, which can include robust cybersecurity protocols, data privacy regulations, explainable AI techniques and adversarial testing to proactively identify and mitigate potential threats (Nandibhatla, 2025).

As AI will be an integral part of Society 5.0, a multidisciplinary approach involving technologists, policymakers, and ethical experts is fundamental to ensure that AI systems benefit society while minimising risks (Medina-Borja, 2017).

1.7.3 Securing Internet of Things (IoT)

Daily routine problems may be solved by something we find growing every day, which are smart devices and mainly smart houses. Almost all appliances we can buy nowadays have a connected version, from the air conditioning to the washing machine but also lights, vacuum cleaners and sound bars, which are called IoT devices. Everything can be connected to Wi-Fi and controlled by a smartphone app or a website. This means that all devices we use are automatically sharing data between themselves and with big tech companies around the world. Therefore, we are now getting more and more dependent on technology than ever, and we find ourselves many times prisoners of it. Society 5.0 pretends to improve this by making technology more helpful and less addicting, solving real-life problems instead of making problems where there are none. The proliferation of IoT poses significant cybersecurity challenges. Many of these connected devices have limited security measures in place, making them attractive targets for cyberattacks, even the less complex ones. In a blockchain-based Society 5.0, data is the lifeblood of the digital ecosystem. The extensive collection and analysis of data, often driven by artificial intelligence and IoT, raise critical privacy concerns. Individuals' personal information, from health records to financial data, is at risk of being exposed in data breaches or misused by malicious actors. Establishing robust data protection regulations and cybersecurity measures, even in the least complex IoT devices, is imperative to safeguard individuals' privacy (Tyagi *et al.*, 2023).

1.7.4 International cybersecurity cooperation

Society 5.0, with its advanced technological integration, promises to usher in an interconnected global landscape where the spectre of cyber threats transcending national borders looms ever larger. In this digital era, the imperative for collaborative endeavours among nations and international organisations becomes paramount. To effectively combat the ever-evolving menace of cybercrime and ensure the security of our interconnected systems, the establishment of consistent and robust cybersecurity standards is not merely a choice but a pure necessity. Only through united efforts and a shared commitment to safeguarding our digital realm can we navigate the complexities of Society 5.0 and harness its transformative potential while minimising the inherent risks (Hüsch & Sullivan, 2023).

1.7.5 Ransomware threats

Ransomware attacks have surged in prevalence, posing a substantial threat to the digital landscape. These attacks involve cybercriminals encrypting valuable data and demanding a ransom for its release. While ransomware is a significant concern, blockchain technology has demonstrated resilience against various security attacks, such as Sybil attacks, selfish mining and Distributed Denial of Service (DDoS) attacks. Firstly, a Sybil attack is characterised by the flooding of a blockchain network with multiple transactions from different fabricated identities. Blockchain's inherent decentralisation and consensus mechanisms make it resistant to Sybil attacks. The network validates transactions through consensus among nodes, making it difficult for malicious actors to manipulate the system by creating fake identities. Secondly, selfish mining is a strategy where miners in a blockchain network attempt to hoard rewards by withholding blocks rather than broadcasting them to the network. Some blockchain protocols, such as Bitcoin's Proof of Work (PoW), have built-in mechanisms to disincentivise selfish mining, maintaining the integrity of the blockchain network. Lastly, Distributed Denial of Service (DDoS) attacks can disrupt blockchain networks by overwhelming them with excessive traffic, rendering them temporarily inaccessible. While DDoS attacks can pose a threat, blockchain networks can implement protective measures, such as traffic filtering and redundancy, to withstand these assaults (Tyagi *et al.*, 2023).

Nevertheless, despite its security advantages, blockchain technology is not immune to trust issues. Decentralisation and the distributed nature of blockchain networks sometimes lead to contentious decisions within the community. For instance, changes in consensus processes or halting new transactions may occur, potentially for the wrong reasons. Trust issues can emerge if users perceive that the blockchain network's actions are not aligned with their interests or values.

In the context of Society 5.0, where data is of paramount importance, ransomware attacks can have devastating consequences. Blockchain technology can provide enhanced data security through features like immutability and cryptographic protection. However, it is essential to develop robust backup and recovery strategies to mitigate the impact of ransomware attacks in blockchain-based systems, which should include regular data backups stored securely off-network, incident response plans, and proactive cybersecurity measures to prevent such attacks. While blockchain technology offers substantial security benefits against various threats, including Sybil attacks and selfish mining, it is not without its trust challenges. The resilience of blockchain networks makes them valuable assets in a data-centric society like Society 5.0, but a comprehensive approach that combines blockchain security with backup and recovery strategies is essential to safeguard valuable data and ensure the continuity of critical services (Wu, 2024).

1.8 SOCIETY 5.0: SOCIAL DEVELOPMENT

Long-term challenge solutions are still not as developed today as daily ones. We can see environmental, humanitarian, financial and social crises becoming more prevalent each day, but no strict strategies have yet been achieved. Humanitarian crises play a big role in developing countries from Asia, Africa and South America, but also in developed countries affected by the huge number of migrants. These humanitarian crises are not a cause by themselves but a consequence of other problems faced by societies all over the world. Wars, poverty, starvation, and natural disasters are among the main causes of migratory crises forcing people to leave their homes because there are no longer conditions to live in there.

When it comes to natural disasters, Society 5.0 can and should minimise their impact. For example, earthquakes cannot be prevented, but their destruction can be reduced. On the one hand, earthquake-proof buildings are essential in risk areas and their structure can be monitored and connected through sensors and systems to calculate its strength in real time and damage after a disaster. For that reason, people will know if the building is still safe to be in or if it is at risk of falling. These connected systems also allow for emergency services to know the dimension of damage in a city, the most affected structures, and the places more important to search and rescue. On the other hand, predicting a natural disaster, such as an earthquake is also a big step towards a safer and more reliable city and society. Even if it can only be predicted thirty seconds before, it can be a big difference in the number of victims, allowing people to take refuge in stronger places, be it under a table, a bed or in the corners of a room (Damaševičius et al., 2023).

Other natural disasters are becoming more and more human activity-related than ever. Hurricanes, forest fires, and floods are among the most usual disasters, whose increasing frequency is intimately related to global warming and climate change. Their impact can also be minimised by better predicting their occurrence. In society 5.0, authorities' alerts can be connected directly to safety systems, allowing, for example, automatically closing windows and doors, preparing areas to flood without damage or closing gas valves to prevent fire propagation.

1.8.1 Sustainability

Nevertheless, it is better and possible to prevent these natural disasters with a concept we now hear every day. Sustainability is one of the most spoken themes in the 21st century. It can be referred to as a society design which can be enabled by a blockchain system and provide a cut in natural resource needs. Today's society has not yet come to that point, and we can see, for example, global warming prevention and mitigation strategies across the world. Even though big steps have been made towards climate change prevention and mitigation, mainly through the compromise of "Paris Agreement", its compliance by limiting global warming to 1.5°C within the deadline of 2050 is still uncertain. However, Society 5.0 may be a way to achieve a carbon-neutral society by 2050.

Firstly, it is important to know that most greenhouse gas emissions are made by urban environments, and that is the first place to make this 5.0 aim a reality. Within cities, buildings are responsible for a major part of total urban emissions, so decarbonising existing building stock, as well as building new carbon-neutral houses, is relevant to cut cities' impact on global warming and, at the same time, reduce global cost for owners.

A carbon-neutral building may be achieved during or after construction, so we can make it in existing and new ones. A large part of lifelong building environmental impact occurs through energy consumption; that is why reducing energy needs is essential to achieve a carbon-neutral place. Air conditioning may be set to a higher temperature in summer and lower in winter, reducing the energy needs of the systems to keep the set temperature. Washing machines, dishwashers and other appliances may be programmed to work during the night, in low demand for electricity when costs are lower, and a bigger part comes from renewables. This is possible today with the technology we have, but it gets a major significance in Society 5.0 with all these systems, appliances and other electricity powered devices connected, which brings us to the concept of a Smart City. Smart cities can control and direct electricity consumption through all devices to reduce the peak needs and keep the total demand stable and as low as possible throughout all day long. This allows to mitigate the problem inherent to renewables, which is the intermittence of power production and many times its displacement of power actual needs (Mishra & Singh, 2023).

1.8.2 Finances

The contemporary world is characterised by the convergence of financial crises with other significant challenges, such as armed conflicts, environmental disasters, and pandemics, giving rise to a multifaceted global landscape.

These phenomena are no longer isolated occurrences, but are intricately intertwined, generating a complex web of challenges that impact societies worldwide. This context underscores the urgent need for a paradigmatic shift, as embodied by the vision of Society 5.0. In the present era, financial crises are inextricably connected to broader social challenges, producing disruptions that are closely correlated with the financial behaviours of individuals. The capacity of individuals to save resources and to make prudent financial decisions becomes instrumental, either in amplifying the ripple effects of financial crises, or in serving as a protective buffer against economic downturns.

Over the past two decades, the global financial system has undergone a profound transformation. The era dominated by cash transactions has receded, giving way to the widespread adoption of debit and credit cards. More recently, the advent of online banking and cryptocurrencies has fundamentally reshaped the way financial transactions are conducted. Cryptocurrencies, for example, have evolved from being merely instruments of payment to becoming powerful investment assets, capturing the interest and investment of individuals across the globe. This evolution highlights the importance of integrating emerging financial systems as essential components of Society 5.0 (Kayani & Hasan, 2024).

At the core of Society 5.0 lies the transformative potential of blockchain-based cryptocurrencies. Owing to their decentralised nature, these technologies are fundamental to a future society in which individuals are emancipated from the constraints imposed by centralised financial institutions. This paradigmatic shift marks a departure from dependence on governments and traditional banking systems, fostering instead a greater reliance on individual agency, discretion, and decision-making. Society 5.0 envisions a financial landscape in which the collective wisdom, choices, and actions of individuals assume a pivotal role in shaping and directing financial and economic systems. It aspires to create an information-centric society wherein individuals operate as key nodes within a human blockchain. Within this visionary framework, global decisions arise from the aggregation of individual actions, collectively determining the trajectory of the economy. The symbiotic relationship between Society 5.0 and blockchain-based cryptocurrencies is thus characterised by a society that evolves from a passive observer to an active participant — consciously shaping its financial and economic destiny in an increasingly complex world, where interconnected challenges demand innovative and adaptive solutions (Richard, 2022).

ACKNOWLEDGEMENTS

A huge thank you to Mrs. Alzira Mendes, Portuguese and English Teacher, for the linguistic review.

REFERENCES

Adel A. (2022). Future of industry 5.0 in society: human-centric solutions, challenges and prospective research areas. *Journal of Cloud Computing (Heidelberg, Germany)*, *11*(1), 40. https://doi.org/10.1186/s13677-022-00314-5

Alderson D. (2019). The future of surgery. *The British Journal of Surgery*, *106*(1), 9–10. https://doi.org/10.1002/bjs.11086

Bader, L., Serror, M., Lamberts, O., Sen, Ö., Van Der Velde, D., Hacker, I., Filter, J., Padilla, E., & Henze, M. (2023). Comprehensively analyzing the impact of cyberattacks on power grids. *arXiv* (Cornell University). https://doi.org/10.48550/arxiv.2305.09390

Barsom, E. Z., Graafland, M., & Schijven, M. P. (2016). Systematic review on the effectiveness of augmented reality applications in medical training. *Surgical Endoscopy*, *30*(10), 4174–4183. https://doi.org/10.1007/s00464-016-4800-6

Bigerna S., Micheli S., Polinori P. (2023). Advantages and disadvantages of Industry 5.0 in the twenty-first century. In Rosak-Szyrocka J., Żywiołek J., Shahbaz M. (Eds.), *Quality management, value creation, and the digital economy* (pp. 20–43). Routledge.

Bradbury, S. (2014). *The evolution of the microscope*. Elsevier.

Channa, M. I., & Ahmed, K. M. (2010). Emergency response communications and associated security challenges. *arXiv* (Cornell University). https://doi.org/10.48550/arxiv.1010.4887

Cyberattacks on Healthcare: A Global Threat That Can't Be Ignored. (2024). UN News. https://news.un.org/en/story/2024/11/1156751

Damaševičius, R., Bacanin, N., & Misra, S. (2023). From sensors to safety: Internet of Emergency Services (IOES) for emergency response and disaster management. *Journal of Sensor and Actuator Networks*, *12*(3), 41. https://doi.org/10.3390/jsan12030041

Deguchi, A., Hirai, C., Matsuoka, H., Nakano, T., Oshima, K., Tai, M., & Tani, S. (2020). What is society 5.0. *Society*, 5(0), 1–24. https://doi.org/10.1007/978-981-15-2989-4

Dinana, H. (2020). Society 5.0 and the Future Economies. Retrieved October 6 from https://www.thecairoreview.com/essays/society-5-0-and-the-future-economies/

Ellis, H. (2019). *The history of surgery*. Boca Raton, FL; London; New York: CRC Press.

Fukuyama, M. (2018). Society 5.0: Aiming for a new human-centered society. *Japan Spotlight*, *1*, 47–50.

Hayashi, Y. (2019). Japanese science and technology basic plan: A perspective of policy process. *Innovation and Development Policy*, *1*(1), 24–38. https://doi.org/10.3724/SP.J.2096-5141.2019.0003

Hüsch, P., & Sullivan, J. (2023). *Global approaches to cyber policy, legislation and regulation: A comparative overview*. Royal United Services Institute.

Ivezic, M. (2020). Cybersecurity and safety in the 5G-enabled smart-everything world. Retrieved October 3, 2023 from https://society5.com/cyber-kinetic-security/cybersecurity-safety-5g-smart-everything/ss

Jessop, Z. M., Al-Sabah, A., Gardiner, M. D., Combellack, E., Hawkins, K., & Whitaker, I. S. (2017). 3D bioprinting for reconstructive surgery: Principles, applications and challenges. *Journal of Plastic, Reconstructive & Aesthetic Surgery: JPRAS*, *70*(9), 1155–1170. https://doi.org/10.1016/j.bjps.2017.06.001

Kayani, U.N., & Hasan, F. (2024). Unveiling cryptocurrency impact on financial markets and traditional banking systems: Lessons for sustainable blockchain and interdisciplinary collaborations. *Journal of Risk and Financial Management*, *17*, 58.

Klslev, E. (2023). Will people fall in love with their chatbot? Retrieved October 4, 2023 from https://www.psychologytoday.com/intl/blog/society-50

Lavanya, V. (2012). Journey of the microscope. *Medica Innovatica*, December, *1* (2) (75–79).

Martelli, N., Serrano, C., van den Brink, H., Pineau, J., Prognon, P., Borget, I., & El Batti, S. (2016). Advantages and disadvantages of 3-dimensional printing in surgery: A systematic review. *Surgery*, *159*(6), 1485–1500. https://doi.org/10.1016/j.surg.2015.12.017

Medina-Borja, A. (2017). How artificial emotion intelligence will change our lives. In Center for Research and Development Strategy, Japan Science and Technology Agency, *Future services & societal systems in Society 5.0* (pp. 223–226) Tokyo, Japan: Japan Science and Technology Agency.

Microscope History. (n.d.). Escola Superior De Agricultura Luiz De Queiroz. https://www.esalq.usp.br/lepse/imgs/conteudo_thumb/Microscope-History.pdf

Mishra, P., & Singh, G. (2023). Energy management systems in sustainable smart cities based on the internet of energy: A technical review. *Energies*, *16*(19), 6903. https://doi.org/10.3390/en16196903

Nandibhatla, A. K. (2025, February 9). OWASP LLM Top 10: Tackling data and model poisoning attacks in AI. *Medium*. https://scrumgit.com/owasp-llm-top-10-tackling-data-and-model-poisoning-attacks-in-ai-d8d873fa9eee

Narvaez-Rojas, C.; Alomia Peñafiel, G.A.; Loaiza Buitrago, D.F.; Tavera Romero, C.A. (2021). Society 5.0: A Japanese concept for a superintelligent society. *Sustainability*, *13*, 6567.

Palacin, V., Ferrario, M.A., Hsieh, G., Knutas, A., Wolff, A., Porras, J. 2021. Human values and digital citizen science interactions. *International Journal of Human-Computer Studies*, *149*, 102605

Richard. (2022, January 5). Society 5.0 — from agile business model to crypto-assets social sharing model. *Medium*. https://medium.com/@richard_12736/society-5-0-from-agile-business-model-to-crypto-assets-social-sharing-model-b67a1322d537

RoX. (2025 February 16). Adversarial attacks: How hackers trick AI & how to test against it. *AICompetence*. https://aicompetence.org/adversarial-attacks-how-to-test-against-it/

Sá, M. J., Serpa, S., & Ferreira, C. M. (2022). Citizen science in the promotion of sustainability: The importance of smart education for smart societies. *Sustainability*, *14*(15), 9356. https://doi.org/10.3390/su14159356

Shiroishi, Y., Uchiyama, K., & Suzuki, N. (2018). Society 5.0: For human security and well-being. *Computer*, *51*(7), 91–95. https://doi.org/10.1109/MC.2018.301104

Sluder, G., & Nordberg, J. J. (2013). Microscope basics. *Methods in Cell Biology*, *114*, 1–10. https://doi.org/10.1016/B978-0-12-407761-4.00001-4

Srilaksmi, N.K., Erkamim, M., Dzakiah, N., Suardana, N.K., & Swambini, N.K. (2023). An examination of cybercrimes leading to vulnerabilities in energy, transportation, and financial systems. *Journal of Digital Law and Policy*, *3*, 38–51.

Tyagi, A. K., Dananjayan, S., Agarwal, D., & Thariq Ahmed, H. F. (2023). Blockchain-Internet of Things applications: Opportunities and challenges for Industry 4.0 and Society 5.0. *Sensors (Basel, Switzerland)*, *23*(2), 947. https://doi.org/10.3390/s23020947

Wu, J., (2024, July 25). Challenges of blockchain technology: Navigating obstacles to adoption and implementation - Blockchain UBC. *Blockchain UBC - Leading the Way in Blockchain Research and Development*. https://blockchainubc.ca/challenges-of-blockchain-technology/

Chapter 2

Navigating the pulse in Industry 6.0

Assessing the resilience of India's healthcare grid

Prerna Reddy Ganga, Kishor Kumar Reddy C, and Kari J Lippert

2.1 INTRODUCTION

In the modern healthcare scene in India, there's been a big change thanks to all the new technology and digital tools. One major shift is how hospitals and doctors are using Electronic Health Records (EHR) systems [1]. These systems are like digital versions of patient files, making it easier for everyone in the healthcare team to share information and keep track of patients' health histories. But these EHR systems aren't just about storing files. They've gotten smarter, too. Now, they can help doctors make better decisions by giving them useful insights and reminders. Telemedicine is another game-changer, especially for people living in remote areas. It lets patients talk to doctors and get medical advice over the phone or through video calls. This means people can get help without having to travel long distances. Then, there's Health Information Exchange (HIE) networks, which help different healthcare places share patient information securely. This makes it easier for doctors to get the info they need to give the best care, no matter where patients go for treatment. When it comes to understanding big health trends, computers are doing a lot of the heavy lifting. They're analysing huge amounts of data to spot patterns and predict things like disease outbreaks. And Artificial intelligence (AI) is making a big impact, too. It's helping doctors read scans more quickly and accurately, which means patients can get diagnosed and treated faster [2].

Virtual health assistants are also becoming more common, giving people health advice and answering their questions online. But it's not all about what doctors are doing. There are also lots of apps and gadgets that help regular folks stay healthy. From fitness trackers to health apps on our phones, there are plenty of ways for people to keep an eye on their health and get support when they need it. Overall, all this new technology is changing the way healthcare works in India. But it's important to make sure everyone can access these tools and that people's private health information stays safe [3]. After all, the goal is to make sure everyone can live their healthiest life possible. In India, technology is making waves in the healthcare sector like never before. From bustling urban centres to remote rural villages,

DOI: 10.1201/9781003643166-2

innovative solutions are revolutionizing the way healthcare services are accessed, delivered, and managed. In this essay, we'll explore how technology is being used in India's healthcare landscape, its impact on various aspects of healthcare delivery, and the challenges and opportunities it presents. Firstly, let's talk about EHR systems. These digital platforms are like digital filing cabinets for patient health information. Instead of relying on piles of paper records, hospitals and clinics can now store and access patient data electronically. This not only saves time but also reduces the risk of errors and makes it easier for healthcare providers to coordinate care across different departments and facilities [4].

One of the most exciting developments in healthcare technology is telemedicine. In a vast country like India, where access to healthcare can be a challenge, telemedicine is a game-changer. Through video calls, patients can now consult with doctors from the comfort of their homes, eliminating the need for long and often arduous journeys to the nearest healthcare facility. Telemedicine also enables the remote monitoring of patients with chronic conditions, allowing doctors to intervene early and prevent complications. HIE networks are another crucial component of India's healthcare technology ecosystem. These networks facilitate the seamless exchange of electronic health data between different healthcare providers and systems. This means that no matter where a patient goes for treatment, their medical records can follow them, ensuring continuity of care and reducing the risk of redundant tests and procedures. The use of data analytics is also transforming healthcare in India. By analysing large datasets of clinical, financial, and operational information, healthcare organizations can identify patterns and trends that can inform decision-making and improve patient outcomes.

Predictive analytics, in particular, holds great promise for early disease detection and prevention, allowing healthcare providers to intervene before conditions worsen. AI is another technology that is making a significant impact on healthcare in India [5]. AI-powered algorithms can analyse medical images, such as X-rays and MRIs, with greater accuracy and speed than ever before, helping doctors diagnose conditions more quickly and accurately. AI-driven virtual health assistants are also becoming increasingly common, providing patients with personalized health advice and support through chatbots and voice interfaces. Mobile health (mHealth) applications and wearable devices are empowering individuals to take control of their health and well-being. With smartphones and smartwatches equipped with health tracking apps and sensors, users can monitor their physical activity, track their sleep patterns, and even measure their heart rate and blood pressure. These tools not only help individuals stay healthy but also enable healthcare providers to remotely monitor patients and intervene when necessary. While technology has the potential to revolutionize healthcare in India, it also presents several challenges [6].

One of the biggest challenges is ensuring that technology is accessible to all segments of society, including those in rural and underserved areas. This requires investments in infrastructure, such as internet connectivity and digital literacy programs, to ensure that everyone can benefit from the advancements in healthcare technology. Data privacy and security are also significant concerns [7]. With the digitization of health records and the widespread use of telemedicine and mobile health apps, protecting patient information from cyber threats is more important than ever. Healthcare organizations must implement robust security measures and adhere to strict privacy regulations to safeguard patient data. Despite these challenges, the potential of technology to transform healthcare in India is immense. By leveraging innovative solutions such as EHR systems, telemedicine, data analytics, AI, and mHealth applications, India has the opportunity to improve access to healthcare, enhance patient outcomes, and reduce healthcare costs [8]. However, realizing this potential will require collaboration between government, industry, and healthcare providers to overcome barriers and ensure that technology is used responsibly and equitably for the benefit of all [9].

2.2 TECHNOLOGY ADVANCEMENTS

Over the years, healthcare in India has seen some big changes thanks to technology. From basic digital systems in the 90s to the latest AI and telemedicine, tech has really transformed how healthcare works. Let's break it down into three phases. In the early days [5], around the 1990s and 2000s, the focus was on getting basic electronic systems in place. This meant digitizing things like billing and scheduling to make hospital administration easier. There were also some early experiments with telemedicine, using satellite tech to connect remote areas with doctors. Then, from the 2010s to the early 2020s, things really took off. Smartphones and the internet became more widespread, and we saw a boom in health apps. These apps let people access healthcare services right from their phones. Plus, networks were set up to help different healthcare providers share data securely. Now, in the mid-2020s and beyond, things are moving even faster [10].

The COVID-19 pandemic pushed telemedicine into the spotlight, with more people than ever relying on remote healthcare. And AI is making a big splash, too. It's being used for things like predicting diseases and personalizing treatment plans. Plus, new tech like blockchain and the Internet of Things (IoT) is helping keep patient data safe and making remote monitoring easier. Back in the 1990s and early 2000s, healthcare in India started to catch up with the digital age. The main goal during this time was to move away from paper and start keeping patient records electronically. Hospitals and clinics began using EHR systems to make this happen. These systems helped them keep track of patient information more easily, cutting down on errors and making it quicker to find what they needed. Alongside this, some basic telemedicine projects were launched [11].

These aimed to connect remote areas with doctors through technology so that people living far from healthcare facilities could still get medical advice and even education about their health. In the 2010s and early 2020s, technology in Indian healthcare really took off. In recent years, the healthcare landscape in India has undergone a profound transformation, thanks to remarkable strides in technology. These advancements aren't just about cutting-edge gadgets and algorithms; they're about enhancing the human experience of healthcare – making it more accessible, personalized, and impactful for individuals and communities across the country [12].

Picture this: Deepak, a farmer in rural Rajasthan, wakes up one morning with a persistent cough. Concerned about his health but hesitant to travel long distances to the nearest city for a consultation, he turns to his smartphone for help [13]. With a few taps on a telemedicine app, Deepak connects with a qualified doctor who listens to his symptoms, asks relevant questions, and provides guidance on potential causes and next steps. Through this virtual consultation, Deepak receives reassurance and actionable advice without ever leaving the comfort of his home. Meanwhile, in a bustling hospital in Bangalore, a doctor begins her rounds, armed not just with a stethoscope and medical charts but also with a tablet loaded with EHRs [14]. As she visits patients in different wards, she seamlessly accesses their medical history, test results, and treatment plans, allowing for more informed decision-making and personalized care. This integration of technology into clinical practice not only enhances efficiency but also improves patient safety and outcomes. In urban centres like Mumbai and Delhi, individuals are embracing wearable health devices and mobile apps to track their fitness goals and monitor their vital signs. Whether it's counting steps, measuring heart rate, or analysing sleep patterns, these digital tools empower users to take charge of their health and make informed lifestyle choices. For busy professionals with hectic schedules, these devices serve as constant reminders to prioritize physical activity and manage stress, thereby reducing their risk of lifestyle-related diseases. Across the country, from remote villages to bustling metropolises, the impact of technology on healthcare is undeniable. It's not just about the latest gadgets or flashy innovations; it's about leveraging technology to address longstanding challenges, bridge gaps in healthcare access, and improve the overall quality of life for millions of people [15].

With more people getting access to the internet and using smartphones, there was a big push to bring healthcare services online. Telemedicine became more popular during this time, with new platforms and apps popping up to connect patients with doctors remotely. People could now get medical advice and even monitor their health from the comfort of their own homes. Plus, HIE networks were set up to make it easier for different healthcare providers to share patient information securely. This meant that no matter where a patient went for treatment, their medical records could follow them [16]. Alongside this, we started seeing more advanced analytics and AI being used in healthcare. These technologies could predict diseases, personalize treatment plans, and even help doctors make better decisions. And let's not forget about mobile health (mHealth) solutions. With fitness trackers, wellness apps, and teleconsultation services, patients were given more control over their health than ever before. In recent years since 2020, we've seen some big changes in healthcare, especially with everything that's been going on with the COVID-19 pandemic. It's like the whole world of healthcare got a big shake-up, and digital solutions have been at the forefront of this transformation. One thing that's really taken off is telemedicine. With all the social distancing and lockdowns, people couldn't always go to the doctor like they used to [17]. So, virtual consultations became the new normal. Now, you can chat with your doctor from your living room. And it's not just talking to doctors online that's become more common.

Artificial intelligence, or AI, has also become a big part of healthcare. Machines are getting better at analysing medical images, which means faster and more accurate diagnoses. Plus, there are now virtual health assistants that can answer your health questions and give personalized advice. Another exciting development is blockchain technology. It's not just for cryptocurrency anymore – it's being explored for its potential to keep patient data safe and make it easier for different parts of the healthcare system to talk to each other. But it's not all about high-tech gadgets and algorithms. Simple things like wearable devices and smartwatches are also changing the way we manage our health. Now, you can keep track of your heart rate, activity levels, and even your glucose levels right from your wrist. Overall, these changes are making healthcare more accessible, personalized, and efficient. But as we embrace all this new technology, it's important to remember to do it in a way that puts people's health and privacy first. After all, the goal is to make sure everyone can get the care they need when they need it, no matter where they are [18].

Rapid technological innovation has been reshaping healthcare in recent years, offering a host of benefits while also posing some significant challenges. Let's delve into the pros and cons of this fast-paced evolution in healthcare technology. On the positive side, one of the most significant benefits of rapid technological innovation is improved access to care. Telemedicine and mobile health (mHealth) solutions have made it possible for patients to access healthcare services remotely, breaking down geographical barriers and ensuring that even those in remote or underserved areas can receive medical attention. This is particularly crucial during times like the COVID-19 pandemic when physical distancing measures are in place [19]. Moreover, technological advancements have led to enhanced efficiency within healthcare systems. Automation of administrative tasks, predictive analytics, and AI-driven decision support systems have streamlined clinical workflows, allowing healthcare providers to optimize resource utilization and focus more on patient care. This not only saves time and resources but also improves the overall quality of care provided. Furthermore, rapid technological innovation has contributed to better patient outcomes. Advanced diagnostics, personalized treatment plans, and remote monitoring tools enable early detection of health issues, timely intervention, and ultimately, improved health outcomes for patients [20]. This proactive approach to healthcare can significantly reduce the burden on healthcare systems and improve patient satisfaction. Cost savings are another significant advantage of embracing technological innovation in healthcare.

Digital health solutions help minimize the need for physical infrastructure, reduce unnecessary hospital visits, and lower healthcare costs for both patients and providers. This is particularly beneficial in resource-constrained settings where healthcare resources are limited. Additionally, the rapid pace of technological innovation fosters an innovation ecosystem within the healthcare sector. Collaboration between healthcare providers, technology companies, researchers, and startups drives continuous improvement and innovation, leading to the development of new solutions and approaches to healthcare delivery. However, alongside these benefits, rapid technological innovation also presents certain challenges and risks that need to be addressed. One of the most pressing concerns is data security and privacy. The digitization of health records and the exchange of sensitive patient information raise concerns about data breaches, privacy violations, and unauthorized access to personal health information. Regulatory challenges also pose a significant hurdle. The rapid evolution of technologies often outpaces regulatory frameworks, leading to uncertainty and compliance issues regarding data protection, medical device regulations, and telemedicine guidelines. This regulatory lag can hinder the widespread adoption of innovative technologies and create barriers to implementation. Moreover, the digital divide remains a significant challenge, particularly in countries like India [21].

Disparities in internet access, digital literacy, and smartphone penetration limit the reach and effectiveness of digital health solutions, leaving many individuals in rural and underserved areas without access to essential healthcare services. Quality of care is another area of concern. While technology can improve efficiency, there is a risk of depersonalization of care and reduced physician-patient interaction. Additionally, reliance on technology may lead to diagnostic errors caused by algorithm biases or technical glitches, undermining patient safety and trust in healthcare systems. Ethical dilemmas also abound in the era of rapid technological innovation. The use of AI in healthcare raises concerns about algorithm transparency, accountability, patient consent, and potential biases in decision-making, shown in Tables 2.1 and 2.2 respectively.

Addressing these ethical considerations is essential to ensuring that technology is used responsibly and ethically in healthcare settings. The evolution of technology in Indian healthcare has brought about significant advancements and benefits, but it also poses challenges that need to be carefully navigated. By addressing issues related to data security, regulatory compliance,

Table 2.1 Growth of telemedicine in India (2010–2024) [18]

Year	*Number of teleconsultations (millions)*	*Telemedicine platforms*
2010	0.1	5
2015	0.5	20
2020	3.0	50
2024	10.0	100

Table 2.2 Adoption of AI in medical imaging analysis [18]

Year	*AI Applications*	*Benefits*
2015	Computer-aided diagnosis	Improved accuracy in detecting lesions
2020	Deep learning algorithms	Faster image analysis
2024	AI-driven radiology solutions	Automated report generation

the digital divide, quality of care, and ethical considerations, stakeholders can harness the power of technology to drive equitable, accessible, and patient-centred healthcare delivery in India. Collaboration between policymakers, healthcare providers, technology developers, and the community is essential to ensure that technological innovation continues to serve the best interests of patients and society as a whole [22].

2.3 DATA SECURITY AND PRIVACY

The healthcare system in India is intricately linked with considerations of privacy and security, forming the bedrock of patient trust and effective healthcare delivery. At its core, the relationship between healthcare and privacy/security underscores the fundamental rights of individuals to control their personal health information while ensuring the integrity and confidentiality of sensitive data. In India, where healthcare access is a critical concern, ensuring robust privacy and security measures is essential to not only protect patient information but also to foster a conducive environment for seeking medical care without fear of unauthorized disclosure or misuse of data [4]. India's healthcare landscape is characterized by a diverse array of healthcare providers, ranging from government-run hospitals to private clinics, each handling vast amounts of patient data on a daily basis. This data encompasses a wide spectrum of information, including medical history, diagnostic reports, treatment plans, and personal demographics.

Given the sensitive nature of this data, ensuring its privacy and security is paramount to upholding patient confidentiality, maintaining trust in the healthcare system, and complying with regulatory requirements. Against this backdrop, India has taken significant strides in establishing regulatory frameworks and standards to govern healthcare data privacy and security.

Table 2.3 Disease burden in India (2019) [19]

Disease	*Number of cases (in millions)*
Hypertension	230
Diabetes	77
Tuberculosis (TB)	2.64 (incidence), 26.9 (prevalence)
Cancer	1.3
Cardiovascular diseases	54.5 (deaths)
Respiratory diseases	11.5 (deaths)
Mental health disorders	197.3 (cases)

The Information Technology Act, 2000, serves as a cornerstone, providing a legal framework for safeguarding EHRs and digital health information. Additionally, guidelines issued by the Ministry of Health and Family Welfare further delineate best practices and standards for protecting healthcare data, ensuring compliance with international norms and standards such as HIPAA (Table 2.3).

Moreover, the impending enactment of the Personal Data Protection Bill is poised to revolutionize the landscape of data privacy in India, offering comprehensive safeguards for personal data across sectors, including healthcare. This legislation is expected to introduce stringent requirements for data protection, including provisions for informed consent, data localization, and penalties for data breaches. By aligning with international best practices and elevating standards for data protection, India's healthcare system can bolster patient trust and enhance the overall quality of care.

Technological advancements have also played a pivotal role in shaping the intersection of healthcare and privacy/security in India. The digitization of health records and the adoption of electronic health information systems have transformed the way patient data is collected, stored, and accessed [6]. While these innovations offer unprecedented opportunities for improving healthcare delivery and clinical outcomes, they also introduce new challenges and vulnerabilities in terms of data security and privacy. To address these challenges, healthcare organizations in India are increasingly investing in robust cybersecurity measures and technological safeguards to protect patient data from unauthorized access, breaches, and cyber threats. Encryption technologies, access controls, and cybersecurity solutions are deployed to fortify digital health systems and ensure the confidentiality and integrity of patient information. Furthermore, training programs and awareness initiatives are implemented to educate healthcare professionals about the importance of data security and privacy best practices, empowering them to safeguard patient data effectively. The healthcare system in India is deeply intertwined with considerations of privacy and security, reflecting a commitment to protecting patient confidentiality and upholding ethical standards of care. Through regulatory frameworks, technological innovations, and organizational initiatives, India's healthcare sector is poised to navigate the evolving landscape of data privacy and security, ensuring that patient data remains sacrosanct and protected against emerging threats in an increasingly digitized healthcare ecosystem.

Data security and privacy are significant considerations in healthcare systems across India, given the sensitivity and confidentiality of the information involved. In navigating this complex landscape, India has established a regulatory framework to safeguard healthcare data, drawing on both domestic legislation and international standards. While the United States' Health Insurance Portability and Accountability Act (HIPAA) serves as a foundational model, India has also crafted its own legal framework, notably the Information Technology Act, 2000. This legislation, along with guidelines issued by the Ministry of Health and Family Welfare, provides a robust foundation for protecting healthcare information. Central to India's approach to data security and privacy in healthcare is the impending enactment of the Personal Data Protection Bill. This comprehensive legislation, when passed, will offer a structured framework for safeguarding personal data across various sectors, including healthcare. Within this evolving legal landscape, the digitization of health records has emerged as a transformative force, offering numerous advantages, such as improved accessibility to patient data and streamlined healthcare delivery. However, this digital shift also raises concerns about data security and privacy, necessitating stringent measures to protect EHRs from breaches and unauthorized access.

Encryption and access controls form the cornerstone of data protection efforts in healthcare systems. Healthcare organizations must implement robust encryption techniques to safeguard sensitive data both during transmission and storage. Additionally, access to this data should be tightly regulated, with stringent authentication measures and role-based access controls ensuring that only authorized personnel can access confidential patient information. Moreover, discussions around data localization requirements underscore the importance of ensuring that healthcare data remains subject to Indian laws and regulations, enhancing data security and privacy by keeping sensitive information within the country's borders. In tandem with technological safeguards, cybersecurity measures play a pivotal role in fortifying healthcare data against evolving threats. Healthcare organizations must invest in cutting-edge cybersecurity solutions to thwart malicious attacks such as ransomware, malware, and phishing attempts. Regular software updates, comprehensive security audits, and ongoing staff training are essential components of a robust cybersecurity strategy.

Furthermore, fostering a culture of cybersecurity awareness among healthcare professionals is paramount, with employees equipped to identify and respond effectively to potential security threats. Central to safeguarding patient data is the principle of informed consent and transparency. Healthcare providers must obtain explicit consent from patients before collecting, storing, or sharing their health information. Patients should be fully informed about how their data will be utilized and who will have access to it, fostering transparency and trust in the healthcare system. Moreover, healthcare organizations must prioritize the development of comprehensive data breach response plans to mitigate the impact of potential security incidents. These plans should outline procedures for identifying, containing, and reporting data breaches in compliance with regulatory requirements, minimizing disruption and preserving patient confidentiality.

Ensuring data security and privacy in healthcare systems in India demands a multi-faceted approach, encompassing regulatory compliance, technological innovation, and organizational vigilance. By adhering to stringent data protection laws and implementing robust security measures, healthcare organizations can effectively safeguard sensitive patient information against unauthorized access and breaches. Moreover, fostering a culture of cybersecurity awareness and transparency is essential in building trust and confidence among patients and stakeholders. As India continues to embrace digital transformation in healthcare, prioritizing data security and privacy will be integral to delivering high-quality, patient-centric care in an increasingly interconnected world. In response to these challenges, healthcare organizations are deploying a range of technical solutions and best practices to safeguard patient data. Encryption technologies, for example, are widely used to protect data both at rest and in transit, ensuring that sensitive information remains secure from unauthorized access. Access controls, such as role-based access control (RBAC) systems, are also employed to restrict access to patient data to authorized personnel only, reducing the risk of data breaches caused by insider threats or external adversaries, as shown in Table 2.4.

In addition to technical measures, healthcare organizations are investing in cybersecurity training and awareness programs to educate staff about the importance of data security and privacy. Employees are trained to recognize and respond to potential security threats, such as phishing emails or social engineering attacks, thereby strengthening the organization's overall security posture. Regular security audits and assessments are conducted to identify vulnerabilities and weaknesses in existing systems and processes, allowing

Table 2.4 Healthcare infrastructure in India [16]

Healthcare facility	*Number*
Hospitals	23,582
Primary health centres	25,650
Community health centres	5,624
Sub-centres	158,378
Diagnostic laboratories	33,000
Pharmacies	1,20,000
Blood banks	2,711
Medical colleges	542

healthcare organizations to proactively address potential security risks before they can be exploited by malicious actors. Furthermore, healthcare organizations are increasingly turning to third-party vendors and service providers to enhance their cybersecurity capabilities. Managed security service providers (MSSPs), for example, offer specialized expertise and resources to help healthcare organizations detect, prevent, and respond to cyber threats. By outsourcing certain aspects of their cybersecurity operations, healthcare organizations can leverage the knowledge and experience of external experts, allowing them to focus on their core mission of delivering high-quality patient care. Despite these efforts, healthcare organizations continue to face significant challenges in protecting patient data from cyber threats.

The healthcare sector remains a prime target for cybercriminals seeking to exploit vulnerabilities in outdated systems, insecure networks, and inadequate security controls. Ransomware attacks, in particular, have emerged as a major threat to healthcare organizations, disrupting operations, compromising patient care, and causing significant financial and reputational damage. In response to the growing threat of ransomware attacks, healthcare organizations are taking proactive steps to strengthen their defences and improve their resilience to cyber threats. This includes implementing multi-layered security controls, such as firewalls, intrusion detection systems, and endpoint security solutions, to detect and block malicious activity. Regular data backups and disaster recovery plans are also essential to ensure that healthcare organizations can recover quickly and effectively from ransomware attacks, minimizing the impact on patient care and organizational operations.

Furthermore, healthcare organizations are collaborating with government agencies, industry partners, and cybersecurity experts to share threat intelligence, best practices, and resources to enhance their collective cybersecurity posture. Information sharing platforms, such as the Healthcare Information Sharing and Analysis Centre (H-ISAC), facilitate collaboration and coordination among healthcare organizations, enabling them to stay ahead of emerging threats and vulnerabilities. Data security and privacy are critical priorities for healthcare organizations in India, given the sensitive nature of the information they handle. By implementing robust technical solutions, adopting best practices, and fostering a culture of cybersecurity awareness, healthcare organizations can mitigate the risks posed by cyber threats and protect patient data from unauthorized access and breaches. However, achieving effective cybersecurity requires a coordinated and collaborative effort involving healthcare providers, government agencies, industry partners, and cybersecurity experts working together to address the evolving challenges and threats facing the healthcare sector.

The Table 2.5 outlines key measures related to regulatory framework, technological safeguards, and organizational vigilance in ensuring data security and privacy in healthcare systems in India. Healthcare management

Table 2.5 Key measures for healthcare data security

Measure	*Key value*
Regulatory framework	Compliance with domestic laws Adherence to international standards Implementation of guidelines Impending Personal Data Protection Bill
Technological safeguards	Encryption for data security Access controls for data access Deployment of cybersecurity solutions Regular software updates
Organizational vigilance	Training for staff awareness Development of data breach response plans Collaboration with vendors Participation in information sharing

in India is a multi-faceted task that involves overseeing a range of activities to ensure the smooth delivery of healthcare services. With a population of over

1.3 billion people, this role is especially crucial. Yet, it comes with its fair share of challenges and opportunities. One of the biggest hurdles in healthcare management is the inadequacy of infrastructure. Shortages of hospitals, medical professionals, and essential resources are widespread, particularly in rural areas where access to healthcare is limited.

Despite increased healthcare spending, there are still disparities in how funds are distributed, leading to unequal access to quality care across different socioeconomic groups. To address these challenges, collaboration among various stakeholders is key. This includes government agencies, healthcare providers, insurers, pharmaceutical companies, and non-profit organizations, all of whom play a role in shaping healthcare policies and practices. Embracing technology and efficient data management practices is crucial for improving healthcare accessibility. Telemedicine, for example, can help bridge gaps in healthcare access, especially in remote areas. Precision medicine, which tailors treatments based on genetic data, holds great promise in revolutionizing healthcare management by making therapies more targeted and effective. However, to truly advance healthcare management in India, capacity-building initiatives are essential. This means investing in infrastructure, providing management training to healthcare professionals, and fostering a culture of continuous learning and innovation. Collaboration between healthcare entities, academic institutions, and industry stakeholders can facilitate the exchange of knowledge and best practices, leading to transformative changes in healthcare delivery.

Policy reforms are also necessary to create an environment conducive to effective healthcare management. This involves reviewing and updating regulations, promoting innovation and research, and ensuring that healthcare services are accessible to all segments of society. While healthcare management in India faces numerous challenges, strategic planning, efficient operations, and effective leadership can help overcome them. However, sustainable progress requires investment in infrastructure, addressing workforce shortages, and enacting policies that promote innovation and equitable access to healthcare services.

2.4 CASE STUDIES

Let's take a closer look at some real-life stories that showcase the incredible impact of healthcare initiatives in India. In one inspiring case, a non-profit organization recognized the unique healthcare needs of a low-income urban neighbourhood. To address these challenges, they trained community members to become health workers. These dedicated individuals delivered essential healthcare services and education right within their communities, promoting preventive care and empowering residents to prioritize their health. Similarly, a ground-breaking public–private partnership took aim at the growing cancer burden in India. By combining resources from both sectors, specialized cancer treatment centres were established in underserved regions. This innovative collaboration not only expanded access to critical treatments but also facilitated early detection and management of the disease, ultimately improving outcomes for patients and families.

Mobile health clinics also emerged as a game-changer in remote villages, bringing primary healthcare services directly to those in need. Staffed with medical professionals and equipped with essential supplies, these clinics tackled geographical barriers and provided much-needed care, all while empowering communities with health education and preventive measures. On the policy front, a government initiative focused on expanding health insurance coverage to marginalized populations proved transformative.

In a remote village in the Indian state of Rajasthan, a pioneering healthcare initiative has transformed the lives of thousands. The Aravind Eye Care System, inspired by the late Dr. G. Venkataswamy, has revolutionized the treatment of cataracts and other eye ailments. Through a network of outreach programs and specialized hospitals, Aravind provides high-quality, affordable eye care to rural communities. By streamlining processes and leveraging economies of scale, they perform thousands of sight-restoring surgeries each year, demonstrating the potential of innovative healthcare delivery models to address pressing public health needs in resource-constrained settings.

In the southern state of Kerala, the Kudumbashree Mission has empowered women to take charge of their communities' health. Through a network of community-based organizations, Kudumbashree trains women as health workers and provides them with essential knowledge and resources to deliver basic healthcare services at the grassroots level. These "Kudumbashree health teams" offer preventive care, health education, and support for maternal and child health, significantly improving health outcomes in underserved rural areas. This grassroots approach to healthcare delivery serves as a model for community empowerment and decentralized healthcare management across India. In the urban sprawl of Mumbai, the Dabbawalas have emerged as unlikely heroes in the fight against malnutrition among schoolchildren. These lunchbox deliverymen, known for their efficient and reliable lunch delivery service, have partnered with local NGOs to launch the "Roti Bank" initiative. Through this program, leftover food from corporate offices is collected, sorted, and redistributed to underprivileged schoolchildren, ensuring they receive nutritious meals every day. The Dabbawalas' logistical prowess and commitment to social responsibility demonstrate the potential for innovative partnerships to address complex healthcare challenges, such as malnutrition and food insecurity, in densely populated urban areas. By offering financial protection against medical expenses, this program ensured that individuals and families could seek healthcare services without fear of financial hardship, leading to improved health outcomes and a more equitable healthcare system. In another success story, a maternal and child health program implemented in collaboration with local NGOs made significant strides in reducing mortality rates in rural areas. Through comprehensive healthcare services and community engagement efforts, this program not only improved health outcomes but also empowered communities to take ownership of their well-being.

These case studies highlight the diverse strategies and partnerships driving healthcare transformation in India. From leveraging technology to fostering community empowerment and expanding health insurance coverage, these initiatives exemplify the innovation and dedication shaping India's healthcare landscape. By embracing a holistic approach and prioritizing equity, accessibility, and quality of care, India is making significant strides towards achieving its healthcare goals and ensuring the well-being of its population.

Basic Healthcare Services (BHS) – A Primary Healthcare Organization in India: BHS, operating in Udaipur, Rajasthan, exemplifies the endeavour to establish a sustainable delivery model for primary healthcare services. This case study underscores the hurdles encountered in setting up and sustaining a privately-led primary healthcare organization in India, emphasizing the imperative of community engagement and trust-building in the context of allopathic healthcare. It also underscores the criticality of devising systems tailored to address community health needs while ensuring financial viability for organizational sustainability. In the state of Tamil Nadu, the "104 Health Information Helpline" has emerged as a lifeline for millions of citizens seeking medical advice and assistance. This toll-free helpline provides round-the-clock access to trained healthcare professionals who offer guidance on health issues, medication, and treatment options.

By leveraging technology to bridge the gap between patients and healthcare providers, the 104 Helpline has become an invaluable resource, particularly for rural and marginalized communities with limited access to healthcare facilities. Its success underscores the importance of telemedicine and remote healthcare services in extending quality healthcare to all corners of India. In the northeastern state of Assam, the "Arogya Nidhi" initiative has transformed the landscape of healthcare financing for low-income families. Administered by the state government in partnership with local healthcare providers, Arogya Nidhi offers subsidized medical treatment and financial assistance to individuals living below the poverty line.

By reducing the financial burden of healthcare expenses, this program has improved access to essential medical services and reduced health disparities among vulnerable populations. Arogya Nidhi serves as a beacon of hope for those grappling with the high costs of healthcare, demonstrating the impact of targeted financial assistance in ensuring equitable access to healthcare for all. In the bustling metropolis of Delhi, the "Mohalla Clinics" initiative has revolutionized primary healthcare delivery for urban slum dwellers. These neighbourhood clinics, strategically located in densely populated areas, offer free or low-cost healthcare services, including consultations, diagnostics, and medications. Staffed by qualified medical professionals, Mohalla Clinics provide timely and affordable healthcare to marginalized communities, reducing reliance on overcrowded hospitals and improving health outcomes for residents. This grassroots approach to healthcare delivery has garnered widespread acclaim for its effectiveness in addressing the healthcare needs of urban poor populations, serving as a blueprint for similar initiatives in other urban centres across India.

2.5 TRENDS AND KEY CHALLENGES

The future of healthcare in India is poised for significant transformation, driven by a convergence of technological advancements, demographic shifts, and evolving healthcare delivery models. Several key trends are expected to shape the landscape of healthcare in India in the coming years, each with profound implications for patients, providers, and policymakers alike. One prominent future trend in Indian healthcare is the widespread adoption of telemedicine and digital health solutions. With the increasing penetration of smartphones and internet connectivity, telemedicine platforms are poised to revolutionize the delivery of healthcare services, especially in remote and underserved areas. These platforms enable patients to consult with healthcare providers remotely, access medical advice, and receive diagnostic evaluations and treatment recommendations without the need for in-person visits. Moreover, digital health solutions such as mobile health apps, wearable devices, and remote monitoring technologies empower individuals to proactively manage their health, track vital signs, and adhere to treatment plans, thereby promoting preventive care and early intervention. Another future trend set to reshape healthcare in India is the rise of precision medicine and personalized healthcare.

Advances in genomics, molecular diagnostics, and data analytics are enabling healthcare providers to tailor medical treatments and interventions to individual patient characteristics, including genetic makeup, lifestyle factors, and environmental influences. By leveraging insights derived from big data and AI, precision medicine holds the promise of optimizing treatment outcomes, minimizing adverse effects, and enhancing the overall efficacy of healthcare interventions.

This paradigm shift towards personalized healthcare represents a departure from the traditional one-size-fits-all approach, offering patients more targeted and effective treatment options based on their unique biological profiles. As we look ahead to the future of healthcare in India, it's not just about medical advancements and policy changes; it's about the stories of the people whose lives are intricately woven into the fabric of the healthcare system, discussed in Table 2.6.

Furthermore, the future of healthcare in India is characterized by the increasing integration of AI and machine learning (ML) technologies into clinical practice. AI-powered algorithms are revolutionizing medical diagnostics, imaging interpretation, drug discovery, and treatment planning, augmenting the capabilities of healthcare providers and improving patient outcomes. For example, AI-driven diagnostic tools can analyse medical images such as X-rays, MRIs, and CT scans to detect abnormalities and assist radiologists in making accurate diagnoses. Similarly, AI-powered predictive analytics models can identify high-risk patient populations, forecast disease outbreaks, and optimize resource allocation, thereby enhancing the efficiency and effectiveness of healthcare delivery.

Additionally, the future of healthcare in India is marked by a shift towards value-based care and outcome-driven reimbursement models. Traditionally, the Indian healthcare system has been characterized by a fee-for-service payment model, incentivizing volume over value and leading to fragmented care and inefficiencies. However, there is growing recognition of the need to transition towards value-based care models that prioritize quality, efficiency,

Table 2.6 Future trends in Indian healthcare

Trend	*Description*
Telemedicine Adoption	Increasing adoption of telemedicine platforms for remote consultations and healthcare delivery.
Artificial Intelligence in Healthcare	Integration of AI algorithms for diagnostics, personalized medicine, and predictive analytics.
Wearable Health Tech	Growth of wearable devices for real-time health monitoring and data collection.
Health Data Analytics	Utilization of big data analytics for population health management and predictive modelling.
Virtual Reality in Healthcare	Use of VR technology for medical training, pain management, and therapeutic interventions.

and patient outcomes. Value-based care emphasizes preventive care, care coordination, and patient engagement, aiming to improve health outcomes while reducing costs and enhancing patient satisfaction. By aligning incentives with quality metrics and patient outcomes, value-based care models incentivize healthcare providers to deliver high-quality, cost-effective care that addresses the holistic needs of patients [12].

The future of healthcare in India is shaped by a confluence of technological innovation, patient-centric care models, and shifting healthcare paradigms. The widespread adoption of telemedicine and digital health solutions, the emergence of precision medicine and personalized healthcare, the integration of AI and ML technologies into clinical practice, and the transition towards value-based care models are all poised to revolutionize the delivery of healthcare services, improve patient outcomes, and enhance the overall quality and accessibility of healthcare in India. As stakeholders across the healthcare ecosystem embrace these future trends and harness the transformative power of technology and innovation, they have the opportunity to usher in a new era of healthcare that is more efficient, effective, and equitable for all [14].

The healthcare sector in India grapples with a multitude of challenges that hinder the delivery of quality healthcare services and impede efforts to improve population health outcomes. One of the most pressing challenges is the limited access to healthcare services, particularly in rural and remote areas where healthcare infrastructure is inadequate. Millions of Indians living in these regions face significant barriers to accessing essential medical care, including long distances to healthcare facilities, transportation challenges, and geographical remoteness.

The lack of healthcare facilities, such as hospitals, clinics, and primary health centres, exacerbates disparities in access to healthcare services, perpetuating inequities in health outcomes across different segments of the population. Compounding the issue of limited access is the severe shortage of healthcare providers in India. The country faces an acute deficit of doctors, nurses, and allied healthcare professionals, with the doctor-to-population ratio falling well below the recommended standards set by international health organizations. This shortage of healthcare professionals strains the existing healthcare infrastructure, leading to overcrowded facilities, long waiting times for appointments, and compromised quality of care. Moreover, the unequal distribution of healthcare professionals, with a concentration in urban areas, further exacerbates disparities in access to healthcare services between rural and urban populations, leaving millions underserved and vulnerable to health risks (Table 2.7).

In addition to workforce shortages, infrastructure deficiencies pose significant challenges to the delivery of healthcare services in India. Many healthcare facilities across the country lack essential amenities and resources, including clean water, sanitation, electricity, and medical equipment. Outdated infrastructure and insufficient hospital beds further strain the

Table 2.7 Challenges in Indian healthcare

Challenge	*Description*
Healthcare Infrastructure	Inadequate infrastructure, including hospitals, medical personnel, and resources, especially in rural areas.
Healthcare Access	Disparities in healthcare access based on geographical location, socioeconomic status, and gender.
Healthcare Affordability	Rising healthcare costs and lack of financial protection, leading to financial burden on patients.
Healthcare Quality	Variations in quality of care, patient safety concerns, and inadequate regulation of healthcare facilities.
Healthcare Information Security	Data privacy concerns, cybersecurity threats, and challenges in maintaining the confidentiality of health records.

healthcare system's capacity to meet the growing demand for medical care. Inadequate infrastructure not only compromises the quality of care provided but also undermines efforts to respond effectively to public health emergencies and disease outbreaks. Addressing infrastructure gaps and investing in modern healthcare facilities are essential steps towards strengthening the resilience and capacity of the healthcare system to meet the evolving needs of the population. Furthermore, healthcare financing remains a critical challenge in India, with high out-of-pocket expenditures and limited access to health insurance coverage.

Many Indians face significant financial barriers to accessing healthcare services, particularly those belonging to low-income and marginalized communities. The high cost of medical care often forces families to choose between seeking treatment and meeting other basic needs, leading to delayed care-seeking behaviour and adverse health outcomes. Moreover, the lack of comprehensive health insurance coverage leaves millions of Indians vulnerable to catastrophic healthcare expenses, pushing them further into poverty and exacerbating socioeconomic disparities in health access and outcomes. In contemplating the future of healthcare in India, it's essential to recognize both the strides made and the hurdles that lie ahead. The trajectory appears promising, with advancements in medical technology, increased access to healthcare services, and growing awareness of preventive care.

However, amidst these positives, significant challenges persist, demanding innovative solutions and concerted efforts from various stakeholders. One prominent trend shaping the future of healthcare in India is the increasing adoption of telemedicine and digital health solutions. With the widespread availability of smartphones and internet connectivity, telemedicine offers a convenient means of accessing healthcare services, particularly for individuals in remote or underserved areas [18].

Virtual consultations, remote monitoring, and mobile health applications are becoming more prevalent, empowering patients to manage their health proactively and facilitating timely interventions. This trend is poised to bridge the gap in healthcare accessibility, especially in regions facing shortages of healthcare professionals and infrastructure. Another notable trend is the rise of personalized medicine and precision healthcare. Advances in genomics, data analytics, and AI are enabling healthcare providers to tailor treatment plans to individual patients' genetic makeup, lifestyle factors, and medical history. Precision medicine holds the promise of more effective therapies, reduced adverse reactions, and improved patient outcomes. As India continues to invest in genomic research and healthcare technology, personalized medicine is likely to become increasingly integrated into clinical practice, revolutionizing disease management and preventive care strategies. Furthermore, the growing emphasis on preventive healthcare and wellness is reshaping the healthcare landscape. Recognizing the importance of early detection and lifestyle modifications in mitigating disease risks, there is a shift towards holistic approaches to health promotion and disease prevention. Government initiatives, corporate wellness programs, and community health campaigns are focusing on raising awareness about healthy living, encouraging regular screenings, and promoting vaccination drives.

By prioritizing preventive measures, India can alleviate the burden of chronic diseases, enhance population health, and reduce healthcare costs in the long run. Despite these promising trends, several challenges loom large on the horizon, threatening to undermine the progress of India's healthcare system. One of the most pressing challenges is the persistent disparity in healthcare access and quality between urban and rural areas. While urban centres boast state-of-the-art medical facilities and specialist care, rural regions often grapple with inadequate infrastructure, healthcare workforce shortages, and limited medical resources. Addressing this urban-rural divide requires targeted investments in rural healthcare infrastructure, incentivizing healthcare professionals to serve in underserved areas, and leveraging technology to deliver telemedicine and mobile health services to remote communities. Another significant challenge is the burden of non-communicable diseases (NCDs) such as diabetes, cardiovascular diseases, and cancer, which are escalating rapidly across India. Factors such as sedentary lifestyles, unhealthy diets, tobacco use, and air pollution contribute to the rise of NCDs, placing immense strain on healthcare systems and economies.

Tackling the NCD epidemic necessitates a multi-faceted approach encompassing health education, lifestyle interventions, early detection programs, and robust healthcare financing mechanisms. Additionally, there is a need for stronger regulation and enforcement of policies targeting tobacco control, food safety, and environmental pollution to curb the prevalence of NCDs and promote public health. Furthermore, India faces ongoing challenges in ensuring healthcare affordability and financial protection for its citizens. Out-of-pocket healthcare expenditures push millions of people into poverty each year, as they struggle to afford essential medical services and medications. While government-sponsored health insurance schemes such as Ayushman Bharat have made significant strides in expanding insurance coverage and reducing catastrophic health expenditures, there is a need for continued efforts to strengthen the healthcare financing system, improve insurance penetration, and enhance the quality and accessibility of healthcare services for all socioeconomic groups.

Moreover, the COVID-19 pandemic has underscored the importance of pandemic preparedness, resilient healthcare systems, and effective crisis management strategies. The pandemic has stretched healthcare infrastructure to its limits, exposing gaps in healthcare delivery, supply chain disruptions, and disparities in access to testing and treatment. Looking ahead, India must invest in strengthening its public health infrastructure, enhancing disease surveillance capabilities, and bolstering healthcare workforce capacity to respond effectively to future health emergencies. Additionally, there is a need for greater collaboration and coordination among government agencies, healthcare providers, private sector entities, and civil society organizations to mount a unified response to pandemics and other health crises. Future of healthcare in India is marked by both promise and challenges.

While advancements in technology, personalized medicine, and preventive healthcare hold immense potential for improving health outcomes and enhancing healthcare delivery,

Alongside the existing challenges, India's healthcare sector suffers from various public health issues, including infectious diseases, non-communicable diseases (NCDs), and emerging health threats. Diseases like tuberculosis, malaria, and HIV/AIDS are still major concerns, especially among vulnerable groups, leading to high rates of illness and death. Antimicrobial resistance adds another layer of complexity to treating these diseases, requiring better surveillance and prevention efforts. Additionally, NCDs like heart disease, diabetes, cancer, and respiratory illnesses are becoming more prevalent due to urbanization, sedentary lifestyles, and poor diets. These diseases strain healthcare resources and call for strategies focusing on prevention, early detection, and management, along with addressing underlying social factors. India also faces emerging health threats like dengue, chikungunya, and the Zika virus, and environmental hazards such as air and water pollution. These challenges demand robust surveillance, emergency response, and collaboration across sectors to safeguard public health. Governance and regulatory issues further impact healthcare quality, safety, and efficiency.

Inconsistent enforcement, lack of guidelines, and fragmented delivery systems result in variations in care and outcomes. Strengthening governance, regulatory oversight, and quality assurance is crucial for patient safety and trust. The healthcare sector also grapples with gaps in health information systems and data collection, hindering evidence-based decision-making and policy formulation. Limited data availability, poor quality, and inadequate infrastructure impede tracking health trends and evaluating programs. Strengthening data collection, analysis capabilities, and sharing mechanisms is vital for informed policymaking and program responses. Addressing these challenges requires collaboration among governments, policymakers, healthcare providers, and civil society. Investments in infrastructure, workforce development, financing, and public health interventions are key to achieving universal health coverage and improving health outcomes for all. By prioritizing these efforts, India can build a more resilient and effective healthcare system that ensures health equity and well-being for its citizens.

The future of healthcare in India is on the brink of a significant transformation, largely propelled by the swift integration of cutting-edge technologies and the growing acceptance of digital health solutions among both healthcare providers and patients. This transformative journey is poised to introduce several emerging trends that carry profound implications for the delivery and accessibility of healthcare services across the nation. One of the noteworthy trends anticipated to gain traction is the widespread adoption of telemedicine services.

Telemedicine has surged in relevance, especially amidst the backdrop of the COVID-19 pandemic, and is projected to sustain its upward trajectory as more individuals opt for remote consultations. Government initiatives and regulatory support are expected to further fuel the expansion of telemedicine services, particularly in remote and underserved areas where access to traditional healthcare facilities remains limited. Simultaneously, the rise of AI is set to revolutionize healthcare delivery in India. AI-powered tools, such as diagnostic algorithms and predictive analytics, hold immense promise in enhancing disease detection, treatment outcomes, and overall patient care. These advancements are poised to usher in an era of precision medicine, where healthcare interventions can be tailored to individual patient profiles, leading to enhanced efficacy and personalized care delivery.

2.6 CONCLUSIONS

From remote patient monitoring to chronic disease management, these innovative solutions hold the potential to address longstanding healthcare challenges and bridge existing gaps in accessibility and delivery. However, amid the promising landscape of technological advancements, several challenges loom large and demand careful consideration and strategic interventions. Chief among these challenges is the lack of robust healthcare infrastructure and interoperability barriers, which impede the seamless exchange of health data among different provider platforms. Furthermore, cybersecurity and data privacy concerns emerge as significant apprehensions, with the healthcare sector increasingly vulnerable to cyber-attacks and breaches. The absence of adequate incentives for providers to digitize healthcare services, coupled with suboptimal internet and smartphone penetration rates, poses additional obstacles to the widespread adoption of digital health solutions.

Moreover, addressing physician attitudes and preferences towards technology adoption, as well as patient preferences for physical interaction, emerge as critical areas requiring attention and intervention. Overcoming these challenges necessitates a comprehensive and collaborative approach involving policymakers, healthcare stakeholders, technology innovators, and regulatory bodies. By addressing these hurdles, India can unlock the full potential of technological advancements and digital health solutions, paving the way for a more accessible, equitable, and resilient healthcare system that meets the evolving needs of its diverse population.

REFERENCES

1. Patel, R., & Gupta, S.K. "Digital Health Adoption in India: Opportunities and Challenges" In *International Conference on Health Informatics (HEALTHINF)*, February 2023.
2. Sharma, A., & Singh, P. "Telemedicine Expansion in Rural India: A Case Study of Implementation Challenges" *Journal of Telemedicine and Telecare*, September 2021.
3. Das, S., & Banerjee, A "Challenges and Opportunities in Mobile Health Adoption in India" *International Journal of Mobile and Blended Learning*, November 2020.
4. Kumar, S., & Verma, R. "AI-Driven Healthcare Solutions for Disease Detection and Management in India" in *Proceedings of the IEEE International Conference on Artificial Intelligence in Medicine and Healthcare (AIMH)*, July 2022.
5. Singh, R., & Kumar, A "Wearable Healthcare Gadgets: Implications for Preventive Medicine in India" In *International Conference on Wearable Sensors and Technologies (WST)*, April 2023.
6. Gupta, N., & Jain, S. "Digital Health Start-ups and Innovations: A Review of Emerging Trends in India" *Journal of Healthcare Innovation*, June 2021.
7. Mishra, V., & Sharma, D. "Cybersecurity and Data Privacy in Indian Healthcare: Current Challenges and Future Directions" *International Journal of Healthcare Information Management*, March 2022.
8. Kumar, A., & Gupta, S. "Government Initiatives and Policy Support for Digital Health in India: A Review of Recent Developments" In *Proceedings of the International Conference on e-Governance and Digital Health (eGADH)*, October 2024.
9. Patel, H., & Shah, K. "Robotics in Healthcare: Applications and Challenges in the Indian Context" *Advances in Robotics for Healthcare*, December 2020.
10. Verma, S., & Singh, M. "AI, Biotech, and Telemedicine: Shaping the Future of Indian Healthcare" In *Conference Proceedings of the International Conference on Biomedical Engineering and Technology (ICBET)*, August 2023.
11. Jain, R., & Sharma, G. "Future Trends in Indian Healthcare: A Perspective on Tech-Enabled Care Delivery" *Journal of Healthcare Technology and Innovation*, October 2021.
12. Gupta, A., & Kumar, R. *"Infrastructure Barriers in Indian Healthcare: A Case Study of Clinical Records Management" In Conference Paper Presented at the International Conference on Health Information Systems (HIS)*, May 2022.
13. Singh, N., & Mishra, A. "Lack of Interoperability Among Provider Platforms: Implications for Indian Healthcare" *Journal of Health Informatics Research*, February 2023.

14. Sharma, S., & Verma, P. "Cybersecurity Threats in Indian Healthcare: Addressing the Vulnerabilities" In *Conference Proceedings of the International Conference on Cybersecurity in Healthcare (CCH)*, September 2021.
15. Kumar, V., & Gupta, D. "Dearth of Provider Incentives for Digitization in Indian Healthcare: A Survey of Challenges and Opportunities" *Journal of Healthcare Management*, January 2024.
16. Patel, M., & Jain, A. "Physician Attitudes and Preferences Towards Technology Adoption in Indian Healthcare" In *Proceedings of the International Conference on Health Psychology and Behavioral Medicine (HPBM)*, November 2020.
17. Anisha, P. R., Reddy, C. K. K., Hanafah, M. M., Murthy, B. R., Mohana, R. M., & Pragathi, Y. V. S. S. "An Intelligent Deep Feature Based Metabolism Syndrome Prediction System for Sleep Disorder Diseases" *Multimedia Tools and Applications*, 83, 51267–51290, 2023. https://doi.org/10.1007/s11042-023-17296-4
18. Yashashwini Reddy, P., Kishor Kumar Reddy, C., & Sithole, N. T. "A Study and Analysis on Diagnosis of Melanoma Cancer with Deep Learning: A Case Study" In A. Hassan, V. K. Prasad, P. Bhattacharya, & P. Dutta (Eds.), *Federated Learning and AI for Healthcare, 5.0*, December 2023, pp:203–218, IGI Global. DOI: 10.4018/979-8-3693-1082-3.ch011
19. Ganga, P. R., Kishor Kumar Reddy, C., Anisha, P. R., & Doss, S., "Validation & Diagnosis of Cystic Fibrosis Prognosis Using Gradient Boost Decision Trees" In *2023 International Conference on Artificial Intelligence for Innovations in Healthcare Industries (ICAIIHI)*, 29–30 December 2023, Raipur, India, DOI: 10.1109/ICAIIHI57871.2023.10489654
20. Mishra, S., & Das, A. "The Role of AI in Chronic Disease Management: Perspectives from Indian Healthcare" *Journal of Artificial Intelligence in Medicine*, April 2023.
21. Sharma, R., & Gupta, V. "Remote Patient Monitoring in Indian Healthcare: Opportunities and Challenges" In *Conference Proceedings of the International Symposium on Telehealth and Remote Monitoring (ISTeRM)*, August 2021.
22. Kishor Kumar Reddy, C. "An Efficient Early Diagnosis and Healthcare Monitoring System for Mental Disorder Using Machine Learning" In B. K. Mishra (Ed.), *Sustainable Science and Intelligent Technologies for Societal Development*, 2023, pp. 22–43, IGI Global.

Chapter 3

Role of artificial intelligence and machine learning in evaluation and assessment

Minirani S, Adithya P, and Minal Shukla

3.1 AI IN EDUCATION

3.1.1 Introduction

In the contemporary landscape of education, the integration of Artificial Intelligence (AI) stands as a pivotal milestone, revolutionizing traditional paradigms of teaching and learning. Harnessing the power of advanced algorithms and machine learning, AI has emerged as a dynamic force capable of personalizing educational experiences on an unprecedented scale. Through intelligent tutoring systems that adapt to individual learning paces, and intricate analytics providing educators with deep insights, AI has the potential to transform classrooms into tailored, interactive environments. As we step into the domain of AI in education, we will examine its diverse applications, discover its countless advantages, tackle the challenges it presents, and envision the profound impact it will undoubtedly have on the future of learning.

AI has integrated into numerous dimensions of education, catalysing transformative shifts in how students learn, educators teach, and institutions operate. This technological revolution has ushered in a new era of personalized learning experiences, with intelligent tutoring systems at the forefront. Powered by advanced algorithms, these systems adeptly assess a student's unique learning style, pace, strengths, and areas requiring improvement. The result is a tailored curriculum that addresses individual needs, ensuring targeted support precisely where it is needed. What's more, AI-driven tutoring systems are available around the clock, offering on-demand assistance regardless of time or location. This real-time accessibility not only addresses the immediate needs of students but also enhances their overall learning experience.

Moreover, AI in education extends its influence to the critical domains of grading and assessment [1]. Here, AI's impact is most pronounced in the realm of Automated Essay Scoring, where algorithms have been refined to assess essays with unparalleled precision. This development accelerates the grading process while ensuring assessments are conducted objectively and

 DOI: 10.1201/9781003643166-3

consistently. Objective assessments, such as multiple-choice and fill-in-the-blank questions, are also swiftly and accurately graded through AI-powered systems. By automating these tasks, educators gain more time to focus on the nuanced aspects of teaching, fostering a more enriching educational environment.

In addition to personalized learning and streamlined assessment, AI lends its prowess to data analysis and predictive analytics, culminating in Early Warning Systems [2]. Leveraging comprehensive student data, AI identifies those at risk of academic challenges, such as falling behind or dropping out, enabling timely interventions. This proactive approach significantly bolsters the chances of student success. Learning Analytics, another facet of AI's impact, equips educators with deep insights gleaned from data analysis [3]. These insights encompass various aspects, including student engagement, learning patterns, and assessment performance. Empowered by this understanding, educators enhance their teaching approaches and develop curricula that are better aligned with individual needs and learning styles [4].

The influence of AI isn't confined to the classroom alone. It has also infiltrated the field of career counselling, transforming the guidance and resources accessible to individuals aspiring for satisfying and prosperous career paths. The analytical strength of AI becomes especially apparent in its capacity to analyse extensive data on industries, job markets, and skill prerequisites. This capability empowers AI-driven career platforms to offer personalized recommendations based on an individual's unique strengths, interests, and goals. Additionally, AI-powered assessments explore an individual's skills and personality traits thoroughly, matching them with appropriate career choices. The accessibility of AI-powered chatbots and virtual advisors ensures individuals have instant, around-the-clock support, especially crucial for those facing critical career decisions. Additionally, AI tracks labour market trends, keeping individuals updated on emerging job opportunities and skill requirements. With these analytical tools, career counsellors can provide more precise and effective guidance, ultimately empowering individuals to make informed and strategic career choices.

In the broader landscape of education, Content Recommendation Systems have undergone a revolution, thanks to AI. These systems, bolstered by advanced algorithms, provide highly personalized content suggestions tailored to individual students' interests and learning levels. By analysing extensive datasets and understanding user behaviour, AI discerns specific preferences, strengths, and areas requiring improvement. This enables the recommendation system to curate a diverse array of educational materials, from articles and videos to interactive exercises, ensuring each student receives content aligned with their unique learning journey. This personalization not only fosters a more engaging and relevant learning experience but also maximizes the efficacy of educational resources, ultimately enhancing the overall quality of education.

AI has also become a cornerstone in the realm of language learning, revolutionizing how individuals acquire new languages. Through advanced algorithms and machine learning, AI offers invaluable assistance in several critical aspects of language acquisition. One of the most prominent features is its capability to facilitate language translation and pronunciation. AI-powered tools adeptly translate text, aiding learners in comprehending and communicating in different languages. Moreover, they provide real-time pronunciation guidance, ensuring learners develop accurate speech patterns. Additionally, AI-driven systems offer instantaneous language correction, offering immediate feedback to enhance accuracy and fluency. Furthermore, language-learning apps powered by AI have taken personalized education to new heights. These applications adapt lessons to cater to individual proficiency levels and learning styles, providing tailored content and exercises. This adaptive approach maximizes the effectiveness of language learning, allowing learners to progress at their own pace. Overall, AI's integration in language learning not only enhances accessibility and convenience but also facilitates more immersive, efficient, and personalized language acquisition experiences for learners worldwide.

AI, plays a pivotal role in transforming personalized learning, customizing educational experiences to cater to the distinct needs of every student. Adaptive Learning Platforms, driven by AI algorithms, play a pivotal role in this transformation. These platforms meticulously analyse students' learning patterns, taking into account their pace, strengths, and preferences. By doing so, they dynamically adjust the curriculum, ensuring that content is delivered in a manner that maximizes comprehension and retention. Additionally, AI-powered systems excel at offering Personalized Recommendations. By continuously monitoring a student's progress and performance, these platforms suggest specific learning resources, ranging from videos and articles to exercises. This curated approach not only keeps learners engaged but also optimizes their learning journey, aligning materials with their current level of comprehension and areas in need of improvement. Through these sophisticated applications of AI, personalized learning transcends a one-size-fits-all approach, ushering in a new era of tailored, effective, and engaging education.

AI is a game-changer in content creation and curation within education. Through advanced algorithms and machine learning, AI has the capability to generate educational content like quizzes, practice exercises, and even entire textbooks, all tailored to specific learning objectives. This not only saves educators valuable time but also ensures that the content is aligned precisely with the curriculum's goals. Additionally, AI-powered algorithms excel at content curation, helping educators sift through the vast expanse of online resources. By analysing relevance, credibility, and effectiveness, AI assists in the discovery and selection of high-quality educational materials, ensuring that students have access to the most valuable and pertinent resources available. This transformative capability of AI in content creation and curation elevates the quality and accessibility of educational materials, enhancing the overall learning experience for students.

AI is a cornerstone of administrative efficiency in educational institutions. It introduces a new level of automation and optimization in various crucial areas. For instance, in admissions and enrolment, AI systems efficiently handle tasks that typically consume significant time and resources. This includes processing applications, managing enrolments, and creating schedules, allowing staff to focus on higher-level decision-making. Furthermore, AI's analytical capabilities prove invaluable in resource allocation. By crunching large volumes of data, AI helps institutions make informed decisions regarding staffing, budget allocation, and facilities management. This ensures that resources are allocated where they are most needed, ultimately leading to a more efficient and cost-effective operation.

The integration of AI into education has undeniably ushered in a new era of dynamic, personalized, and efficient learning experiences. From intelligent tutoring systems that adapt to individual learning styles, to AI-driven content creation and curation, the impact of this technology is far-reaching. The transformative power of AI is most evident in its ability to assess and address the unique needs of each student, revolutionizing traditional teaching methods. Moreover, in the realm of assessments, AI's proficiency in automated grading ensures fairness and consistency, liberating educators to focus on fostering enriched learning environments. The implementation of Early Warning Systems and Learning Analytics, facilitated by AI, empowers educators with data-driven insights, enabling timely interventions and refined teaching strategies.

Beyond the classroom, AI extends its influence to career counselling, offering personalized guidance based on an individual's strengths and aspirations. This analytical prowess aids in making informed career decisions, aligning interests with opportunities in the job market. Additionally, AI optimizes administrative processes, from admissions to resource allocation, streamlining operations and ensuring resources are allocated where they are needed most.

Furthermore, AI's contribution to language learning, with tools for translation, pronunciation, and real-time correction, fosters a more immersive and efficient language acquisition process. This not only enhances accessibility but also accelerates proficiency, enabling learners to communicate with confidence in diverse linguistic settings.

In conclusion, the integration of AI in education signifies a monumental shift towards a more adaptive, inclusive, and data-informed learning environment. As AI continues to evolve, its potential to revolutionize education further is boundless. Through personalized learning experiences, streamlined administrative processes, and enhanced language acquisition, AI is poised to shape the future of education, unlocking new levels of potential and paving the way for a more equitable and effective educational landscape. With AI as a catalyst, the pursuit of knowledge becomes not only enriching but also uniquely tailored to the needs and aspirations of each individual learner.

3.1.2 EIDP: Proactive approach

In the realm of education, the quest for academic success and retention is a critical endeavour. As institutions strive to provide quality education, ensuring that students not only enrol but also persist and thrive is paramount. The advent of technology, particularly the integration of AI, has paved the way for a proactive approach towards early intervention and dropout prediction. This transformative shift in educational practices empowers educators to identify at-risk students and implement timely interventions, ultimately fostering a more supportive and inclusive learning environment.

Early Intervention and Data-Driven Practices (EIDP) are transformative approaches that empower educational institutions to proactively address the challenges in teaching-learning. Early intervention in education refers to the strategic efforts made by educators and institutions to identify students who may be at risk of academic challenges or potential dropout at an early stage. The goal is to provide targeted support and resources to address their specific needs before these challenges escalate, ensuring that students have every opportunity to succeed.

AI, with its capacity for data analysis and predictive analytics, stands as a formidable tool in early intervention strategies. By leveraging comprehensive student data, AI algorithms can detect patterns and trends indicative of potential academic struggles or disengagement. This data-driven approach allows educators to identify at-risk students in a timely and accurate manner, thus enabling proactive intervention measures. Predictive analytics, a key component of early intervention strategies, harnesses the power of AI to forecast student outcomes based on historical data, behaviour patterns, and various academic indicators. Through this method, educators gain valuable insights into the potential challenges a student may face, allowing for the implementation of tailored support measures. AI-enabled systems meticulously examine a range of factors that may serve as early warning signs of academic difficulties. These may include attendance records, grades, engagement levels, participation in extracurricular activities, and even social interactions. By analysing this wealth of information, AI can flag students who exhibit signs of struggling or disengagement, prompting educators to take action before the situation escalates [5].

One of the most profound advantages of leveraging AI for early intervention is the ability to provide highly personalized support to at-risk students. With detailed insights into each student's strengths, weaknesses, and learning styles, educators can craft interventions that are specifically tailored to address their unique needs. This individualized approach not only increases the likelihood of success but also fosters a sense of belonging and support, crucial factors in preventing dropout. Armed with the insights gleaned from

AI-driven early intervention strategies, educators can implement a range of targeted interventions. These may include additional tutoring, mentoring programs, counselling services, or modifications to the curriculum. By addressing the specific challenges faced by at-risk students, educators can create an environment that is conducive to their academic growth and success.

While the benefits of early intervention through AI are profound, it is essential to navigate the ethical considerations surrounding student privacy and consent. Institutions must establish transparent policies and protocols for data collection, storage, and usage. Clear communication with students and their families about the purpose and potential outcomes of data analysis is imperative to ensure trust and compliance.

In addition to early intervention, AI-driven dropout prediction models have emerged as a powerful tool in safeguarding students' educational journeys. These models employ sophisticated algorithms to analyse a wide range of factors, including academic performance, attendance, socio-economic background, and even extracurricular involvement. By synthesizing this information, AI can generate accurate predictions about which students may be most vulnerable to dropping out. Armed with the insights garnered from dropout prediction models, institutions can implement targeted strategies to mitigate the risk. These may include personalized counselling sessions, academic support programs, mentorship initiatives, and financial aid opportunities. By addressing the underlying factors that contribute to dropout risk, educators can create a more conducive environment for student success.

A paradigm shift in education has occurred with the proactive strategy of early intervention and dropout prediction, fuelled by AI. By utilizing AI's analytical capabilities, instructors may precisely identify at-risk children and provide individualized guidance. This improves academic performance while also fostering an inclusive and supportive culture within educational institutions. The ability for early intervention and dropout prediction will also advance with technology, ensuring that every student has the chance to flourish and achieve in their academic endeavours.

3.1.3 The challenge of dropout rates

The issue of high dropout rates in educational institutions is a persistent challenge that undermines the potential of countless students and poses a significant obstacle to achieving educational equity. Dropout rates not only hinder individual progress but also have broader societal and economic implications. To tackle this issue, education systems worldwide are increasingly turning to EIDP as a powerful tool to identify at-risk students and implement targeted interventions. In this section, we discuss some of the challenges posed by dropout rates and the pivotal role of EIDP in mitigating this issue.

High dropout rates have far-reaching consequences, affecting students, communities, and the economy. Students who leave school prematurely are often faced with limited job prospects, reduced earning potential, and a higher likelihood of engaging in risky behaviours. Moreover, dropout rates perpetuate cycles of poverty and inequality, further exacerbating social disparities.

Addressing the challenge of dropout rates in educational institutions is a complex endeavour fraught with numerous hurdles. One of the primary difficulties lies in identifying students who are at risk of dropping out. This task is multifaceted, considering factors such as academic performance, attendance, socio-economic background, and personal circumstances, all of which contribute to the risk profile. Without systematic tools and processes in place, educators may find themselves struggling to identify these students in a timely manner. Moreover, educational institutions often grapple with resource constraints that impede their ability to implement comprehensive dropout prevention programs. These limitations encompass financial constraints, understaffing, and a lack of adequate training for educators in effective intervention strategies. This dearth of resources further complicates efforts to proactively address dropout rates.

Additionally, the absence of data-driven strategies poses a significant challenge for many institutions. Without established practices for monitoring and analysing student performance and behaviour, educators may inadvertently overlook crucial indicators of potential dropout risk. This information vacuum hampers the institution's ability to intervene promptly and effectively (Figure 3.1).

Another critical factor in dropout prevention is parental or guardian involvement. Engaging parents and guardians in a collaborative effort is often instrumental in preventing dropout. However, various factors, such as parental work schedules, language barriers, and limited access to resources, can impede their ability to provide the necessary support.

Overcoming these challenges requires a multifaceted approach, encompassing the implementation of systematic tools for student identification, addressing resource limitations, establishing data-driven practices, and finding innovative ways to engage parents and guardians in the dropout prevention process. Only through concerted efforts and a commitment to proactive strategies can educational institutions hope to make substantial progress in reducing dropout rates and ensuring the success of all students.

By leveraging advanced technology and data analytics, EIDP offers a systematic framework for identifying at-risk students, tailoring interventions, and monitoring progress.

Through the employment of data analytics, EIDP conducts comprehensive assessments, delving into students' academic performance, attendance records, and behavioural patterns. This panoramic view equips educators with a nuanced understanding, enabling them to discern potential dropout risks at their earliest stages. Concurrently, predictive analytics, a cornerstone of EIDP, harnesses the power of advanced modelling techniques to forecast which students are most susceptible to facing dropout risk. This foresight

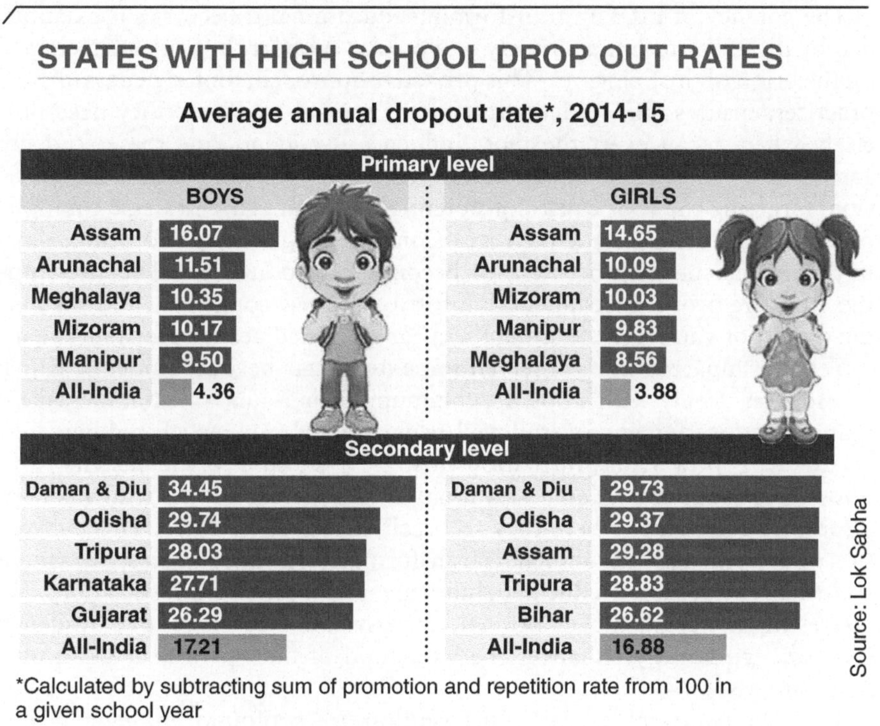

Figure 3.1 Dropout rates in India [6].

not only empowers educators with a proactive stance but also allows for the optimal allocation of resources, ensuring that interventions are targeted towards those who stand to benefit the most.

With the arsenal of data-driven insights at their disposal, educators can craft interventions that are tailored to address the specific challenges faced by at-risk students. These multifaceted interventions encompass an array of approaches, from personalized tutoring to mentorship programs, counselling services, and an assortment of academic support initiatives.

Moreover, EIDP facilitates ongoing progress monitoring, ensuring that interventions remain finely attuned to the evolving needs of each student. By continuously tracking academic performance, attendance, and levels of engagement, educators are poised to adapt their strategies as necessary, assuring that students remain on a trajectory towards sustained academic success. In this intricate interplay of data-driven practices and EIDP, educators are not only equipped with a powerful toolkit to combat high dropout rates but also empowered to foster a more inclusive, supportive educational environment. Ultimately, this concerted effort culminates in a tangible reduction in dropout rates, unlocking increased opportunities for students to realize their full academic potential and secure a brighter future.

The potency of EIDP in transforming educational trajectories is exemplified by its ability to initiate timely and precise support for students teetering on the brink of dropping out. This proactive approach, rooted in data-driven practices, ensures that students receive the critical assistance they need precisely when it can exert the most influence. By identifying these students early, before their challenges escalate, educators can intervene with precision, targeting the root causes of potential dropout. This strategic timing is pivotal in thwarting academic regression, enabling students to bridge gaps in their understanding before they become insurmountable. This pre-emptive measure proves instrumental in fortifying their confidence and sustaining their motivation, positioning them for sustained academic achievement.

Yet, the impact of early intervention extends far beyond academic gains. It resonates deeply with students, communicating a fundamental message – that their education and overall well-being are not only recognized but cherished. This powerful affirmation fosters a culture of inclusivity and unwavering support within educational institutions. It imbues students with a sense of belonging, where they feel seen, heard, and valued. This cultivation of a nurturing environment is transformative, as it creates a space where students are not merely educated, but truly nurtured. This holistic approach, addressing the academic, emotional, and psychological facets of a student's journey, significantly enhances the likelihood of their persistence and ultimate success.

The reverberations of early intervention are profound, influencing not only the individual students but also the broader educational ecosystem. It contributes to higher retention rates, as students are more likely to remain engaged and committed to their educational path. Moreover, it sets a powerful precedent, demonstrating that educational institutions are not merely reactive entities but proactive advocates for each student's potential. In essence, the power of early intervention is a testament to the transformative impact that timely and targeted support can have on the holistic development and success of every student. It represents an ethos of care, commitment, and belief in the potential of every learner, ultimately shaping a brighter, more inclusive future for education.

The issue of high dropout rates must be addressed by educational institutions all across the world. EIDP integration is a breakthrough strategy for tackling this complicated problem. EIDP enables educators to early identify at-risk kids, carry out focused interventions, and track progress by leveraging the power of data analytics and predictive modelling. With the help of these preventative measures, educational institutions may build a welcoming environment that not only discourages dropouts but also promotes inclusivity and academic achievement. The potential for EIDP to have a significant impact on dropout rates and academic outcomes is limitless as technology develops. We have the chance to open the door for a more inclusive and equitable educational environment with coordinated efforts and a dedication to data-driven practices.

3.2 APPLICATION OF AI IN EDUCATION SYSTEM

3.2.1 AI in early intervention

Embedded within the realm of AI-driven early intervention and dropout prediction is a comprehensive reservoir of data, painstakingly harvested from multifarious dimensions of a student's academic odyssey. This treasure trove encompasses a kaleidoscope of information, spanning from granular academic performance records to nuanced attendance patterns, encapsulating a mosaic of demographic particulars and subtle behavioural cues. Through an arduously detailed process of data collection and fastidious curation, AI systems lay down the groundwork, akin to a masterful artisan crafting a finely tuned instrument, all in the pursuit of distilling profoundly meaningful, prescient, and actionable insights for the betterment of a student's educational trajectory.

The initial stage of handling raw data, often fraught with inconsistencies, redundancies, and superfluous details, necessitates a meticulous refining process before it can be deemed suitable for rigorous analysis. This pivotal phase, aptly referred to as data preprocessing, entails a comprehensive endeavour to cleanse and reshape the data, a task of paramount importance to ensure its utmost quality and pertinent applicability. Through the judicious removal of extraneous elements and the harmonization of formats, AI systems engage in a transformative act, akin to skilled artisans sculpting raw material into a refined masterpiece. This dedicated endeavour readies the data for its subsequent journey through the analytical pipeline, setting the stage for more profound and insightful examinations.

Inside the complex web of data, there are specific details or factors, each with its own meaning. In the area of predicting students' needs and foreseeing dropouts, these important details include careful attendance records, thoughtful test scores, how students engage in class, and background information. AI, acting like a smart judge of this information, uses advanced methods to identify and separate these crucial details very precisely, similar to how musicians finely tune their instruments for harmony. The resulting organized information, now containing the most important parts, is ready for further analysis. This careful process makes the system very good at understanding and helping students. It greatly improves the learning experience for every student involved.

The choice of machine learning algorithms is contingent upon the intrinsic characteristics of the specific problem under consideration. Within the domain of early intervention and dropout prognosis, a discerning selection process is undertaken, wherein algorithms including but not limited to decision trees, support vector machines, logistic regression, and neural networks find common application. These algorithms, each distinguished by its unique mathematical underpinnings and computational intricacies, furnish a diverse array of analytical methodologies (Figure 3.2).

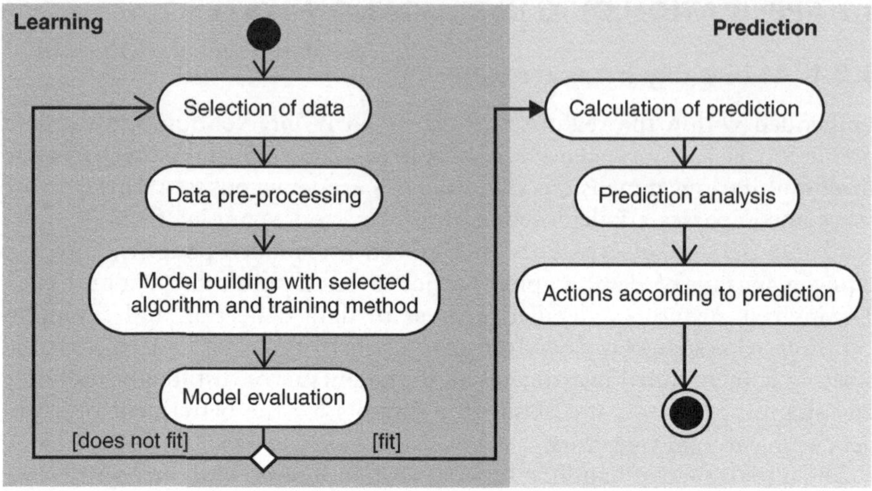

Figure 3.2 Basic working of ML and DL algorithms in Prediction [7].

Each algorithm possesses unique strengths that enhance predictive capabilities in complementary ways. Decision trees provide intuitive, interpretable models ideal for identifying hierarchical relationships within data. Support vector machines excel at delineating complex decision boundaries, and logistic regression offers a probabilistic framework for classification. In contrast, neural networks, with their adaptability and ability to model intricate non-linear relationships, represent deep learning at its best. By skilfully combining these algorithms, a sophisticated and multifaceted approach to prediction emerges. This strategic blend of their strengths significantly amplifies the effectiveness of predictive modelling efforts, showcasing the nuanced artistry in machine learning model selection, especially in fields like early intervention and dropout prediction.

In the training phase, the chosen algorithm delves into a vast dataset rich with annotated information. For dropout prediction, this dataset comprises extensive historical student records, juxtaposed with their academic outcomes – either successful graduation or premature discontinuation. Within this structured environment, the algorithm embarks on an iterative process of computational assimilation and understanding. Through this methodical exposure, the algorithm identifies discernible patterns and intricate relationships latent within the data. This learning process involves the algorithm internalizing relevant features and attributes, refining its internal parameters to approximate an optimal state. Consequently, the algorithm gains an enhanced capacity to generate precise and well-informed predictions regarding the likelihood of student dropout.

Following the culmination of the training phase, the model is subjected to a stringent evaluation process, a pivotal undertaking aimed at ascertaining its precision and dependability. This evaluative scrutiny is conducted leveraging an independent cohort of data, denominated as the testing dataset, which the model has not hitherto encountered. Through this deliberate segregation of data, an unexplored sample space is introduced, affording a measure of the model's adaptability and generalizability. The assessment hinges on the model's performance in the context of novel and previously unseen instances. This empirically grounded evaluation serves as a litmus test for the model's efficacy within authentic, real-world scenarios. It is a critical phase in affirming the model's proficiency to furnish reliable prognoses, thereby underpinning its utility in practical educational applications.

Endowed with a duly trained and rigorously validated model, the AI system stands prepared to undertake predictive operations predicated upon contemporaneous data streams. Within the ambit of early intervention, this signifies the analytical scrutiny of extant student data in order to expeditiously identify nascent indicators of potential risk. Such discernments, should they materialize, function as a clarion call, promptly apprising educators and administrative stakeholders of exigent circumstances, thereby instigating a proactive and pre-emptive response. This process epitomizes the real-time operationalization of the predictive framework, wherein the model's prescient capacities are deployed in an ongoing manner to scrutinize current data, thus providing timely insights into evolving circumstances. The ensuing alerts, grounded in empirical analysis, offer an invaluable toolset for educational practitioners and decision-makers, equipping them to undertake judicious interventions conducive to the amelioration of academic outcomes and the cultivation of a conducive learning environment.

AI systems, far from static entities, are conceived as entities characterized by perpetual learning and adaptive capacities. They are expressly engineered to accommodate an ongoing process of refinement and evolution. This dynamic comportment is predicated upon the infusion of novel data, which, upon integration, facilitates the recalibration and retraining of the model. This iterative undertaking serves to incrementally augment the model's precision, acuity, and contextual pertinence over successive iterations.

The dynamic proclivity of AI systems thus ensures their sustained alignment with the evolving exigencies and complexities inherent to the educational milieu. This inherent adaptability serves as a testament to the model's capacity to remain attuned to emergent trends, shifts in pedagogical paradigms, and advancements within the educational landscape. As a result, the AI system stands poised to offer ever more pertinent and discerning insights, rendering it a valuable and continually refined asset within the educational domain. Within the operational framework of the AI system, educators and administrators assume a central and instrumental position in the feedback loop. Empowered with the discerning insights furnished by the system, they undertake the strategic implementation of interventions, concurrently assuming the responsibility for vigilant monitoring of their efficacy. This iterative process engenders a potent symbiosis between human agency and

AI, synergistically converging the analytical aptitude intrinsic to AI with the nuanced cognitive acumen and empathic sensibilities embodied by educators [8]. This collaborative interplay, characterized by a reciprocal exchange of information, feedback, and action, establishes a formidable amalgam poised for effectual transformative influence.

The augmentation of AI-driven analytical capabilities with the human element not only amplifies the discerning depth of interventions but also ensures their alignment with the contextual exigencies and socio-emotional dimensions of the educational landscape. Consequently, this concerted fusion of human intelligence and artificial cognitive capacities constitutes a catalytic force for constructive pedagogical metamorphosis. The application of AI in the realm of early intervention and dropout prognostication unfolds as a cyclical and iterative progression. With each successive iteration, an augmented corpus of data is amassed, enabling a more comprehensive contextualization of student trajectories. Concurrently, a broader array of interventions is deployed, their outcomes meticulously scrutinized. Within this iterative paradigm, the AI system engages in a continuous process of learning and adaptation, refining its predictive capacities based on empirical feedback loops.

This iterative modality serves as the crucible wherein the efficacy of AI-driven interventions undergoes meticulous refinement and enhancement. The cumulative effect of this iterative process is the progressive fine-tuning and optimization of the system's prognostic capabilities. This iterative paradigm, characterized by its feedback-rich nature, stands as a linchpin in the iterative development of AI-driven educational interventions, serving to fortify their applicability and impact within the dynamic educational landscape.

3.2.2 How student data can be used

Let us discuss the intricate process of selecting and curating datasets tailored for early intervention and dropout prediction in education, highlighting key considerations and best practices.

A robust dataset for early intervention and dropout prognostication necessitates the inclusion of a diverse array of student records, thereby engendering a comprehensive portrayal of their academic trajectory. This compilation spans a spectrum of essential metrics, encompassing scholastic performance indicators, longitudinal attendance records, standardized assessment results, rates of coursework fulfilment, and instances of disciplinary involvement. This multifaceted compilation of information collectively affords a holistic vantage point from which to scrutinize a student's educational odyssey, thereby furnishing discerning perspectives on potential risk factors.

This encompassing dataset not only captures quantitative facets of academic progression but also encompasses pertinent qualitative dimensions, which together engender a panoramic representation of a student's scholastic journey. The combination of these diverse records forms an integral component of the analytical infrastructure for early intervention and dropout prediction, facilitating nuanced prognostications that are informed by a comprehensive understanding of the student's academic trajectory. This holistic approach is instrumental in cultivating a discerning and effective framework for educational support and intervention.

The inclusion of demographic data emerges as an indispensable facet within datasets designed for the prognostication of dropout rates and the facilitation of timely interventions. This contextual information encompasses pivotal determinants including gender, ethnicity, socio-economic stratum, linguistic proficiency, and special education designation. The nuanced comprehension of the demographic composition within a student cohort assumes a critical role in discerning potential disparities and furnishing the basis for tailored intervention strategies, bespoke to the distinctive attributes of specific subgroups. This deliberate and stratified approach proves instrumental in mitigating potential educational challenges and inculcating an environment conducive to equitable academic attainment across a diverse spectrum of demographics (Figure 3.3).

Behavioural indicators constitute a reservoir of invaluable insights into a student's cognitive involvement, motivational disposition, and overarching socio-emotional well-being. This encompassing category encompasses metrics pertaining to classroom involvement, punctuality, participation in supplementary educational pursuits, as well as instances of disciplinary engagement. The integration of behavioural data augments the dataset's granularity, affording a discerning lens through which to perceive incipient indicators of potential concern that may not be readily discernible through the sole prism of academic metrics. The incorporation of behavioural dimensions serves to enrich the comprehensive analytical framework, providing a more holistic assessment of a student's scholastic experience. This multifaceted approach is predicated on the recognition that a student's academic journey is inherently entwined with their behavioural manifestations, thus warranting an integrated analytical perspective. Consequently, the amalgamation of behavioural indicators with conventional academic metrics culminates in a robust dataset, poised to yield nuanced insights and facilitate early intervention strategies tailored to the multifaceted dimensions of a student's educational trajectory.

The inclusion of data pertaining to the social and emotional well-being of students assumes paramount importance in the construction of a comprehensive dataset for early intervention purposes. This entails the integration of indicators germane to student motivation, self-perception, interpersonal relationships, and mental health status. Profound insights into the social and emotional milieu afford the capacity for tailored interventions, which, in turn, can ameliorate underlying factors impinging upon a student's scholastic advancement. The incorporation of social and emotional dimensions engenders an enriched analytical framework, acknowledging the intricate

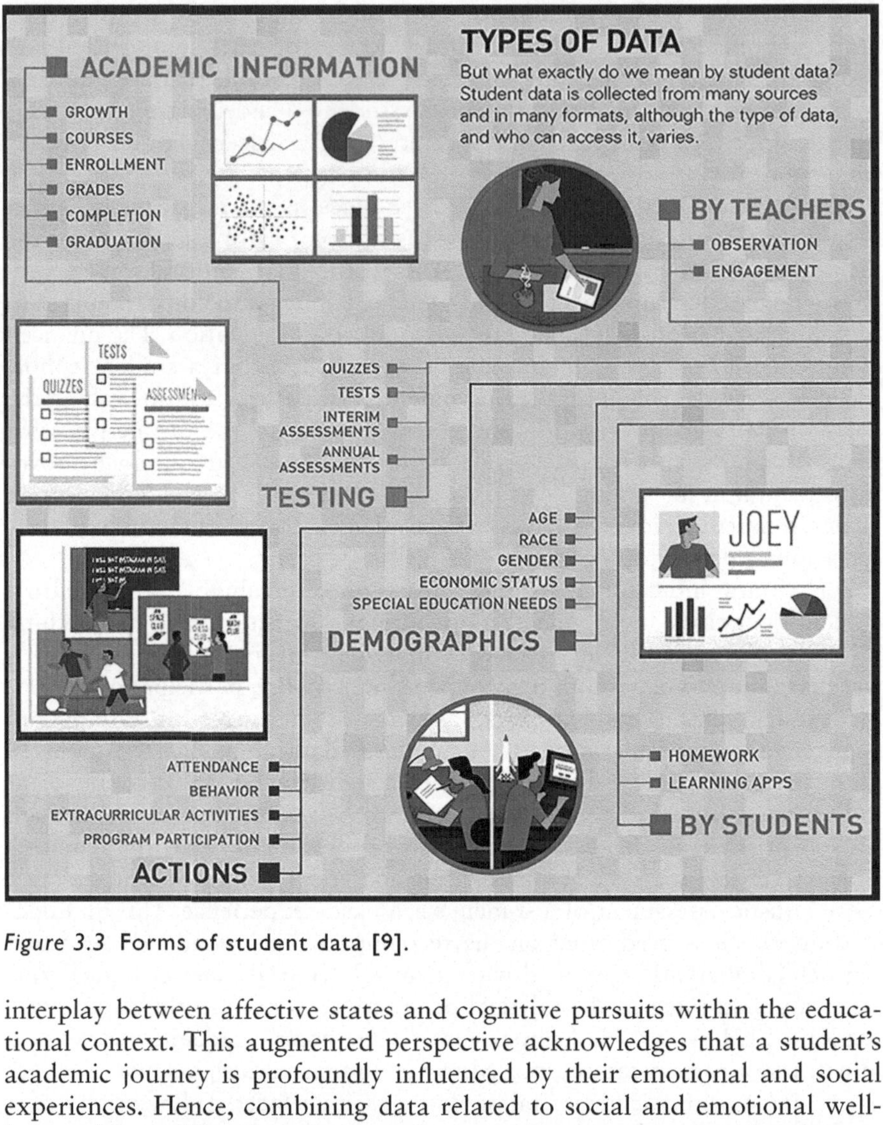

Figure 3.3 Forms of student data [9].

interplay between affective states and cognitive pursuits within the educational context. This augmented perspective acknowledges that a student's academic journey is profoundly influenced by their emotional and social experiences. Hence, combining data related to social and emotional wellbeing with traditional academic measures creates a comprehensive dataset. This dataset is well-equipped to provide insightful observations and support tailored early intervention strategies that address the diverse aspects of a student's educational journey effectively.

Longitudinal data, chronicling the progressive trajectory of a student's academic journey over time, constitutes an indispensable resource for early intervention and dropout prognosis. This retrospective panorama facilitates the discernment of discernible trends and recurrent patterns, which may serve as indicative markers of declining academic performance or an augmented susceptibility to premature educational disengagement. The examination of this temporal continuum empowers educators to strategically intercede at pivotal junctures, potentially effecting a transformative impact on a student's educational trajectory. The inclusion of longitudinal data underscores the acknowledgement that academic development is inherently dynamic, subject to a plethora of multifaceted influences and fluctuations over time. This longitudinal perspective, therefore, assumes a pivotal role in enhancing the comprehensive scope of the analytical framework, providing a reflective lens through which to identify salient turning points. Consequently, the integration of longitudinal data with extant datasets amplifies the discerning capacity for early intervention, ultimately fostering a more responsive and adaptive educational milieu.

The integration of data regarding a student's familial background and community environment enriches the dataset with substantive contextual depth. Variables encompassing parental educational attainment, household stability, and the accessibility of community resources wield noteworthy influence over a student's educational trajectory. These extrinsic determinants hold the potential to significantly shape the educational experience. A judicious understanding of these external influences affords the capacity to design interventions that address broader systemic challenges. The assimilation of information pertaining to family and community context serves to broaden the analytical framework, recognizing the pivotal role that external environments play in moulding educational outcomes. This contextual perspective acknowledges the nuanced interplay between familial and communal dynamics and a student's academic development. Consequently, the amalgamation of such contextual data with prevailing academic metrics augments the dataset's capacity for insightful early intervention, facilitating a more tailored and effective response to the diverse challenges encountered within the educational milieu.

Contextual variables encapsulate facets pertinent to the educational milieu per se. This encompasses attributes such as the scale of the educational institution, its urban or rural designation, the availability of educational resources, and the prevailing teacher-student ratios. The discerning appreciation of the contextual backdrop within which education transpires confers indispensable insight for the discerning interpretation of student data, whilst concurrently enabling the bespoke calibration of interventions germane to distinct educational settings. The incorporation of contextual variables engenders a heightened level of granularity within the analytical framework, recognizing the influential role wielded by the educational environment in shaping scholastic outcomes. This contextual acumen illuminates the intricate interplay between the structural attributes of an educational establishment and the academic progress of its student cohort. Consequently, the amalgamation of these contextual parameters with extant academic metrics amplifies the dataset's capacity for astute early intervention, cultivating a more refined and targeted approach attuned to the diverse exigencies prevalent across distinct educational contexts.

In the pursuit of tailored interventions, the inclusion of specialized data tailored to specific demographic subsets emerges as an imperative. For instance, within the cohort of English Language Learners (ELL), assessments gauging linguistic proficiency assume pivotal significance. Similarly, students necessitating special education provisions may possess Individualized Education Plans (IEPs), housing indispensable information essential for the formulation and refinement of intervention strategies. This deliberate integration of specialized data for distinct population cohorts manifests an earnest recognition of their distinctive academic and developmental requisites. It is predicated upon the principled acknowledgement that precise interventions hinge upon a comprehensive understanding of the particular challenges and exigencies encountered by these delineated subgroups. Consequently, the incorporation of specialized data confers the capacity for interventions that are not only circumspect but also cogently aligned with the nuanced requirements of individuals within these demarcated populations. This bespoke approach underscores a dedicated commitment to inclusivity and equity within the educational paradigm.

Safeguarding the integrity and reliability of data constitutes a paramount undertaking. This imperative necessitates the implementation of rigorous validation protocols, meticulous data cleansing procedures, and judicious normalization processes. These efforts are directed towards the expurgation of outliers, rectification of errors, and the standardization of formats. A dataset marred by inaccuracies and irregularities poses a substantial risk, as it has the potential to engender erroneous prognostications and, by extension, render interventions ineffectual. The meticulous validation process involves a systematic examination of the dataset's coherence, wherein aberrant or anomalous entries are identified and subsequently addressed. Concurrently, the data cleansing endeavour entails a discerning and systematic eradication of any erroneous or superfluous information, culminating in a refined and reliable dataset. Additionally, the normalization process serves to establish uniformity in data representation, a pivotal step in harmonizing the diverse elements within the dataset.

It is incumbent upon practitioners in the field of early intervention and dropout prediction to exercise scrupulous diligence in these data refinement processes. The repercussions of neglecting data quality can reverberate significantly, potentially compromising the efficacy of interventions and the accuracy of prognostications. As such, a steadfast commitment to data quality assurance stands as an indispensable pillar in fortifying the robustness and dependability of predictive models in the educational domain.

Respecting the privacy rights of students and adhering to ethical guidelines is an indispensable mandate when compiling datasets for early intervention and dropout prediction, particularly within the Indian context. It is imperative that data is anonymized and aggregated to shield individual identities. Additionally, strict compliance with pertinent privacy laws and regulations, such as the Personal Data Protection Bill in India, is of paramount importance. The process of anonymization entails the deliberate removal of personally identifiable information, ensuring that individual identities remain concealed within the dataset. Simultaneously, data aggregation serves to amalgamate information in a manner that precludes the identification of individual students, thereby fortifying privacy safeguards. This twofold approach is crucial in maintaining the confidentiality and integrity of the data.

Adherence to established privacy laws and regulations, such as the forthcoming Personal Data Protection Bill in India, is crucial in safeguarding the rights of students and instilling public trust in educational institutions. This legislation establishes robust protocols for the handling, storage, and dissemination of student data, thereby providing a legal and ethical framework that reinforces privacy protections. In essence, the conscientious consideration of student privacy concerns and unwavering adherence to ethical guidelines are foundational tenets that underlie the ethical compilation of datasets for early intervention and dropout prediction within the Indian educational landscape.

3.2.3 ML model used

Let us understand how a DFFNN (Deep FeedForward Neural Network) works in Early Intervention of Dropout and Prediction.

Multilayer perceptrons, or MLPs for short, are a powerful family of artificial neural networks specialized in pattern recognition and regression tasks. They are sometimes referred to as Deep Feedforward Neural Networks. Their value is especially evident when it comes to early intervention dropout prediction. With each layer sending its output to the next layer, these networks are made up of several layers of interconnected nodes. The complex relationships found in the data can be understood by them due to its architectural design.

First, a vector of features about a student's behaviour, attendance, academic performance, and socio-economic status is sent to the input layer. The network's comprehension is based on these characteristics. After that, the information flow is mediated by one or more hidden layers. These layers' neurons calculate the weighted total of their inputs. The strength of the connections between neurons is indicated by these weights. Then, an activation function is used. Non-linearity is introduced, which is required for the network to learn and approximate complicated, non-linear correlations in the data. Rectified Linear Unit (ReLU), Sigmoid, and Tanh activation functions are often used. This procedure of weighted sum calculation and activation function application is repeated iteratively across the network, which is known as forward propagation.

The data then reaches the output layer. This layer is critical to the final prediction. A single neuron is sufficient for binary classification problems (with or without dropout). Multiple neurons are used for multi-class classification. The output layer employs problem-specific activation functions, such as sigmoid for binary classification and SoftMax for multi-class classification. Next, a loss function is used to compare the network's output to the actual target. The difference between the expected and actual numbers is quantified by this function. For binary classification, binary cross-entropy and for multi-class classification, categorical cross-entropy are common loss functions.

Backpropagation is used to improve the model. This procedure entails computing the gradient of the loss function with respect to each weight and then adjusting the weights using gradient descent or one of its variants. In order to identify the ideal set of weights that minimize prediction error, the network is trained to minimize the loss function across a number of epochs. A subset of the data is reserved for validation throughout training. This makes it possible to track how well the model performs with unknown data. After training, the model's performance is assessed on an independent test set to gauge how well it generalizes. Early stopping can be used to avoid overfitting, a condition in which the model becomes overly tailored to the training set. This means keeping an eye on the validation loss and stopping training if it begins to rise, which indicates that the model is overfitting.

After training successfully, the model is prepared for use. Through the input of relevant student attributes, the network generates an output that represents the expected probability of dropout. It's crucial to remember that proper implementation calls for consideration of features, preparation of the data, hyper parameter adjustment, and possibly the addition of other methods to improve performance, such as batch normalization, dropout regularization, and model ensemble (Figure 3.4).

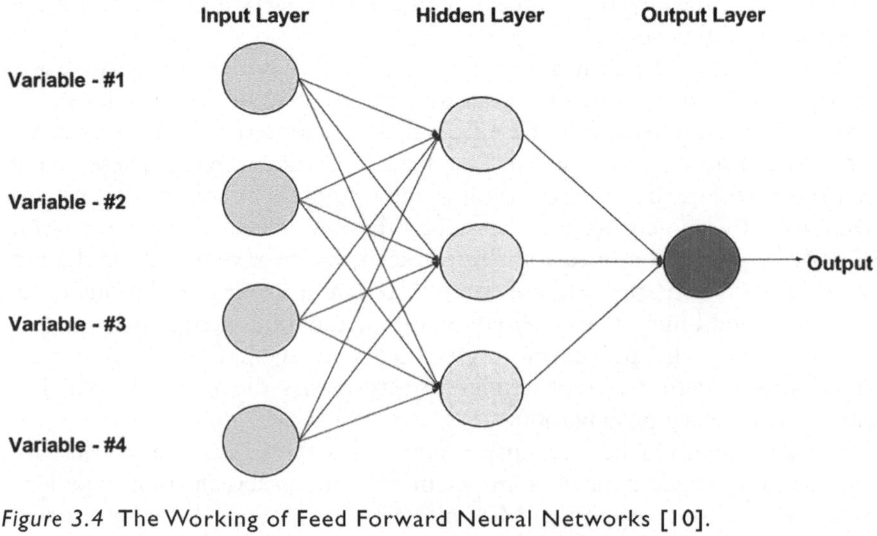

Figure 3.4 The Working of Feed Forward Neural Networks [10].

When it comes to assessing the effectiveness of deep feedforward neural networks for early intervention dropout prediction, a variety of performance indicators are essential. These metrics provide numerical assessments of the model's predictive ability and enable a thorough analysis over the training and testing datasets. First, there is accuracy, which is a basic indicator that represents the percentage of correctly identified cases in all observations. By dividing the total number of forecasts by the number of correct guesses, this is computed. Even while accuracy offers a simple evaluation, it might not always be the most relevant metric, especially when the dataset is unbalanced and one class (such as dropout) greatly dominates the other.

However, when datasets are unbalanced, precision and recall become important measurements. Positive prediction accuracy is quantified by precision. The calculation involves dividing the total number of true positive and false positive forecasts by the number of true positive predictions. Recall measures how well the model detects positive instances. It is sometimes referred to as sensitivity or true positive rate. The ratio of true positives to the total of false negatives and true positives is used to compute it. As a composite metric that represents the harmonic mean of recall and precision, the F1-Score takes shape. This score, which is particularly useful in situations when there is an unequal class distribution, provides a balanced assessment of recall and precision. It provides a thorough evaluation of the model's performance by skilfully integrating the two measures.

The Confusion Matrix emerges as a valuable tool, presenting a detailed breakdown of the model's predictions. It delineates true positives, true negatives, false positives, and false negatives, providing a granular understanding of where the model may be making errors.

When combined, these performance indicators provide an invaluable instrument for assessing a deep feedforward neural network's efficacy in early intervention dropout prediction. A comprehensive evaluation of the model's performance may be obtained by customising the metrics used to the particularities of the problem and dataset. This empowers data scientists and educators to work together to enhance student results.

3.3 CHALLENGES AND ETHICAL CONSIDERATIONS

EIDP, in general, brings forth a host of ethical considerations that must be carefully navigated to ensure the well-being and equitable treatment of all students involved.

Transparency and explainability are crucial in ensuring that the AI models and algorithms used in early intervention are comprehensible to all stakeholders. Students, parents, and educators should have a clear understanding of how the system functions and makes its recommendations. This transparency builds trust and allows for informed decision-making. Guarding against biases in both data and algorithms is paramount. AI systems should be rigorously tested to ensure they do not disproportionately disadvantage or advantage any particular group based on factors such as race, gender, socio-economic status, or other sensitive attributes. Addressing and mitigating biases is essential for providing fair and equal opportunities to all students. Handling student data with utmost care is a critical ethical consideration. Compliance with data protection regulations and the implementation of robust security measures are imperative to safeguard sensitive information. Students and their families should feel confident that their personal information is being handled securely and with respect for their privacy. Obtaining informed consent from students and their parents or guardians is a fundamental ethical principle. They should be fully aware of how their data will be used and how the intervention system will operate. This ensures that stakeholders have agency over their information and understand the purpose and potential impact of the intervention.

While AI can provide valuable insights, it should never replace human judgement and expertise. There should always be a human element involved in decision-making, and automated systems should complement and support, rather than override, the knowledge and experience of educators. Human oversight ensures that the unique circumstances and needs of individual students are taken into account. Clearly defining who is responsible for the outcomes of the AI-based intervention is essential. This may involve a combination of educators, administrators, AI developers, and other stakeholders. Establishing clear lines of accountability helps ensure that any challenges or issues that arise are addressed promptly and effectively. Regular assessment of the performance of the AI system is crucial in ensuring that it achieves its intended goals without causing unintended harm. If biases or other ethical concerns emerge, they should be identified and rectified promptly. Ongoing monitoring allows for adjustments to be made to the intervention approach as needed. Ensuring that all students, regardless of their background or circumstances, have equal access to the benefits of early intervention using AI is a core ethical principle. Efforts should be made to remove barriers and provide support to those who may be at risk of dropping out, regardless of their socio-economic status or other demographic factors.

Consideration of the potential long-term effects of using AI for early intervention is crucial. It is important to track and evaluate whether this approach leads to improved educational outcomes for students. Ethical responsibility extends beyond immediate intervention and requires a commitment to the long-term success and well-being of each student. Recognizing and respecting cultural differences that may affect a student's educational experience is essential. AI interventions should be culturally sensitive and inclusive, taking into account the diverse backgrounds and perspectives of students and their families. The implementation of interventions should be carried out with care to avoid stigmatizing students who receive extra support. The process should be discreet and conducted with respect for students' dignity, ensuring that they feel valued and empowered in their educational journey. Maintaining open lines of communication with all stakeholders about the purpose, process, and outcomes of the AI-based intervention is critical. Regular feedback loops provide an opportunity for continuous improvement and ensure that the intervention remains aligned with the ethical principles and goals set forth.

Early intervention in dropout prevention using AI holds great promise for supporting students in their educational journey. However, navigating the ethical considerations is paramount to ensure that the well-being and rights of students are protected. Transparency, fairness, privacy, and ongoing evaluation are among the key ethical principles that should guide the implementation of AI-based interventions. By upholding these principles, we can create a supportive and inclusive educational environment that empowers all students to thrive.

ACKNOWLEDGEMENT

We are grateful to the anonymous reviewers for their valuable suggestions that have helped us to improve our work.

REFERENCES

1. Shute, V. J., & Becker, B. J. (2010). "Innovations in the Use of Computerized Testing." *Review of Educational Research*, 80(3), 392–417.
2. Adnan, M., Habib, A., Ashraf, J., Mussadiq, S., Raza, A., Abid, M., Bashir, M., & Khan, S. (2021). Predicting at-Risk Students at Different Percentages of Course Length for Early Intervention Using Machine Learning Models. *IEEE Access*, 9, pp. 7519–7539. 10.1109/ACCESS.2021.3049446
3. Siemens, G., & Baker, R. S. (2012). "Learning Analytics and Educational Data Mining: Towards Communication and Collaboration." In *Proceedings of the 2nd International Conference on Learning Analytics and Knowledge*, 252–254.
4. Perera, D., Kay, J., Koprinska, I., & Yacef, K. (2009). "Clustering and Sequential Pattern Mining of Online Collaborative Learning Data." *IEEE Transactions on Knowledge and Data Engineering*, 21(6), 759–772.
5. Yudelson, M. V., Koedinger, K. R., & Gordon, G. J. (2013). "Individualized Bayesian Knowledge Tracing Models." In *Proceedings of the 6th International Conference on Educational Data Mining*, 82–89.
6. Times of India. (2018, January 3). *School Dropout Rates in India*. Times of India. https://timesofindia.indiatimes.com/india/school-dropout-rates-in-india/articleshow/62333614.cms
7. Vitolina, I. (2015). E-inclusion modeling for blended e-learning course. *Procedia Computer Science*, 65, 744–753, 10.1016/j.procs.2015.09.019

8. Koedinger, K. R., & Corbett, A. T. (2006). “Cognitive Tutors: Technology Bringing Learning Science to the Classroom.” In K. Sawyer (Ed.), *The Cambridge Handbook of the Learning Sciences*, Cambridge University Press, 61–78.
9. Getting Smart. (2015, July 9). Infographic: What Is Student Data? Getting Smart. https://www.gettingsmart.com/2015/07/09/infographic-what-is-student-data
10. Learn OpenCV. (n.d.), “Understanding Feedforward Neural Networks.” *Learn OpenCV*, https://learnopencv.com/understanding-feedforward-neural-networks/

Chapter 4

Resume analysis and generation using LLM

Chittaranjan Pradhan, Biswaroop Nath, Nayanika Chatterjee, Aniruddha Banerjee, Anuraag Banerjee, and Abhranil Chattopadhyay

4.1 INTRODUCTION

The creation of an effective resume is a critical step for job seekers, as it involves distilling years of education, experience, and unique skills into a concise document that stands out in a competitive job market. With the rise of automated applicant tracking systems, candidates face the added challenge of optimizing their resumes to meet the specific requirements of each job posting. This process can be arduous and time-consuming, often requiring a strategic use of keywords and a focus on relevant experiences and achievements.

In response to this challenge, the proposed tool, Resume Analyzer, aims to streamline the resume customization process by leveraging the capabilities of Large Language Models (LLMs). These advanced models have shown remarkable proficiency in following instructions and performing a variety of complex tasks, which suggests their potential utility in assisting job applicants with tailoring their resumes efficiently. By inputting a general-purpose resume into Resume Analyzer, candidates could significantly reduce the time spent on modifying their resumes for each job application, potentially enhancing their chances of success in the job market.

Resume Analyzer is a tool designed to assist job applicants in creating personalized resumes tailored to specific job postings. The key features of Resume Analyzer include:

- *Personalized Resume Generation*: Users can input their detailed resume and the job posting they are interested in, and Resume Analyzer will generate a personalized resume tailored to that job posting.
- *Language Understanding and Information Extraction*: Utilizing state-of-the-art language models like Google's Gemini, Resume Analyzer can extract relevant details from both the job description and the user's resume.
- *Role-Specific Refinement*: The tool refines the user's resume by focusing on role-specific details, ensuring that the final document aligns closely with the job requirements.

DOI: 10.1201/9781003643166-4

- *Time Efficiency*: Resume Analyzer significantly reduces the time required to tailor a resume for a specific job, completing the process in a matter of seconds.
- *Ease of Use*: The tool is user-friendly and does not require any fine-tuning by the end-user, making it accessible to a wide audience.
- *Quality Control*: Novel task-specific evaluation metrics are proposed to control for alignment and hallucination, ensuring the quality of the generated resumes.
- *Accessibility*: Resume Analyzer is available for use, providing a practical solution for job applicants facing the challenge of resume customization.

These features aim to streamline the resume tailoring process, making it more efficient and less prone to human error, thereby enhancing the chances of applicants passing through automated filtering systems.

Resume Analyzer is designed to assist job seekers in tailoring their resumes to specific job postings. The process begins with the user uploading their detailed resume and the job description they are targeting. The tool then utilizes a Large Language Model (LLM) to extract relevant data from both the resume and the job description. With this information, Resume Analyzer generates a personalized resume that aligns with the job's requirements, optimizing the candidate's chances of passing through automated filtering systems and catching the attention of potential employers. This streamlined approach aims to save applicants significant time and effort in the job application process.

4.2 LITERATURE REVIEW

The field of natural language processing [1] and text mining [2] has seen significant advancements, leading to innovative applications across various sectors. One such application is the analysis and categorization of resumes, coupled with the extraction of information pertinent to job seekers. Studies have delved into techniques for parsing resumes to extract key information [3, 4], inferring skills from the content of resumes [5], and aligning candidates with job positions [6]. However, there is a noticeable absence in research on the automated creation of resumes that are tailored to specific job descriptions.

Resumes [7, 8] are structured documents summarizing an individual's professional experience, education, skills, and qualifications. Key sections typically include contact information, a professional summary, work experience, education, skills, and certifications or awards. These sections collectively provide a comprehensive view of a candidate's suitability for a job.

Resume analysis [9] involves the automated extraction and examination of relevant information from resumes, enabling recruiters to efficiently identify suitable candidates. Key tasks include information extraction, entity recognition, skills matching, and relevance scoring. These tasks streamline the evaluation [10] of resumes, making the recruitment process faster and more accurate.

Research has suggested methodologies for extracting information from resumes in two stages, starting with the identification of distinct blocks of information [11]. Yet, the concept of generating customized resumes for particular job openings has not been directly tackled. The use of large language models (LLMs) [12] as assistants capable of following instructions has emerged as a focal point in recent studies. Large Language Models (LLMs) are deep learning models trained on vast text datasets using transformer architectures. These models employ self-attention mechanisms to understand and generate coherent text, capturing long-range dependencies and contextual relationships. LLMs are pre-trained on large corpora to learn language patterns and fine-tuned on specific tasks to enhance their performance in applications such as resume analysis. These models have been employed for retrieving information [13, 14], aiding in educational pursuits [15, 16], and analyzing textual data [17, 18].

In resume parsing, LLMs can identify and extract key sections and information using techniques like tokenization, named entity recognition (NER), and pattern matching. They facilitate the extraction and normalization of information, ensuring consistency and accuracy. For instance, LLMs can extract job titles and descriptions, normalize skills to a standardized format, and match them with job requirements.

Despite their capabilities, LLMs face challenges such as data privacy, bias, and contextual understanding. Ensuring compliance with privacy regulations, mitigating biases in training data, and improving contextual comprehension are ongoing challenges.

In one instance, study [19] utilized ChatGPT for the generation of resumes in both structured and unstructured formats. The focus of this study, however, was on augmenting data for subsequent classification tasks rather than on crafting resumes that meet specific job criteria. This highlights a gap in the current literature and underscores the potential for future research to bridge this divide by developing methods for the automatic generation of targeted, job-specific resumes.

4.3 BASIC CONCEPTS

In this section, we discussed the model used and the performance metrics used in this paper.

4.3.1 Models

Large Language Models (LLMs) are advanced artificial intelligence models trained on vast amounts of text data to understand and generate human-like language. These models are typically built using deep learning techniques, especially transformer architectures [20]. LLMs have become increasingly popular and influential in NLP tasks due to their ability to handle various language-related tasks such as text generation, translation, summarization, and more.

LLMs operate by processing input text through multiple layers of neural networks, which encode and transform the text representation to capture semantic meaning and context. One of the key features of LLMs is their ability to generate coherent and contextually relevant text based on the input they receive.

Some well-known examples of LLMs include OpenAI's GPT series [21] (e.g., GPT-3), Google's Gemini-Pro [22], and Facebook's RoBERTa [23]. These models demonstrated amazing performance on a wide range of NLP tasks and have significantly advanced the state of the art in the field.

LLMs have numerous applications across various domains, including chatbots, content generation, language translation, and more. They continue to be an active research area with ongoing efforts to improve their capabilities, efficiency, and interpretability.

4.3.2 Performance metrics

This study uses two performance metrics as follows:

4.3.2.1 Overlap coefficient

The overlap coefficient [24] is used to compute the similarity between two sets, often used in comparing the similarity of two binary or categorical data sets. It calculates the proportion of elements that are common to both sets comparative to the total number of elements in the smaller set. Mathematically, the overlap coefficient is defined as:

$$OC(A,B) = (|A \cap B|) / \min(A,B) \tag{4.1}$$

The overlap coefficient ranges from 0 to 1, where:

- $OC = 0$ indicates no overlap between the sets.
- $OC = 1$ indicates complete overlap, meaning that all elements in the smaller set are also present in the larger set.

The overlap coefficient is commonly used in various fields, such as bioinformatics, information retrieval, and pattern recognition, to quantify the similarity between data sets. It provides a simple and intuitive measure of overlap that is particularly useful when dealing with binary or categorical data.

4.3.2.2 Cosine similarity

Cosine similarity [25] is used to find out how similar two vectors are in terms of their orientation, regardless of their magnitude. It is commonly used in various fields, including NLP, information retrieval, and machine learning, to compare the similarity between documents, text embeddings, or any other high-dimensional data represented as vectors.

Cosine similarity is used to evaluate the similarity between two documents expressed in a high-dimensional space. It is computed between two vectors A and B by means of the cosine angles between them as:

$$CS\,(A,B) = (A.\ B)/\|A\|\ \ \|B\| \tag{4.2}$$

Where:

- $(A.B)$ represents the dot product of vectors A and B.
- $\|A\|$ and $\|B\|$ represent the Euclidean norms (magnitudes) of vectors A and B, respectively.

Cosine similarity ranges from −1 to 1:

- A value of 1 indicates that the vectors are perfectly aligned, implying that the documents are identical or highly similar.
- A value of −1 indicates that the vectors are diametrically opposed, implying that the documents are dissimilar.
- A value of 0 indicates orthogonality, implying that the documents have no similarity.

Cosine similarity is advantageous because it is independent of the magnitude of the vectors and focuses solely on the direction, making it robust to differences in document length or scale. It is widely used in tasks such as document retrieval, clustering, and recommendation systems for its effectiveness in capturing semantic similarity.

4.4 PROPOSED APPROACH

In this section, we offer an in-depth exploration into the operational architecture of our groundbreaking Resume Analyzer tool, providing a nuanced depiction of each constituent element. Illustrated in Figure 4.1, Resume Analyzer embodies our envisioned pipeline, enriched by the facilitation of LLM, dedicated to the nuanced and personalized generation of resumes.

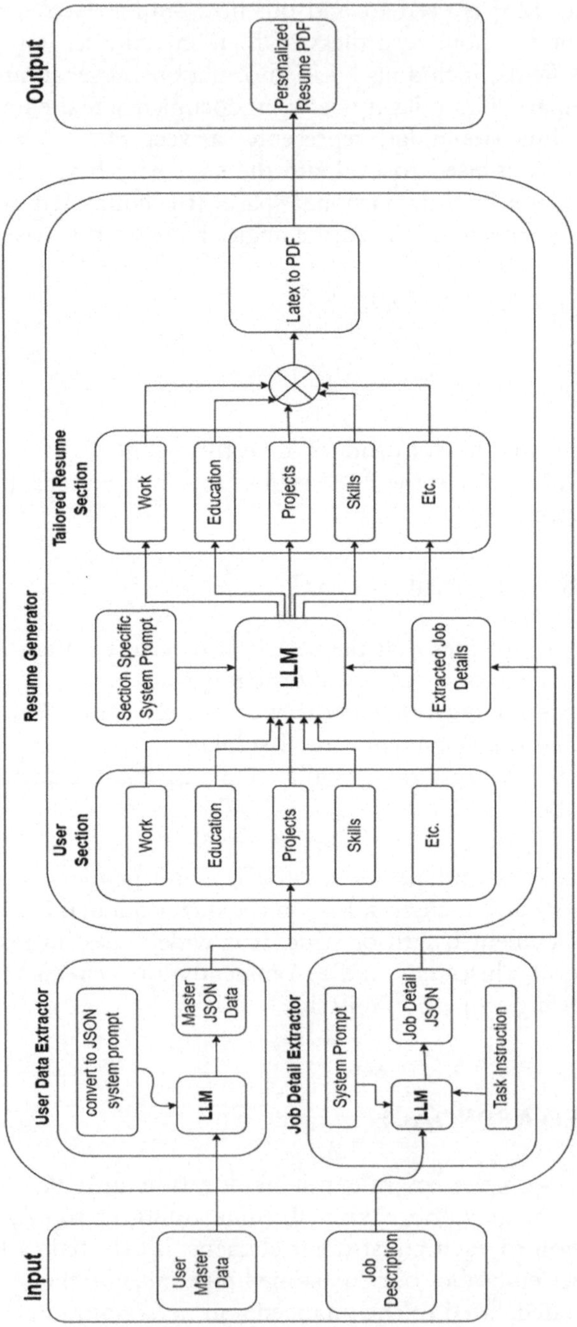

Figure 4.1 Workflow of proposed approach.

The system comprises three foundational components, each intricately interwoven to orchestrate the resume creation process:

- *User Data Extractor*: Serving as the cornerstone of Resume Analyzer, the User Data Extractor module assumes the critical task of seamlessly transforming user-provided baseline resumes, typically sourced in PDF format, into a meticulously structured JSON representation. By employing sophisticated parsing algorithms and data standardization techniques, this module not only captures the textual content of the resume but also discerns semantic meaning and contextual nuances. Through this process, it lays the groundwork for subsequent processing stages, ensuring that the resume data is amenable to advanced analysis and manipulation.
- *Job Details Extractor*: Operating synergistically with the User Data Extractor, the Job Details Extractor module fulfills the indispensable role of dynamically parsing and analyzing job descriptions sourced directly from users. Through a streamlined interface that allows users to simply copy and paste the job posting text, this module utilizes advanced natural language processing (NLP) algorithms to dissect the content, extracting key keywords, requirements, and contextual cues. The extracted information is then encapsulated in a structured JSON format, poised to inform and guide the subsequent stages of the resume generation process. By seamlessly integrating user-provided job details, this module ensures that the generated resumes are finely tuned to meet the specific demands and expectations of prospective employers.
- *Resume Generator*: Positioned at the nexus of the pipeline, the Resume Generator module represents the culmination of Resume Analyzer's processing prowess. Drawing upon the meticulously structured user data and the rich contextual insights gleaned from the job requirements, this module embarks on the intricate task of synthesizing personalized resumes tailored to each user's unique profile and aspirations. Leveraging a sophisticated ensemble of machine learning models, NLP techniques, and heuristic algorithms, the Resume Generator meticulously crafts each resume, optimizing both content and formatting to resonate harmoniously with the intended job roles and industry norms. Through iterative refinement and adaptive learning mechanisms, it continually refines its output, striving to deliver resumes of unparalleled quality and relevance.

In dissecting each component of the pipeline, we will explore not only the technical intricacies but also the human-centric considerations embedded within Resume Analyzer. From leveraging NLP algorithms to parse job descriptions and candidate resumes with precision to incorporating machine learning models for personalized recommendation engines, every facet is meticulously designed to optimize both efficiency and effectiveness. Furthermore, we'll examine the seamless integration of user feedback loops, ensuring continuous improvement and adaptability to evolving industry trends.

4.4.1 User data extractor

In today's competitive job market, efficiently matching candidates with suitable positions is crucial for both job seekers and employers. Recognizing the importance of this challenge, our tool incorporates a sophisticated user data extractor as a pivotal component. This extractor seamlessly processes resumes submitted in PDF format, regardless of their structure or format, along with the corresponding job description. By leveraging advanced Language Models (LLMs) chosen by the user, the extractor accurately extracts key information from resumes, aligning candidate skills and experiences with job requirements. The structured output, presented in a JSON format, not only enhances readability but also facilitates seamless integration with other tools and systems. With careful design and implementation, the user data extractor streamlines the job application process, empowering users to make informed decisions and enhancing their competitiveness in the job market. The key features and benefits of this extractor:

- *Purpose Clarity*: The user data extractor is designed to serve as a crucial component within a larger tool aimed at streamlining the job application process. Its primary purpose is to automate the extraction of key information from resumes provided by users, enhancing efficiency and accuracy in matching candidates with job descriptions.
- *Input Flexibility*: The extractor accommodates both structured and unstructured resumes in PDF format, allowing users to seamlessly submit their resumes regardless of their organization or formatting. This flexibility ensures accessibility for users with varying resume styles.
- *Integration with Job Description*: By prompting users to input the job description of the position they are applying for, the extractor aligns extracted resume data with specific job requirements. This integration enables tailored analysis, ensuring that highlighted skills and experiences directly correspond to the job's demands.
- *Customization with LLMs*: Users have the option to select a Language Model (LLM) of their choice for the extraction process. This customization empowers users to leverage LLMs that align with their preferences or specific language understanding needs, enhancing the accuracy and relevance of the extracted information.
- *Off-the-Shelf Implementation*: The extractor utilizes the chosen LLM in an off-the-shelf manner, eliminating the need for additional training or fine-tuning. This approach simplifies implementation while maintaining robust performance, as the LLM is deployed through carefully structured prompts designed to facilitate precise information extraction.

- *Structured Output*: Extracted information is organized into a structured JSON format, providing a standardized representation of the user's resume data. This structured output enhances readability and compatibility with downstream processes, facilitating seamless integration with other tools or systems.
- *Enhanced Understanding through Prompting*: The extraction process is guided by a prompt that frames the task within a relevant context, such as designing a resume parsing application. This contextualization helps the LLM comprehend the intent behind the extraction task, enhancing its ability to accurately identify and extract key sections and information from the resume text.
- *Terminology Clarification*: The term "UserData" is introduced to refer to the structured dictionary of information extracted from the user's resume. This terminology simplifies communication and understanding within the tool's documentation and discussions, facilitating clarity and ease of reference.
- *Ease of Interpretation*: By encapsulating extracted information within a structured dictionary, the extractor enhances the interpretability of resume data for users and downstream processes alike. This clear representation of key details facilitates informed decision-making and personalized recommendations for job applications.
- *Continuous Improvement*: While the extractor operates effectively in its current state, there remains room for iterative enhancements and refinements. Feedback from users and ongoing monitoring of performance metrics can inform future iterations, ensuring that the extractor evolves to meet changing user needs and technological advancements.

In this process, we extract sections and critical details from the user's resume using the selected Language Model (LLM) without the need for additional training. This is achieved through structured prompts that guide the information extraction process. For example, a prompt might be: "You are tasked with developing a system that processes text-formatted resumes, converting them into structured JSON format". Through this method, a users' information structured dictionary is being generated by LLM. This dictionary, referred to as UserData, serves as a convenient reference for subsequent sections.

4.4.2 Job details extractor

This tool accepts textual job descriptions. To utilize it, you simply need to copy and paste the job description from your chosen job listing into the tool's text box.

Our approach aims to extract the key information from the job description, such as the position title and required qualifications. We employ two carefully crafted prompts for this purpose using the LLM: the task prompt, which gathers essential job details, and the system prompt, which prompts the LLM to adopt the persona of an experienced CV writer. Specifically, we craft a character for the system prompt that mirrors the focus, tone, and style commonly found in professional resumes, such as: "Your career is well-seasoned".

The task prompt guides the LLM in extracting vital job information from the provided job description and presenting it in an organized JSON format. By combining specific questions, the LLM adeptly collects, structures, and delivers crucial data in a JSON format. This precise extraction is crucial for the subsequent stage of the pipeline, which involves customizing resumes.

The extracted data typically encompasses key components such as the job title, keywords, objectives, duties, required and preferred qualifications, company name, and other corporate details. Throughout this document, we will denote this JSON representation of job data as "JobDetails".

4.4.3 Resume generator

At the heart of our architecture lies the process of resume customization, wherein a tailored version of the user's resume is generated to suit the specific job requirements. This component takes inputs from the previous phases: UserData and JobDetails.

The UserData is segmented into sections for processing, aligning with the typical components of a resume, such as Personal Details, Education, Work Experience, Skills, Achievements, Projects, and so forth. Recognizing that resumes naturally adhere to a structured framework, we opt to process them in sections. This approach ensures that the context length limit manageable by LLMs is not exceeded, thereby facilitating smoother processing.

Furthermore, studies have shown that LLMs often overlook information buried within lengthy contexts and struggle to extract data from excessively long passages. To address this, processing one section at a time resolves these issues effectively.

Here's how we handle each UserData section iteratively: Initially, the personal details section is extracted exactly as it appears, preserving the accuracy of information such as name, address, and phone number. Then, we proceed through the UserData sections systematically. For each section, we provide the selected LLM with both the JobDetails and the UserData section data, along with a question tailored to that specific section. The LLM's task is to reconstruct each section of the resume after analyzing the user's existing resume and retaining relevant information.

The system prompt serves as a guide for crafting a resume tailored to the job description provided in JobDetails, drawing from best practices endorsed by career counselors and expert resume writers. LLMs adhere strictly to the following principles:

- *Relevance*: Emphasize accomplishments that are directly pertinent to the job description.

- *Honesty*: Prioritize truthful and unbiased language.
- *Specificity*: Give precedence to job-specific relevance over generic achievements.
- *Design*:
  - Voice: Prefer the active voice whenever possible.
  - Proofreading: Ensure all grammar and spellings are correct.

These carefully constructed instruction-style prompts are designed to mirror the expertise and tone of seasoned CV writers, leveraging the instructional comprehension capabilities of state-of-the-art LLMs. Our tool uses the Gemini Pro LLM from Google's Deepmind. We employed a parsing function that we developed to convert the generated output into a JSON format. Ultimately, we utilize a LATEX engine to format the final resume output.

4.5 RESULTS ANALYSIS

This section presents the results provided by the proposed tool.

As in Figure 4.2, we can see the simple UI of our tool, with a entry box for Job description text, a place for uploading the resume file of user, containing

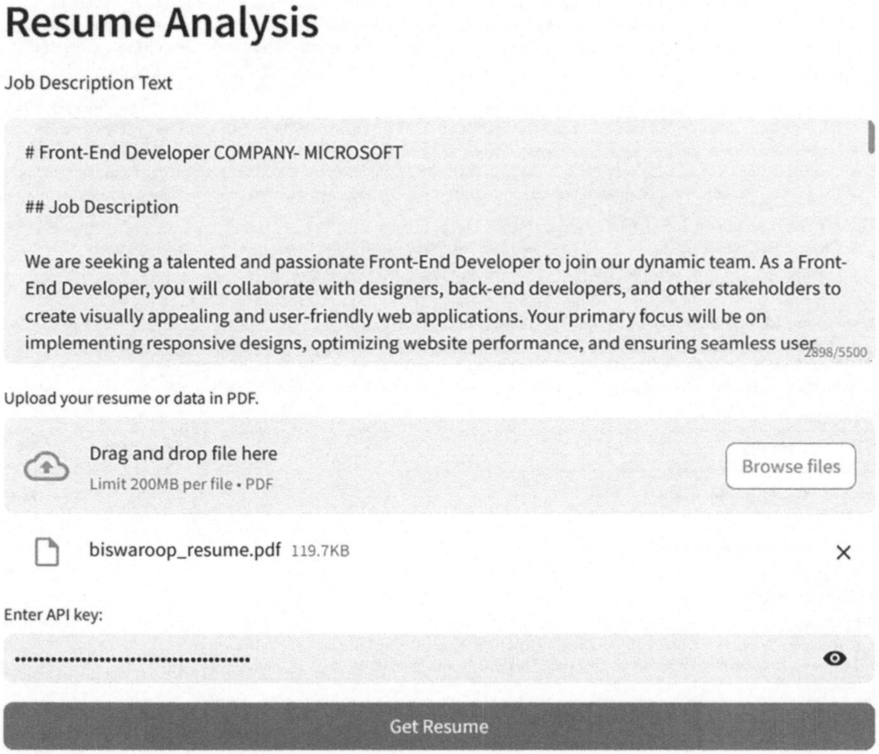

Figure 4.2 UI of proposed tool.

the bio-data, and another textbox for API key for Google Gemini Pro, which has a hide function to provide confidentiality to the API key of the user. In Figure 4.2, we have already provided a sample "Front-end developer" job description and a sample resume. The sample resume in the input is provided in Figure 4.3.

BISWAROOP NATH

35 Subhasnagar, Dumdum, Pin 700034

123-456-7890 biswaroopnath2001@gmail.com linkedin.com/in/biswaroop github.com/biswaroop

With over 2+ years of experience as a Data Engineer specializing in Front end technologies and data pipelines for large-scale distributed database systems, I bring both expertise and enthusiasm for handling complex challenges to your team.

Work Experience

Electronics Company **May 2026 – August 2027**
Software Engineer Intern *City, State*

- Developed a service to automatically perform a set of unit tests daily on a product in development in order to decrease time needed for team members to identify and fix bugs/issues.
- Incorporated scripts using Python and PowerShell to aggregate XML test results into an organized format and to load the latest build code onto the hardware, so that daily testing can be performed.
- Utilized Jenkins to provide a continuous integration service in order to automate the entire process of loading the latest build code and test files, running the tests, and generating a report of the results once per day.
- Explored ways to visualize and send a daily report of test results to team members using HTML, Javascript, and CSS.

Startup, Inc **May 2024 – August 2025**
Front End Developer Intern *City, State*

- Assisted in development of the front end of a mobile application for iOS/Android using Dart and the Flutter framework.
- Worked with Google Firebase to manage user inputted data across multiple platforms including web and mobile apps.
- Collaborated with team members using version control systems such as Git to organize modifications and assign tasks.
- Utilized Android Studio as a development environment in order to visualize the application in both iOS and Android.

Projects

Gym Reservation Bot | *Python, Selenium, Google Cloud Console* **January 2021**

- Developed an automatic bot using Python and Google Cloud Console to register myself for a timeslot at my school gym.
- Implemented Selenium to create an instance of Chrome in order to interact with the correct elements of the web page.
- Created a Linux virtual machine to run on Google Cloud so that the program is able to run everyday from the cloud.
- Used Cron to schedule the program to execute automatically at 11 AM every morning so a reservation is made for me.

Ticket Price Calculator App | *Java, Android Studio* **November 2020**

- Created an Android application using Java and Android Studio to calculate ticket prices for trips to museums in NYC.
- Processed user inputted information in the back-end of the app to return a subtotal price based on the tickets selected.
- Utilized the layout editor to create a UI for the application in order to allow different scenes to interact with each other.

Transaction Management GUI | *Java, Eclipse, JavaFX* **October 2020**

- Designed a sample banking transaction system using Java to simulate the common functions of using a bank account.
- Used JavaFX to create a GUI that supports actions such as creating an account, deposit, withdraw, list all acounts, etc.
- Implemented object-oriented programming practices such as inheritance to create different account types and databases.

Education

KIIT University **Sep. 2020 – May 2024**
Bachelor of Technology in Information Technology *CGPA, 9.01/10*

Aditya Academy Secondary **2019 – 2020**
CBSE Class 12th *Percentage, 92 %*

Aditya Academy Secondary **2017 – 2018**
CBSE Class 10th *Percentage, 95 %*

Technical Skills

Languages: Python, Java, C, HTML/CSS, JavaScript, SQL
Developer Tools: VS Code, Eclipse, Google Cloud Platform, Android Studio
Technologies/Frameworks: Linux, Jenkins, GitHub, JUnit, WordPress

Figure 4.3 Sample input resume.

As in Figure 4.3, we can see it contains sample bio-data of the user, with a lot of description of the projects and is indeed a cluttered resume.

Figure 4.4, shows what happens after pressing the "Get Resume" button, and we have included a toast notification system that notifies the progress of the resume-generation pipeline. This gives the user a more transparent way of knowing if everything is going as planned by the program and notifies the user if any error occurs in between.

In Figure 4.5, it shows a preview of the generated resume, we have also provided a download button so that users can conveniently download the resume. The resume generated by the LLM, is provided in Figure 4.6, where we can see the resume looks much cleaner, the projects are written more professionally, and the details such as Educational Qualifications and contact details of the user are intact and exactly the same as provided by the original bio-data.

In Figure 4.7, it shows a preview of the performance metrics, as discussed in the previous sections. We can clearly see the Overlap co-efficient between various data – 0.894 (resume and user-data), signifying that the resume is

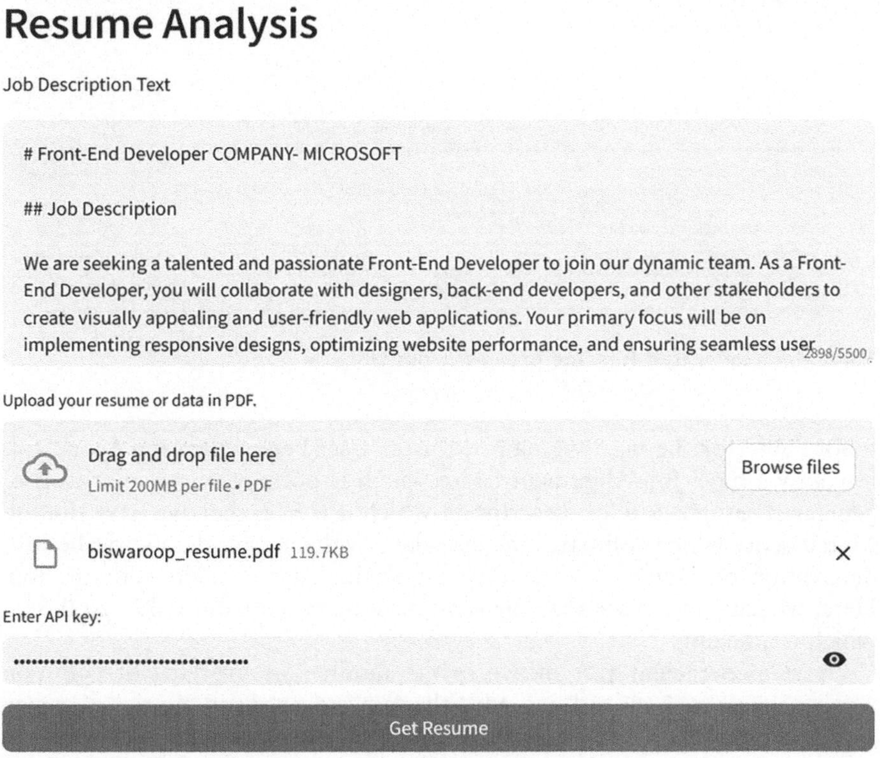

Figure 4.4 Toast notifications – showing progress of proposed tool.

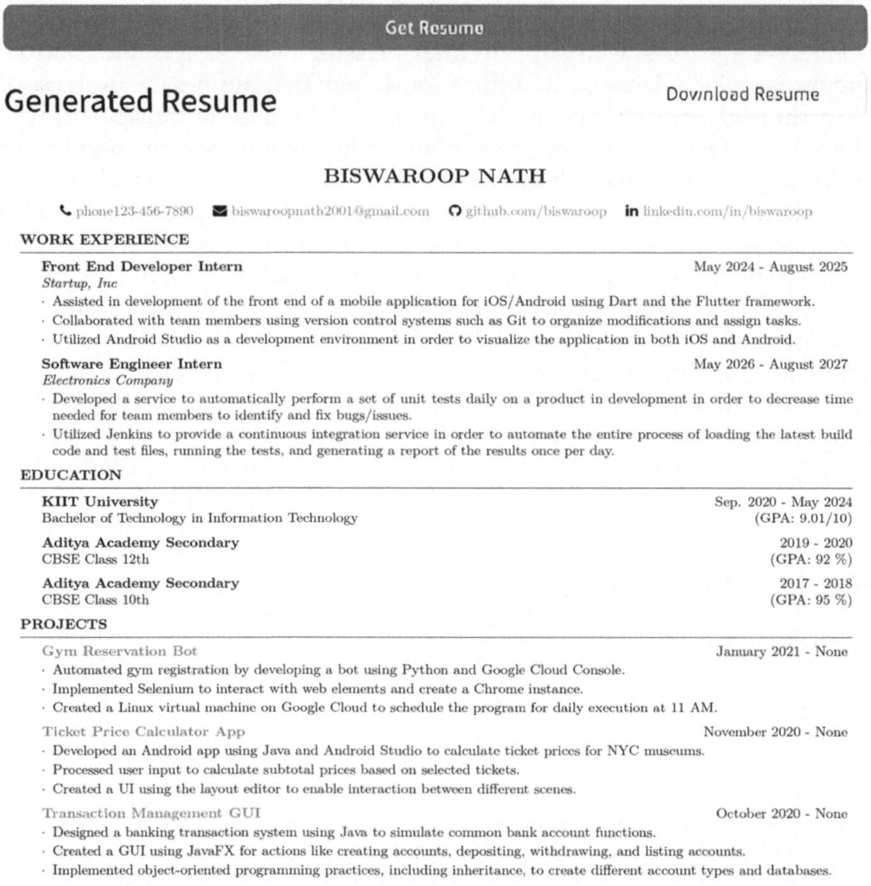

Get Resume

Generated Resume

Download Resume

BISWAROOP NATH

phone123-456-7890 biswaroopnath2001@gmail.com github.com/biswaroop linkedin.com/in/biswaroop

WORK EXPERIENCE

Front End Developer Intern May 2024 - August 2025
Startup, Inc
- Assisted in development of the front end of a mobile application for iOS/Android using Dart and the Flutter framework.
- Collaborated with team members using version control systems such as Git to organize modifications and assign tasks.
- Utilized Android Studio as a development environment in order to visualize the application in both iOS and Android.

Software Engineer Intern May 2026 - August 2027
Electronics Company
- Developed a service to automatically perform a set of unit tests daily on a product in development in order to decrease time needed for team members to identify and fix bugs/issues.
- Utilized Jenkins to provide a continuous integration service in order to automate the entire process of loading the latest build code and test files, running the tests, and generating a report of the results once per day.

EDUCATION

KIIT University Sep. 2020 - May 2024
Bachelor of Technology in Information Technology (GPA: 9.01/10)

Aditya Academy Secondary 2019 - 2020
CBSE Class 12th (GPA: 92 %)

Aditya Academy Secondary 2017 - 2018
CBSE Class 10th (GPA: 95 %)

PROJECTS

Gym Reservation Bot January 2021 - None
- Automated gym registration by developing a bot using Python and Google Cloud Console.
- Implemented Selenium to interact with web elements and create a Chrome instance.
- Created a Linux virtual machine on Google Cloud to schedule the program for daily execution at 11 AM.

Ticket Price Calculator App November 2020 - None
- Developed an Android app using Java and Android Studio to calculate ticket prices for NYC museums.
- Processed user input to calculate subtotal prices based on selected tickets.
- Created a UI using the layout editor to enable interaction between different scenes.

Transaction Management GUI October 2020 - None
- Designed a banking transaction system using Java to simulate common bank account functions.
- Created a GUI using JavaFX for actions like creating accounts, depositing, withdrawing, and listing accounts.
- Implemented object-oriented programming practices, including inheritance, to create different account types and databases.

Figure 4.5 Generated Resume preview from UI.

tailor-made for the user, we named this as User-Personalization Score, following we have Job Alignment Score which is nothing but comparison of generated resume and job-description which is 0.522, then we have the Job Match Score which compares the user-data or the master-data with the Job-description providing an idea how the original data is suited for the job. Here, we can clearly see that the score has increased from 0.377 to 0.522, which is amazing.

Overlap-coefficient just measures the number of overlaps of common words, hence, we have also included the same comparison metrics as cosine similarity, which is done after converting the textual data in a vector form, thus negating different biases. In this case, too, we can see in Figure 4.7, the User-personalization score is high at 0.886, and the Job match score increased from 0.250 to 0.303, which is again an improvement.

BISWAROOP NATH

phone123-456-7890 biswaroopnath2001@gmail.com github.com/biswaroop linkedin.com/in/biswaroop

WORK EXPERIENCE

Front End Developer Intern May 2024 - August 2025
Startup, Inc

- Assisted in development of the front end of a mobile application for iOS/Android using Dart and the Flutter framework.
- Collaborated with team members using version control systems such as Git to organize modifications and assign tasks.
- Utilized Android Studio as a development environment in order to visualize the application in both iOS and Android.

Software Engineer Intern May 2026 - August 2027
Electronics Company

- Developed a service to automatically perform a set of unit tests daily on a product in development in order to decrease time needed for team members to identify and fix bugs/issues.
- Utilized Jenkins to provide a continuous integration service in order to automate the entire process of loading the latest build code and test files, running the tests, and generating a report of the results once per day.

EDUCATION

KIIT University Sep. 2020 - May 2024
Bachelor of Technology in Information Technology (GPA: 9.01/10)

Aditya Academy Secondary 2019 - 2020
CBSE Class 12th (GPA: 92 %)

Aditya Academy Secondary 2017 - 2018
CBSE Class 10th (GPA: 95 %)

PROJECTS

Gym Reservation Bot January 2021 - None

- Automated gym registration by developing a bot using Python and Google Cloud Console.
- Implemented Selenium to interact with web elements and create a Chrome instance.
- Created a Linux virtual machine on Google Cloud to schedule the program for daily execution at 11 AM.

Ticket Price Calculator App November 2020 - None

- Developed an Android app using Java and Android Studio to calculate ticket prices for NYC museums.
- Processed user input to calculate subtotal prices based on selected tickets.
- Created a UI using the layout editor to enable interaction between different scenes.

Transaction Management GUI October 2020 - None

- Designed a banking transaction system using Java to simulate common bank account functions.
- Created a GUI using JavaFX for actions like creating accounts, depositing, withdrawing, and listing accounts.
- Implemented object-oriented programming practices, including inheritance, to create different account types and databases.

Figure 4.6 Generated Resume by LLM.

Resume Metrics

Overlap Coefficient

User Personlization Score	Job Alignment Score	Job Match Score
0.894	0.522	0.377
↑ [resume,master_data]	↑ [resume,JD]	↑ [master_data,JD]

Cosine Similarity

User Personlization Score	Job Alignment Score	Job Match Score
0.886	0.303	0.250
↑ [resume,master_data]	↑ [resume,JD]	↑ [master_data,JD]

Figure 4.7 Various metrics of Resume from proposed tool.

4.6 CONCLUSION

In this paper, we present Resume Analyzer, a user-friendly pipeline designed to assist job seekers in tailoring their resumes to specific job postings. The process is simple: users provide their resume and the corresponding job description to our pipeline, which is facilitated by advanced Large Language Models (LLMs). Currently, our tool supports Gemini LLM. However, we envision future enhancements to incorporate additional LLMs, particularly those that are open-source.

While LLM-powered systems like Resume Analyzer offer significant benefits, it's essential for users to be aware of potential challenges, such as the phenomenon of hallucination. Hallucination occurs when the model generates content that may be contextually relevant but not necessarily accurate or grounded in reality. To address this, future iterations of our tool could explore techniques like retrieval-augmented generation and generation using knowledge graphs.

Retrieval-augmented generation involves incorporating external knowledge sources, such as databases or pre-existing text, to enhance the generation process. By leveraging additional information, we can improve the relevance and accuracy of the generated content. Similarly, generation using knowledge graphs involves structuring knowledge in the form of graphs and using graph-based algorithms to generate text. This approach can help ensure that generated content is grounded in factual information and reduces the likelihood of hallucination.

By integrating these emerging concepts into our pipeline, we aim to further enhance the accuracy and effectiveness of Resume Analyzer, providing job seekers with a valuable tool to optimize their resume-writing process and improve their chances of success in the competitive job market.

REFERENCES

1. Prakash M. Nadkarni, Lucila Ohno-Machado, and Wendy W. Chapman. "Natural language processing: an introduction", *Journal of the American Medical Informatics Association* 18.5: 544–551, 2011.
2. Ramzan Talib, et al. "Text mining: techniques, applications and issues", *International Journal of Advanced Computer Science and Applications* 7.11: 414–418, 2016.
3. XiaoWei Li, Hui Shu, Yi Zhai, and ZhiQiang Lin. "A method for resume information extraction using bert-bilstm-crf", In *21st International Conference on Communication Technology (ICCT)*, IEEE, 1437–1442, 2021.
4. VV Satyanarayana Tallapragada, V Sushma Raj, U Deepak, P Divya Sai, and T Mallikarjuna. 2023. "Improved resume parsing based on contextual meaning extraction using BERT", In *7th International Conference on Intelligent Computing and Control Systems (ICICCS)*, IEEE, 1702–1708, 2023.

5. Kameni Florentin Flambeau Jiechieu and Norbert Tsopze. "Skills prediction based on multi-label resume classification using CNN with model predictions explanation", *Neural Computing and Applications* 33, 5069–5087, 2021.
6. Yi Xing, Allan James, and Croft W. Bruce. "Matching resumes and jobs based on relevance models", In *Proceedings of the 30th Annual International ACM SIGIR Conference on Research and Development in Information Retrieval*, 809–810, 2007.
7. Peg Thoms, et al. "Resume characteristics as predictors of an invitation to interview", *Journal of Business and Psychology* 13: 339–356, 1999.
8. Hans Jürgen Eysenck. "The measurement of personality [Resume]", *Proceedings of the Royal Society of Medicine*, XL, 75–80, 1946.
9. Kun Yu, Guan Gang, and Zhou Ming. "Resume information extraction with cascaded hybrid model", In *Proceedings of the 43rd Annual Meeting of the Association for Computational Linguistics (ACL)*, 2005.
10. Cathryn L. Cox, and William H. Glick. "Resume evaluations and cosmetics use: when more is not better", *Sex Roles* 14: 51–58, 1986.
11. Chen Jie, Zhang Chunxia, Niu Zhendong, et al. "A two-step resume information extraction algorithm", *Mathematical Problems in Engineering*, 1, 5761287, 2018.
12. Yang Jingfeng, Jin Hongye, Tang Ruixiang, Han Xiaotian, Feng Qizhang, Jiang Haoming, Zhong Shaochen, Yin Bing, and Xia Hu. "Harnessing the power of LLMs in practice: A SURVEY on ChatGPT and beyond", *ACM Transactions on Knowledge Discovery from Data*, 18, 6, 2024. https://doi.org/10.1145/3649506
13. Wei Xiang, Cui Xingyu, Cheng Ning, Wang Xiaobin, Zhang Xin, Huang Shen, Xie Pengjun, Xu Jinan, Chen Yufeng, Zhang Meishan, et al. "Zero-shot information extraction via chatting with chatgpt", *arXiv preprint arXiv:2302.10205*, 2023.
14. Zhu Yutao, Yuan Huaying, Wang Shuting, Liu Jiongnan, Liu Wenhan, Deng Chenlong, Dou Zhicheng, and Wen Ji-Rong. "Large language models for information retrieval: a survey", *arXiv preprint arXiv:2308.07107*, 2023.
15. Ibrahim Adeshola and Adeola Praise Adepoju. "The opportunities and challenges of ChatGPT in education", *Interactive Learning Environments*, 32, 1–14, 2023.
16. Chung Kwan Lo. "What is the impact of ChatGPT on education? A rapid review of the literature", *Education Sciences*, 13(4), 410, 2023.
17. G. Meyer Jesse, J. Urbanowicz Ryan, C.N. Martin Patrick, Karen O'Connor, Li Ruowang, Peng Pei-Chen, J. Bright Tiffani, Tatonetti Nicholas, Won Kyoung Jae, Gonzalez-Hernandez Graciela, et al. "ChatGPT and large language models in academia: opportunities and challenges", *BioData Mining*, 16(1), 20, 2023.
18. Pal Soumen, Bhattacharya Manojit, Lee Sang-Soo, and Chakraborty Chiranjib. "A domain-specific next-generation large language model (LLM) or ChatGPT is required for biomedical engineering and research", *Annals of Biomedical Engineering*, 52, 1–4, 2023.
19. Skondras Panagiotis, Psaroudakis George, Zervas Panagiotis, and Tzimas Giannis. "Efficient resume classification through rapid dataset creation using ChatGPT", In *14th International Conference on Information, Intelligence, Systems & Applications (IISA)*. IEEE, 1–5, 2023.

20. Tianyang Lin, et al. "A survey of transformers", *AI Open*, 3, 111–132, 2022.
21. Introducing ChatGPT, (n.d.). https://openai.com/index/chatgpt/
22. AI for Every Developer, (n.d.). https://ai.google.dev/
23. Yinhan Liu, et al. "Roberta: a robustly optimized bert pretraining approach", *arXiv preprint arXiv:1907.11692*, 2019.
24. Henry F. Inman and Edwin L. Bradley Jr "The overlapping coefficient as a measure of agreement between probability distributions and point estimation of the overlap of two normal densities", *Communications in Statistics-theory and Methods*, 18(10), 3851–3874, 1989.
25. Alfirna Rizqi Lahitani, Permanasari Adhistya Erna, and Setiawan Noor Akhmad. "Cosine similarity to determine similarity measure: Study case in online essay assessment", In *4th International Conference on Cyber and IT Service Management*. IEEE, 2016.

Chapter 5

Role of artificial intelligence and machine learning in drug discovery, personalised treatment, and simplifying medical treatment

Rinku Manvani, Harsh Purohit, Bhumika Choksi, Chita Ranjan Sahoo, and Sejal Shah

5.1 INTRODUCTION

The term AI (Artificial Intelligence), also called Machine Intelligence, was coined by MacCarthy in 1956. However, the thought was first raised in the 1940s when computer pioneers Alan Turing (the developer of the Turing test) and John Von Neumann worked on examining how machines could "think" to differentiate humans from machines. They worked on the thought that there is a possibility that Machines can think and mimic human behaviour. A milestone was achieved in this field when, in 1956 researchers proved that a machine becomes able to solve a problem if they are allowed to access a vast amount of memory and GPS (General Problem Solver) came into existence. Hence, AI is defined as the ability of the machine to learn, mimic, enhance, and replicate the behaviour of humans by experience through reasoning and learning. The capacity of AI to evaluate large volumes of data gives it an advantage over humans and experiences much faster in comparison to humans (Fleming, 2018; Kumar, 2019). However, AI has become feasible in recent years because of advancements in the computing world and an increase in data.

AI is connected to us in many ways and plays a critical role in our lives. AI has been used in computer programs for years. Today AI is found practically everywhere, not only targeted to business but also in education, cybersecurity, and our daily lives in the form of Personal Assistants – Siri, Cortana, Alexa, Google Assistant, Google Lens, etc., Manufacturing robots, Automated Finance investing, Automated Mass Transport, Virtual Travel booking agents, etc. With AI, now Myntra is aware of the things consumers want to purchase throughout different seasons, Hotstar now understands which shows and movies consumers enjoy watching and Google knows which medications users look up. Still, experts believe that more innovation is yet to come using AI.

AI is based on the advantages of multiple subjects, including linguistics, computer science, and mathematics, etc. Here, computer science provides tools for designing algorithms, and mathematics will provide tools for

DOI: 10.1201/9781003643166-5

problem-solving (Chan et al., 2019). AI encompasses several domains like logic, probability theory, algorithms, economics, solution searches, and knowledge representation, as well as machine learning as its basic model. (Dara et al., 2022). AI is based mainly on 2 components.

1.1 Machine Learning.
1.2 Artificial Neural Networks.

The goal of machine learning (ML) algorithms is to find patterns in previously classed data sets. An area of machine learning called "deep learning" (DL) makes use of artificial neural networks, or ANNs. (Tyagi, 2020).

5.1.1 Machine learning

Machine Learning makes the computer learn from experience, data, and algorithms for which it uses probability theory and statistics. Supervised and unsupervised machine learning are the two categories. Supervised algorithms may apply what they have learnt in the past to new sets. Unsupervised are the ones that draw the inference from the new data set. The algorithm of Machine Learning establishes linear and non-linear relationships in a given data set. Statistical methods are used to train the algorithms to predict the data set (Dara et al., 2022).

Machine learning is a deep learning subset that is represented by a neural network consisting of three or more layers (Dara et al., 2022). These neural networks "learn" from enormous amounts of data by attempting, but failing miserably, to imitate the way the human brain works. Deep learning is a crucial technological advancement for autonomous vehicles that allows them to distinguish between stop signs and lampposts. It is imperative to have voice control on consumer gadgets, such as tablets, TVs, cellphones, and hands-free speakers. It yields results that were previously impossible. A computer model that has undergone deep learning can do categorization tasks straight from photos, text, or audio. Neural network topologies and a significant amount of labelled data are used to train models. This benefits newly created applications or programs with numerous output categories (Chan et al., 2019).

5.1.2 Artificial neural network

The "Biological Neurons" in the human brain serve as both an inspiration and a point of overlap for the concept. It consists of three layers' worth of neural, or node, components.

1. An Input Layer.
2. Hidden layer.
3. Output layer.

Neurons have mathematical functions to process the incoming data and predict the output based on it. ANN learns from instances. Healthcare suffers because of the rising medical costs and sometimes because of inefficient procedures (Secinaro et al., 2021). However, AI was just added to the field of Medicine where it could be beneficial to mankind by speeding up the process of treatment, achieving greater accuracy, supervisory activities, and developing intensive care and has opened new dimensions for overall health care. These advancements provide quick, cost-efficient, more accurate, and better solutions to healthcare problems. AI is providing new paths to the healthcare sector and giving it a much-needed makeover. AI is proving to be a game changer in the area of health care. It is improving the industrial aspects from Robotic surgeries to securing private records against cyber criminals. The virtual assistance of AI has decreased unnecessary hospital visits and is proven to be a time-saver for medical staff as well as patients. AI has the potential to enhance structure design, evaluate therapeutic targets, and identify hit and lead compounds.

5.2 AI IN DRUG DISCOVERY

The pharmaceutical industry has significantly increased the digitization of its data in recent years. With increased digitization may come the challenge of collecting, evaluating, and implementing that information to report serious healthcare issues (Holmes et al., 2004). The wide chemical space, which contains over 10^{60} molecules, promotes the production of many distinct medicinal compounds (Mak & Pichika, 2019). However, the medicine-creation procedure is hampered by an absence of cutting-edge technology, making it a slow and costly work that may be rectified by utilizing AI (Mishra, 2018). AI has the power to speed up the validation of therapeutic targets, optimize structural design, and identify hit and lead compounds (Mak & Pichika, 2019; Sellwood et al., 2018). Pharmaceutical businesses may have data sets containing millions of molecules to develop new drugs, which makes them difficult to handle with standard ML techniques (Ayers et al., 2022). A computational model consisting of the quantitative structure-activity relationship (QSAR) makes considerable volumes of chemicals or basic physicochemical properties, like log P or log D, easy to forecast. However, these models still need to go far to be able to anticipate biologically complicated qualities like the effectiveness of a medication or its severe side effects (Marr, 2021; Paul et al., 2021).

A machine learning system learns the syntax and principles of chemistry from an existing collection of molecules to create new, innovative compounds. The extension and modification of existing chemical structures no longer require the use of hard-coded rules and heuristics, as was the case with earlier de novo design methodologies. Many drug development strategies are based on the visual characteristics reflected in photographs captured under a microscope. A drug discovery team looks for connections between the phenotypic response and the chemical structure in these situations. It used to be necessary to analyse these images manually or to create specialized software to find certain features in the images. Deep learning methodologies have made this procedure simpler and made new kinds of analysis possible (Ramsundar et al., 2019). Chemical building blocks are used to create drugs and other organic molecules in a process that has long been considered to be both an art and a science. There hasn't been much progress in automating the creation of pathways from readily accessible chemicals to complicated compounds. Large datasets of chemical processes are being evaluated using deep learning, and these analyses are utilized to suggest workable pathways for the synthesis of complicated compounds (Walters & Barzilay, 2021).

A few years ago, only research laboratories could use certain methods, but today, the whole drug discovery pipeline uses them. Although we aren't yet able to claim that an "AI-discovered medicine" has been discovered, ML techniques are being used in the fields described here as well as several others, from target discovery through clinical trial design. By displaying molecular distributions and features, the enormous virtual chemical space represents a molecular topography map (Paul et al., 2021). The purpose of chemical space visualization is to locate bioactive chemicals by gathering positional data on molecules within the space. Consequently, virtual screening (VS) helps identify appropriate compounds for additional investigation. Several open-access chemical databases include ChemBank, DrugBank, PubChem, and ChemDB. In addition to ligand and structure-based techniques, other in silico methods for virtual compound screening enable more effective removal of non-target compounds, enhanced profile analysis, and more cost-effective therapeutic molecule selection (Mak & Pichika, 2019). Molecular fingerprint identification and coulomb matrices are two examples of drug design techniques that consider the physical, chemical, and toxicological aspects of choosing a lead molecule.

Medicine combination discovery relies heavily on deep learning. In the SARS-CoV-2 pandemic, AI, machine learning, and deep learning aided in the creation of vaccinations and pharmaceuticals. The challenging process of finding new drugs could be made easier, faster, and less complex with the assistance of deep learning. The algorithms of Deep learning can create compounds with the required properties and anticipate pharmacological properties and drug-target interactions. Deep learning algorithms may perform well on molecular, clinical, and population data, and a range of toolkits can be employed to search for important patterns in the data. Thanks to machine learning and deep learning, researchers can now discover protein structures more quickly through molecular modelling and predictive analytics. Figure 5.1 shows the drug discovery process using AI.

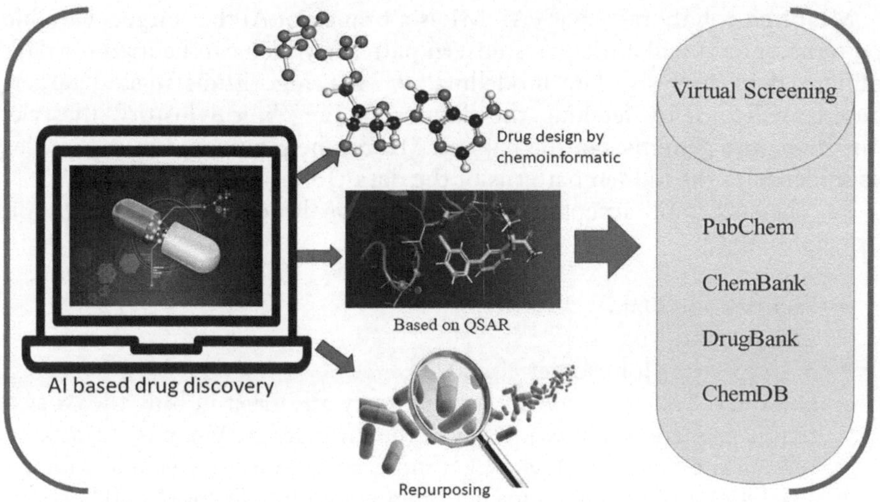

Figure 5.1 AI-based drug discovery.

5.3 AI IN PERSONALISED TREATMENT

Medicines are chemical compounds that are used to cure or prevent diseases and have been proven to be a key to saving lives. Often, the patients are treated with similar medication or treatment as those diagnosed with the same symptoms. But, a single drug or treatment cannot be effective against all individuals of the same kind. One particular drug could be effective against one age group of people but could show different effectiveness against another age group with similar clinical symptoms. Personalized medicine can be an alternative to this obstacle. All the body systems vary clinically, so it is better suggested that the medication must be customized according to the clinical condition of the individual (Schork, 2019). This approach is often called precision /individual /customized medication/ patient-specific treatment. Personalized treatment becomes key when the patients show less response to the generalized treatment or require a unique treatment for an uncommon disease.

In the past 10 years, there has been a remarkable and significant acceptance of AI in various domains (Fleming, 2018). AI is becoming a promising tool in all sectors of the world including personalized treatment, drug discovery, maternal health, patient health, and managing medical data in the healthcare field. AI in personalized treatment is a promising approach to revolutionizing healthcare. The progression of AI in the medical field is pumped by the digitalization of health-related information and advancement in technology. (Topol, 2019; Hashimoto et al., 2018; Mayo & Leung, 2018).

ML often collaborates with AI. ML is a branch of AI that targets variable patterns of data and anticipates unseen patterns which can be used for data mining, data analysis, data modelling, etc. ML algorithms suggest target-based medicines by fetching the data from the clinical history, lifestyle, nutrition, and genomics of the patient. ML can help prevent future diseases as it identifies the hidden patterns of the data (Johnson et al., 2021)

For the successful acceptance of AI in the health care sector, three main principles apply –

- Security and Data
- Insights and Analytics
- 3.3Expertise (Johnson et al., 2021)
- Data and Security – means transparency and trust in how the AI system is trained and how it is given the knowledge. We must be able to rely on the work of AI as we get more and more comfortable with it.
- Analytics and Insights – means "augmented intelligence" and "actionable insights" must be in favour of what humans do and not against it. AI gets information from multiple structured and unstructured sources, it can use computer vision and multimodal applications, and it can help medical professionals in constructing more meticulous conclusions in the area of health care. (Nurse making a care plan for a pregnant woman, Physician prescribing a drug to a patient).
- Expertise means – the AI system is trained and created by human professionals, which requires expertise in the field to lead the workforce change and always requires new skills. Developing a dazzling AI model to meet the needs of the rising globe and produce high-quality commercial apps needs qualified specialists with expert knowledge of the newest technology.

5.3.1 Naïve Bayes Classifier

The Naive Bayes Classifier is among the best and easiest classification algorithms currently available. It enables the rapid development of ML models with accurate prediction capabilities. Some often used Spam filtering, sentiment analysis, and article classification, which are all done using Naive Bayes algorithms. Naive Bayes models come in three varieties: Bernoulli, Multinomial, and Gaussian. Data mining's categorization function classifies objects based on how similar they are to one another. These approaches to Data classifiers are utilized in various real-time applications to categorize data. Before any new pharmaceuticals in the pharmaceutical industry may be sold, they must first receive FDA approval. The FDA uses only human labour, which adds time to the approval process for new drugs. The FDA uses a classification technique to simplify the approval process for new treatments. One way to classify unknown drugs is by looking at their relative molecular mass, hydrogen bond donors/acceptors, number of flexible rotation keys, polar surface area, and hydrophobic constant, amongst other chemical characteristics. By doing so, the FDA can reduce the time it takes to approve new treatments. To create a new database, these six attributes and the drug category were acquired from the US National Library of Medicine. Using the gathered data, the Naive Bayes algorithm may accurately forecast the target class of unknown medications (Daley, 2023).

5.4 AI APPLICATIONS

5.4.1 AI in diagnostic assistance

AI was once misinterpreted as a competitor of medical experts but has now got the true recognition of being a second helping hand of professionals. AI is now considered a tool that complements and assists human work instead of replacing physicians and healthcare staff. It is a strong belief that AI has the ability to enhance all parts of the healthcare system, from diagnosis to therapy. AI, with its efficiency and accuracy, has proved to be a wonderful support for overloaded healthcare professionals and helps to reduce their pressure (Lee & Yoon, 2021). AI supports healthcare personnel with various tasks, including data collection and documentation, administrative work, patient outreach and monitoring, device automation, etc. There is evidence that AI algorithms outperform humans in the processing of medical imagery, identifying biomarkers and symptoms from Electronic Medical Records, Characterization, and prognostic reports of disease (Miller & Brown, 2018).

The need for healthcare utility is constantly in trend, creating a shortage of healthcare facilities and physicians in many countries. With the rise of smartphones and other wireless technologies, a new model of healthcare delivery has emerged, allowing for the provision of services like health tracking applications and search platforms through remote interactions that are accessible at any time and from any place (Reddy et al., 2019). These services are potential tools for underserved places that lack specialists. Along with that, such technologies minimize the exposure of patients to the clinic, keeping them away from unnecessary clinic exposure to infections (Combi et al., 2016)

Telehealth services are crucial for regions with limited access to healthcare and a shortage of specialists. They can help to reduce costs and minimize the risk of exposure to transmissible diseases at clinics. In developing nations, telehealth technology is particularly helpful since it allows for the growth of healthcare systems and the creation of infrastructure to match the present demands of patients (Bohr & Memarzadeh, 2020). Although the concept is simple, these solutions nonetheless necessitate thorough independent evaluation to guarantee the efficacy and safety of patients.

Deep learning models can decode images from medical scans, such as MRIs, CT scans, and X-rays, to establish a diagnosis (Bhattacharjee, 2022).

In medical images, anomalies and risks can be identified by algorithms. Deep learning is heavily utilized in the diagnosis of cancer. Significant progress in computer vision has recently been made possible by machine learning and deep learning. Medical imaging allows for a quicker diagnosis of diseases, which simplifies treatment (Fleming, 2018).

5.4.1.1 ANN (artificial neural network)

A network of associated nodes or vertices that resemble the structure of neurons in a real brain is called an artificial neuron. Each edge or connection can communicate with other nodes or neurons. Artificial neurons can communicate with connected neurons by processing the signals received. As learning continues, the weight of nodes and edges (neurons and connections) can change, which affects the intensity of the connection's signal by either increasing or decreasing it. Layers of neurons are aggregated, with each layer modifying its inputs in different ways. Signals travel through the layers, potentially passing through them multiple times, from the input layer to the output layer. A neural network is frequently trained by figuring out how much the planned output differs from the projected output. This disparity contains the error. The network then uses this error value, along with a learning technique, to adjust its weighted associations. As adjustments are made over time, the neural network will provide output that progressively resembles the intended output. Before the training can be discontinued under specific circumstances, these adjustments can be made multiple times. ANN first tried to leverage the architecture of the human brain to perform tasks that were hard for conventional algorithms to execute properly. Neurons are linked together in a variety of ways to allow for the conversion of some inputs into outputs. The network produces an ordered, weighted graph. An ANN consists of a set of simulated neurons.

In medication delivery, pharmaceutical research, and Blood-Brain Barrier (BBB) permeability, ANN is frequently employed as a classifier. ANNs are frequently employed in various medical specialties, particularly cardiology. The fields of radiology, diagnostics, electronic signal analysis, and radiological image analysis have all made extensive use of ANNs (Bhattacharjee, 2022). ANN can potentially speed up and lower the cost of finding new antimicrobial agents while assisting drug development research.

5.4.2 AI in patient care

5.4.2.1 Random forest

Supervised learning is a popular machine learning approach that includes the Random Forest algorithm. When it comes to machine learning, supervised learning is sufficiently adaptable to handle issues involving both classification and regression. Using ensemble learning, the approach improves model performance and tackles complicated problems by combining several classifiers. Instead of relying on only one decision tree, the random forest takes into account the majority of predictions and uses their votes to determine the outcome. It's done using forecasts from each tree. Having more trees in the forest improves accuracy by decreasing the likelihood of overfitting. It predicts outcomes with reasonable accuracy, especially for the vast dataset. It requires less training time compared to other algorithms. It can remain accurate even when a substantial amount of data is missing. These are the key features.

The first step in running Random Forest is to build the forest by connecting N decision trees; the second step is to generate forecasts for each of those trees. This technique can be used to identify illness patterns and associated risks.

5.4.3 AI in maternal care

A thriving society requires maternal and infant health. The major issues associated with maternal health include – miscarriage, ectopic pregnancy, preeclampsia, and failure in the progression of labour pain, which results in C-sections (Saaka & Hammond, 2020; Fernández et al., 2019; Murali et al., 2020; Phipps et al., 2019; Padilla et al., 2021). Iron deficiency is another factor regarding maternal health that occurs due to postpartum and antepartum hemorrhage, retained placenta, and vaginal infection during delivery (Patrick et al., 2020; Yockey et al., 2016; Polivka et al., 1997).

The major issues associated with the neonatal or newly born baby are uterine growth retardation, birth asphyxia during delivery shoulder dystocia, septicemia, premature delivery, or any other congenital abnormalities (Gulzar Ahmad et al., 2022).

The complications related to the mother as well as the baby need to be detected in time so healthcare professionals can take action against them. Advancements in technology aid in improving the medical sector with the use of AI, Sensing, and Computing. It has made possible the early detection of complications so that action can be taken on an urgent basis.

Sensing technology has been demonstrated to be an effective tool for enhancing mother and newborn health care, patient monitoring during pregnancy, and live monitoring of several diseases (Runkle et al., 2019). ML strategies are another tool used for the diagnosis of abnormalities in a mother during the early stages of pregnancy, as they have the potential to predict a baby's health by detecting brain and general body growth. Advancements in developing ML algorithms detect complications like preterm birth risk, wild stress detection, prenatal risk, congenital heart disease, and postpartum depression. (King et al., 2019; Woolery & Grzymala-Busse, 1994; Veena & Aravindhar, 2021; Zhang et al., 2021; Chu et al., 2020).

The currently used algorithms for best pregnancy outcomes include – SVM (Support Vector Machine), ANN (Artificial Neural Networks), and RA (Regression Analysis)(Shahid et al., 2019, Davidson & Boland, 2021).

5.4.4 Managing data

5.4.4.1 SVM (support vector machine)

A well-known supervised learning algorithm, Support Vector Machine (SVM), is proficient in handling both regression and Machine Learning Classification tasks. Finding the best line or decision boundary to divide n-dimensional space into classes is the primary objective of the support vector machine (SVM) approach, which will enable subsequent data points to be classified quickly (Johnson et al., 2021). SVM chooses the extreme points and vectors to aid in the creation of the hyperplane.

The SVM approach is based on support vectors, which stand in for these extreme cases. When two different cases have the same features, the SVM algorithm is used to model the one that can differentiate the two datasets. We first train our model with the number of images of two different datasets so it can learn about the difference, and then we test it. The support vector constructs a decision boundary between these two datasets. When an extreme boundary case comes, it uses its learning. On the basis of support vectors, it will give the correct decision. For linearly separable data, or data that can be split into two groups with a single straight line, linear support vector machines (SVM) can be utilized to classify the data. Linear SM classifier is the name of the classifier used for such data. Non-linear SVM is utilized for data that is not linearly separated. It is possible to use a Non-Linear SVM classifier to sort datasets that can't be sorted with a straight line. These datasets are called "non-linear data".

Currently, SVMs rank among the most effective methods for predicting chemical and biological properties and for the computational discovery of active substances. Their usage in drug development is predicted to grow even more. This will also involve the creation of SVM-based meta-classifiers that mix several techniques and use their unique advantages and complementarities. It has shown excellence in resolving classification issues across several biomedical domains, particularly in bioinformatics.

SVM modelling is a potential classification technique for identifying individuals with diseases that are common worldwide, such as diabetes and pre-diabetes. Recent research suggests that individuals with pre-diabetes can avoid becoming diabetic by changing their lifestyle or taking medication. Therefore, early screening and diagnosis form the cornerstone of effective prevention programs. To recognize individuals who are at increased risk of developing diabetes or who already have pre-diabetes, various risk scores and prediction formulas have been developed. These equations include prevalent risk variables such as body mass index (BMI) and familial history of diabetes. Further research on this method in other

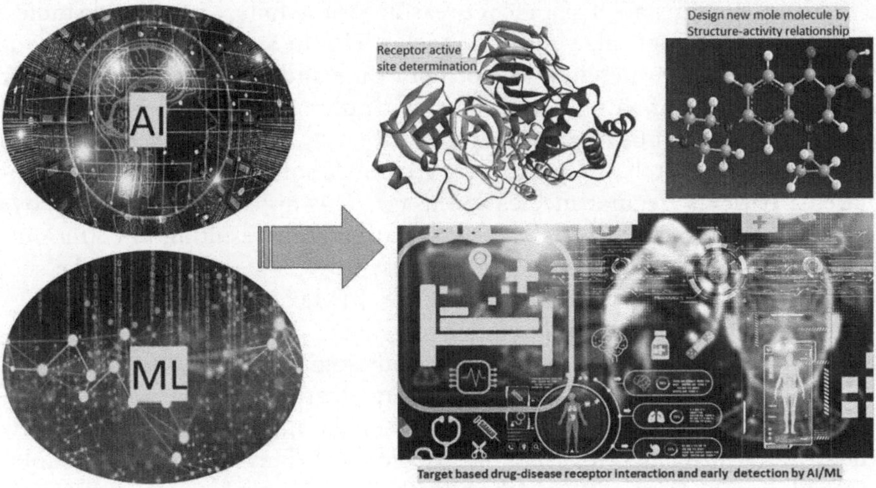

Figure 5.2 AI and ML applications.

complex diseases with comparable components is necessary. Figure 5.2 illustrates the various uses of ML and AI.

5.5 ROLE OF AI IN CLINICAL TRIALS

Clinical trials are a crucial step in evaluating the safety and effectiveness of medical treatments for specific health problems in human subjects. These trials typically last for 6–7 years and require significant financial investment. However, only 1 out of every 10 compounds that undergo clinical trials are ultimately successful, leading to significant losses for the industry (Hay et al., 2014). These failures might result from insufficient infrastructure, inadequate technology needs, or poor patient selection. Due to the abundance of digital medical data that exists at present, there are numerous opportunities for analysis, research, and improvement in the healthcare industry, these issues can be mitigated with the assistance of AI (Harrer et al., 2019). Patient enrollment consumes one-third of the time necessary for clinical research. Enrolling eligible people can help to reduce the 86% of clinical studies that would otherwise be completed (Fogel, 2018). AI can assist in identifying a particular group of sick individuals for participation in Phase II and III clinical trials by analysing the patient's specific genome-exposome profile. This study can assist in the early identification of potential treatment targets in the subjects chosen (Mak & Pichika, 2019).

The discovery of potential drug candidates in preclinical trials is a crucial step in identifying lead compounds that are likely to pass clinical trials. AI-powered techniques like predictive Machine Learning can help predict lead compounds early on, taking into account the specified patient group. AI can help with other reasoning techniques as well (Harrer et al., 2019).

Patient dropouts are a significant reason for the failure of clinical studies, with nearly 30% of trials failing due to this. This necessitates further recruitment efforts and results in a loss of both time and money. To prevent this, close monitoring of patients and support in adhering to the intended clinical trial procedure can be of great help.

In Phase II research, AiCure developed mobile software to monitor schizophrenia patients' regular medication usage. This approach helped improve adherence to the clinical trial procedure, ultimately resulting in better outcomes. The patient adherence rate increased by 25%, resulting in the successful completion of the clinical experiment (Mak & Pichika, 2019).

AI may be used in medical imaging to generate imaging biomarkers automatically. It may also be used to aid the radiologist in reading the studies. AI may detect worrisome patches due to image features that are suggestive of cancerous tissue (Weiss et al., 2020). This might allow radiologists to study the research while first focusing on specific areas. Furthermore, the radiologist will benefit from ensuring that no places to investigate are overlooked. The pathological treatment of a disease may be assisted by the application of AI technology, which may also be utilized to construct multivariate models based on clinical trial data (Pianykh et al., 2020).

AI has a significant impact on medical image processing since it allows for the automation of formerly human operations such as structure, tissue, and organ segmentation (Savadjiev et al., 2019). By avoiding human involvement, AI eliminates the subjectivity and unpredictability that are inherent in every process that relies on humans. A crucial aspect of clinical trials is the improvement of objectivity since it enables the homogeneity of processes like picture segmentation (Li et al., 2020). It speeds up reading times and enhances the radiologist's workflow. This may be done by developing a CAD system that can categorize photos automatically. A screening will be conducted by the CAD to distinguish between healthy and pathological photographs. The radiologists' time efficiency will be increased by prioritizing those imaging studies that were labelled as abnormal during the radiological reading (Oakden-Rayner, 2019).

Clinical studies are complex and costly. Scientists can use machine learning and deep learning to gather data from multiple sources and identify potential candidates for clinical trials. With the utilization of deep learning, it is also possible to monitor the progress of these trials while minimizing human errors and interference continuously. Scientists can use machine learning and deep learning to gather data from multiple sources and identify potential candidates for clinical trials (Fultinavičiūtė, 2022).

5.5.1 Linear regression

Regression analysis is a set of statistical methods used to evaluate, examine, or estimate the relationship between two or more variables. Regression analysis comes in various forms. To describe the relationship between one or more explanatory aspects and a scalar response, machine learning algorithms like linear regression are utilized. The methodology is known as multiple linear regression when dealing with many explanatory variables and as simple linear regression when dealing with one. When all other predictor variables remain constant, a fitted linear regression model can determine the connection between a single predictor variable and the response variable under certain conditions. The following are examples of popular linear regression estimation methods: Least-squares, Maximum-likelihood, Bayesian, Quantile, Principal component, and Least angle. In Health Care, Regression can be used to predict medical information (Bhattacharjee, 2022). For example, it can be used to predict the length of stay at the hospital, the prediction of total surgical process time to enable efficient use of OT, and it is also being used to predict healthcare expenses of individuals based on some variables. Regression becomes helpful in forecasting information, analysing results, and correcting errors, but there are also some limitations. It is assumed that the relationship between the variables remains unchanged (Girnyak, 2022). This assumption may not always hold and may lead to misleading results, which is the drawback

5.6 SIMPLIFICATION OF TREATMENT

AI and ML approaches could effectively assist in the collection, conversion, analysis, and understanding of data at each stage of the development and use of novel therapeutics. AI/ML modelling approaches have a broad range of utilization, such as replacing, reducing, and improving the usage of model animals in preclinical development. AI/ML systems can aid in identifying individuals with specific illness features or clinical conditions during clinical trials. They help facilitate data collecting and processing, which is later submitted to regulatory agencies for marketing authorization procedures. AI and ML technologies have unparalleled capacities in interpreting intricate biological data, forecasting molecular interactions, and recognizing novel potent candidates. These technologies enable academics to analyse extensive datasets with enhanced speed and accuracy compared to previous methods. AI systems can efficiently analyse enormous databases of organic compounds to find molecules that possess the needed features, thereby greatly accelerating the first phases of drug discovery (Selvaraj et al. 2021; Xu et al. 2021; Paul et al. 2021; Vatansever et al. 2021).

An essential obstacle in the area of developing drugs is the process of identifying and confirming appropriate therapeutic targets. AI and ML algorithms have the ability to analyse genetic, genome-wide, and proteomic data to recognize potential targets for diseases. Through the identification of patterns and correlations in biological data, AI can forecast the probability of a target's effectiveness (Paul et al. 2021; Vora et al. 2023). This empowers researchers to make well-informed decisions before undertaking time-consuming and expensive experimental procedures. The methods of screening new drug candidates include assessing their effects on biological systems. AI and ML models can forecast the actions of substances in intricate biological settings, simplifying the process of choosing compounds for further experimentation. By employing this predictive methodology, valuable time and resources are conserved since only the most auspicious individuals go to further phases of advancement. AI/ML-driven computational simulations anticipate molecule-target protein interactions, transforming drug design. Simulated drug design improves specificity, potency, and safety (Dara et al. 2022). Thus, AI-guided rational drug design accelerates lead compound optimization, promoting precision medicine. AI/ML can increase patient recruitment, forecast responses, and optimise trial designs in clinical studies. These technologies can identify participants, predict outcomes, and customize therapy regimens using patient data. More efficient trials, lower costs, and higher success rates result.

Integration of AI/MI technologies into drug research has the potential to transform the field, but it also poses risks and problems that must be considered: data bias and quality, lack of interpretability, overfitting, and generalization, ethical and regulatory concerns, dependency on data quantity, unintended consequences, technical challenges, human expertise and judgment, long-term safety and efficacy (Ahmed et al. 2020). AI/ML adoption is rising throughout drug development phases and therapeutic domains, according to the FDA. Pharmaceutical and biologic applications have increasingly included AI/ML components. These contributions involve drug discovery, clinical trials, post-market safety monitoring, and advanced pharmaceutical production. The Pharmaceuticals Agency acknowledged the rapid development of AI for the need for a regulatory approach to enable safe and effective research, regulation, and implementation of human and animal medication.

5.7 LIMITATIONS

Various AI techniques, such as unsupervised as well as supervised learning approaches, reinforcement learning, and evolutionary and rule-based algorithms, have the potential to assist in resolving these difficulties. These strategies usually rely on the examination of extensive data sets that can be utilized in many ways. These methodologies can forecast the effectiveness and harmfulness of new therapeutic molecules more accurately and efficiently than older methods. Proteins or genetic pathways implicated in diseases can be discovered by AI algorithms, opening up new avenues for drug development. This has the potential to broaden the range of drugs discovered beyond the constraints of traditional methods and could ultimately result in the creation of innovative and more efficient therapies. Consequently, conventional approaches to pharmaceutical research have had moderate success in previous times, but they are constrained by their dependence on experiment and their incapacity to precisely forecast the actions of novel bioactive chemicals. Conversely, AI-based methods has the capacity to augment the efficacy and precision of drug discovery procedures, hence facilitating the creation of more potent treatments.

Now, medicinal chemistry procedures extensively on a trial-and-error approach and extensive experimental techniques. These methods entail scrutinizing a vast quantity of possible medicinal molecules to pinpoint those that possess the desired characteristics. Nevertheless, these approaches might be characterized by their sluggishness and high expenses and frequently produce outcomes with limited precision. Furthermore, their scope may be constrained by the accessibility of appropriate test substances and the challenge of precisely forecasting their actions within the organism.

Although AI has an opportunity to bring about significant advantages in drug research, it is crucial to acknowledge and address certain obstacles and restrictions. An important obstacle is the accessibility of appropriate data. AI-based methodologies usually need a significant amount of data to be used for training purposes. Oftentimes, the accessibility of data may be restricted, or the data itself may possess subpar quality or lack consistency, hence impacting the precision and dependability of the outcomes. Another obstacle arises from ethical considerations, as AI-based methods have the potential to generate questions regarding fairness and bias. For example, if the data utilized to train a machine learning system exhibits bias or lacks representativeness, the consequent predictions may be erroneous or unjust. Developing novel medicinal chemicals ethically and equitably is crucial. Various methodologies and techniques can be employed to surmount the challenges encountered by AI in the domain of chemical medicine. Data augmentation is a method that involves creating artificial data to enhance existing datasets. By augmenting the amount and variety of data utilized to train machine learning algorithms, the accuracy and dependability of the outcomes can be enhanced. Another approach is to employ explainable AI techniques, which aim to deliver understandable and transparent justifications for the predictions generated by ML algorithms. By doing so, it is possible to mitigate issues regarding discrimination and equity in AI-based methods and gain a more comprehensive comprehension of the basic processes and assumptions that drive the predictions.

Contemporary AI-driven methodologies do not serve as a replacement for conventional experimental techniques, nor can they supplant the proficiency and knowledge possessed by human researchers. AI is limited to generating predictions just from the existing data, and it is essential for human researchers to subsequently verify and understand the outcomes. Moreover, incorporating AI alongside conventional experimental techniques can further optimize the drug development process. Integrating the prognostic abilities of AI with human researchers' proficiency and knowledge can enhance drug discovery efficiency and expedite the advancement of novel treatments.

REFERENCES

Ahmed Z, Mohamed K, Zeeshan S, Dong X. (2020). Artificial intelligence with multi-functional machine learning platform development for better healthcare and precision medicine. *Database*, 2020, baaa010.

Ayers, M., Jayatunga, M., Goldader, J., & Meier, C. (2022, March 23). Adopting AI in drug discovery. *BCG Global*. Retrieved July 24, 2023, https://www.bcg.com/publications/2022/adopting-ai-in-pharmaceutical-discovery

Bhattacharjee. (2022, February 2). Artificial intelligence (AI) and its assistance in medical diagnosis | ImpactQA. *ImpactQA*. Retrieved July 24, 2023, from https://www.impactqa.com/blog/artificial-intelligence-ai-and-its-assistance-in-medical-diagnosis/

Bohr, A., & Memarzadeh, K. (2020). The rise of artificial intelligence in healthcare applications. In *Artificial intelligence in healthcare* (pp. 25–60). Academic Press.

Chan, H. S., Shan, H., Dahoun, T., Vogel, H., & Yuan, S. (2019). Advancing drug discovery via artificial intelligence. *Trends in Pharmacological Sciences*, 40(8), 592–604.

Chu, R., Chen, W., Song, G., Yao, S., Xie, L., Song, L., ... & Kong, B. (2020). Predicting the risk of adverse events in pregnant women with congenital heart disease. *Journal of the American Heart Association*, 9(14), e016371.

Combi, C., Pozzani, G., & Pozzi, G. (2016). Telemedicine for developing countries. *Applied Clinical Informatics*, 7(04), 1025–1050.

Daley. (2023, March 24). Artificial intelligence in healthcare: 39 Examples improving the future of medicine. *Built In*. Retrieved July 24, 2023, from https://builtin.com/artificial-intelligence/artificial-intelligence-healthcare

Dara, S., Dhamercherla, S., Jadav, S. S., Babu, C. M., & Ahsan, M. J. (2022). Machine learning in drug discovery: a review. *Artificial Intelligence Review*, 55(3), 1947–1999.

Davidson, L., & Boland, M. R. (2021). Towards deep phenotyping pregnancy: A systematic review on artificial intelligence and machine learning methods to improve pregnancy outcomes. *Briefings in Bioinformatics*, 22(5), bbaa369.

Fernández, A. D. R., Fernández, D. R., & Sánchez, M. T. P. (2019). A decision support system for predicting the treatment of ectopic pregnancies. *International Journal of Medical Informatics*, 129, 198–204.

Fleming, N. (2018). How artificial intelligence is changing drug discovery. *Nature*, 557(7706), S55.

Fogel, D. B. (2018). Factors associated with clinical trials that fail and opportunities for improving the likelihood of success: A review. *Contemporary Clinical Trials Communications*, 7(11), 156–164. doi: 10.1016/j.conctc.2018.08.001

Fultinavičiūtė, U. (2022, April 20). AI benefits in patient identification and clinical trial recruitment has challenges in sight. *Clinical Trials Arena*. https://www.clinicaltrialsarena.com/features/ai-clinical-trial-recruitment/

Girnyak, M. (2022, March 18). What are the major limitations of machine learning algorithms? – PostIndustria. (n.d.). *What Are the Major Limitations of Machine Learning Algorithms? – PostIndustria*. https://postindustria.com/what-are-the-major-limitations-of-machine-learning-algorithms/

Gulzar Ahmad, S., Iqbal, T., Javaid, A., Ullah Munir, E., Kirn, N., Ullah Jan, S., & Ramzan, N. (2022). Sensing and artificial intelligent maternal-infant health care systems: A review. *Sensors*, 22(12), 4362.

Harrer, S.; Shah, P.; Antony, B.; Hu, J. (2019). Artificial intelligence for clinical trial design. *Trends in Pharmacological Sciences*, 40(8), 577–591. https://doi.org/10.1016/j.tips.2019.05.005

Hashimoto, D. A., Rosman, G., Rus, D., & Meireles, O. R. (2018). Artificial intelligence in surgery: promises and perils. *Annals of Surgery*, 268(1), 70.

Hay, M., Thomas, D., Craighead, J. et al. (2014). Clinical development success rates for investigational drugs. *Nature Biotechnology*, 32, 40–51.

Holmes, J., Sacchi, L., & Bellazzi, R. (2004). Artificial intelligence in medicine. *The Annals of The Royal College of Surgeons of England*, 86, 334–338.

Johnson, K. B., Wei, W. Q., Weeraratne, D., Frisse, M. E., Misulis, K., Rhee, K., ... & Snowdon, J. L. (2021). Precision medicine, AI, and the future of personalized health care. *Clinical and Translational Science*, 14(1), 86–93.

King, Z. D., Moskowitz, J., Egilmez, B., Zhang, S., Zhang, L., Bass, M., ... & Alshurafa, N. (2019). Micro-stress EMA: A passive sensing framework for detecting in-the-wild stress in pregnant mothers. *Proceedings of the ACM on Interactive, Mobile, Wearable and Ubiquitous Technologies*, 3(3), 1–22.

Kumar, S. (2019, December 12). Advantages and disadvantages of artificial intelligence. *Medium*. https://towardsdatascience.com/advantages-and-disadvantages-of-artificial-intelligence-182a5ef6588c

Lee, D., & Yoon, S. N. (2021). Application of artificial intelligence-based technologies in the healthcare industry: Opportunities and challenges. *International Journal of Environmental Research and Public Health*, *18*(1), 271.

Li, G. M., Tsoutsos, N. G., Gupta, N., Karri, R. (2020). Computer aided design (CAD) model search and retrieval using frequency domain file conversion, *Additive Manufacturing*, 36. https://doi.org/10.1016/j.addma.2020.101554

Mak, K. K., & Pichika, M. R. (2019). Artificial intelligence in drug development: present status and future prospects. *Drug Discovery Today*, 24(3), 773–780.

Marr, B. (2021, July 2). What are the negative impacts of artificial intelligence (AI)? | Bernard Marr. *Bernard Marr*. https://bernardmarr.com/what-are-the-negative-impacts-of-artificial-intelligence-ai/

Mayo, R. C., & Leung, J. (2018). Artificial intelligence and deep learning–Radiology's next frontier?. *Clinical Imaging*, 49, 87–88.

Miller, D. D., & Brown, E. W. (2018). Artificial intelligence in medical practice: the question to the answer?. *The American Journal of Medicine*, 131(2), 129–133.

Mishra, V. (2018). Artificial intelligence: The beginning of a new era in pharmacy profession. *Asian Journal of Pharmaceutics (AJP)*, 12(02), 7276.

Murali, S., Miller, K., & McDermott, M. (2020). Preeclampsia, eclampsia, and posterior reversible encephalopathy syndrome. *Handbook of Clinical Neurology*, 172, 63–77.

Oakden-Rayner, L. (2019). The rebirth of CAD: How is modern AI different from the CAD we know? *Radiology: Artificial Intelligence* May 29; 1(3):e180089. doi: 10.1148/ryai.2019180089

Padilla, Abir, Zakowski, & Carvalho. (2021, August 9). How AI could help doctors reduce maternal mortality. *Harvard Business Review*. Retrieved July 24, 2023, from https://hbr.org/2021/08/how-ai-could-help-doctors-reduce-maternal-mortality

Patrick, H. S., Mitra, A., Rosen, T., Ananth, C. V., & Schuster, M. (2020). Pharmacologic intervention for the management of retained placenta: A systematic review and meta-analysis of randomized trials. *American Journal of Obstetrics and Gynecology*, 223(3), 447.e1–447.e19.

Paul, D., Sanap, G., Shenoy, S., Kalyane, D., Kalia, K., & Tekade, R. K. (2021). Artificial intelligence in drug discovery and development. *Drug Discovery Today*, 26(1), 80.

Phipps, E. A., Thadhani, R., Benzing, T., & Karumanchi, S. A. (2019). Pre-eclampsia: pathogenesis, novel diagnostics and therapies. *Nature Reviews Nephrology*, 15(5), 275–289.

Pianykh, O. S., Langs, G., Dewey, M., Enzmann, D. R., Herold, C. J., Schoenberg, S. O., Brink, J. A. (2020). Continuous learning AI in radiology: Implementation principles and early applications. *Radiology*, 1, 6–14. doi: 10.1148/radiol.2020200038

Polivka, B. J., Nickel, J. T., & Wilkins III, J. R. (1997). Urinary tract infection during pregnancy: a risk factor for cerebral palsy?. *Journal of Obstetric, Gynecologic, & Neonatal Nursing*, 26(4), 405–413.

Ramsundar, B., Eastman, P., Walters, P., & Pande, V. (2019). *Deep learning for the life sciences: applying deep learning to genomics, microscopy, drug discovery, and more*. O'Reilly Media.

Reddy, S., Fox, J., & Purohit, M. P. (2019). Artificial intelligence-enabled healthcare delivery. *Journal of the Royal Society of Medicine*, 112(1), 22–28.

Runkle, J., Sugg, M., Boase, D., Galvin, S. L., & C. Coulson. (2019). Use of wearable sensors for pregnancy health and environmental monitoring: Descriptive findings from the perspective of patients and providers. *Digital Health*, 5, 20552 07619828220

Saaka, M., & Hammond, A. Y. (2020). Caesarean section delivery and risk of poor childhood growth. *Journal of nutrition and metabolism*, 24, 6432754.

Savadjiev, P., Chong, J., Dohan, A., Vakalopoulou, M., Reinhold, C., Paragios, N., Gallix, B. (2019). Demystification of AI-driven medical image interpretation: Past, present and future. *European Radiology*, 29(3), 1616–1624. doi: 10.1007/s00330-018-5674-x

Schork, N. J. (2019). Artificial intelligence and personalized medicine. *Precision Medicine in Cancer Therapy*, 178, 265–283.

Secinaro, S., Calandra, D., Secinaro, A., Muthurangu, V., & Biancone, P. (2021). The role of artificial intelligence in healthcare: a structured literature review. *BMC Medical Informatics and Decision Making*, 21, 1–23.

Sellwood, M. A., Ahmed, M., Segler, M. H., & Brown, N. (2018). Artificial intelligence in drug discovery. *Future Medicinal Chemistry*, 10(17), 2025–2028.

Selvaraj, C., Chandra, I., Singh, S.K. (2021). Artificial intelligence and machine learning approaches for drug design: Challenges and opportunities for the pharmaceutical industries. *Molecular Diversity*, 26, 1–21.

Shahid, N., Rappon, T., & Berta, W. (2019). Applications of artificial neural networks in health care organizational decision-making: A scoping review. *PloS One*, 14(2), e0212356.

Topol, E. J. (2019). High-performance medicine: the convergence of human and artificial intelligence. *Nature Medicine*, 25(1), 44–56.

Tyagi, N. (2020, April 24). 6 Major branches of artificial intelligence (AI) | Analytics steps. https://www.analyticssteps.com/blogs/6-major-branches-artificial-intelligence-ai

Vatansever S, Schlessinger A, Wacker D, Kaniskan HÜ, Jin J, Zhou MM, Zhang B. (2021). Artificial intelligence and machine learning-aided drug discovery in central nervous system diseases: State-of-the-arts and future directions. *Medicinal Research Reviews*, May; 41(3):1427–1473.

Veena, S., & Aravindhar, D. J. (2021). Remote monitoring system for the detection of prenatal risk in a pregnant woman. *Wireless Personal Communications*, 119, 1051–1064.

Vora LK, Gholap AD, Jetha K, Thakur RR, Solanki HK, Chavda VP. (2023). Artificial intelligence in pharmaceutical technology and drug delivery design. *Pharmaceutics*, 15(7), 1916.

Walters, W. P., & Barzilay, R. (2021). Critical assessment of AI in drug discovery. *Expert Opinion on Drug Discovery*, 16(9), 937–947.

Weiss, J. et al. (2020). Artificial intelligence-derived imaging biomarkers to improve population health. *The Lancet Digital Health*, 2(4), e154–e155.

Woolery, L. K., & Grzymala-Busse, J. (1994). Machine learning for an expert system to predict preterm birth risk. *Journal of the American Medical Informatics Association*, 1(6), 439–446.

Xu Y, Liu X, Cao X, Huang C, Liu E, Qian S, Liu X, Wu Y, Dong F, Qiu CW, Qiu J. (2021). Artificial intelligence: A powerful paradigm for scientific research. *The Innovation*, 2(4), 1–20.

Yockey, L. J., Varela, L., Rakib, T., Khoury-Hanold, W., Fink, S. L., Stutz, B., ... & Iwasaki, A. (2016). Vaginal exposure to Zika virus during pregnancy leads to fetal brain infection. *Cell*, 166(5), 1247–1256.

Zhang, Y., Wang, S., Hermann, A., Joly, R., & Pathak, J. (2021). Development and validation of a machine learning algorithm for predicting the risk of postpartum depression among pregnant women. *Journal of Affective Disorders*, 279, 1–8.

Chapter 6

Analyzing trust relationships within social networks

A comprehensive study

Manoj Kumar Srivastav, Somsubhra Gupta, Subhranil Som, and Biswaranjan Acharya

6.1 INTRODUCTION

In today's online world, social network analysis (SNA) helps to understand how people connect and interact on social networks platform [1, 2]. These networks are really important to make communication and exchange knowledge [3]. Many social networks platforms have been developed on the Web, such as Facebook andInstagram. In those networks, it may be the case that a lot of the end-users are usually physically unknown with each other [27]. In this case, if two unknown participants wish to communicate with each other for various reasons, the evaluation of their trustworthiness along a certain trust path between them within the social network is mandatory [4]. But, the level of trustworthiness may vary and it may be multidimensional [33], and it is sometimes subjective and depends on the person's specific role within the network. It is not an easy task to define trust, as trust cannot be easily defined through mathematical formulas or algorithm procedures [5]. Trust may rely on several factors, from psychological and sociological factors to computer security factors. Trust is essential for building connections and effective communication in social networks, and it depends on various factors. Direct factors like how long people have known each other, how often they interact, and the type of interactions they have play a role. Referenced factors such as influence score, number of followers [42], posts, total likes, and engagement rate also influence trust levels. Understanding these factors is crucial in SNA, especially in online spaces where trust is vital for communication and collaboration. By considering these factors, we can better understand how trust operates in online communities and foster positive relationships and collaboration [35].

Generally, Trustworthiness can vary based on several factors (Table 6.1):

The paper examines trust dynamics in social networks and explores the drawbacks of using binary trust labels. It proposes a fuzzy approach to trust modeling. It highlights the distinctions between direct factors, relying on personal interactions, and referenced factors, incorporating external metrics. The study further outlines various calculation methods, including machine learning techniques, a linear combination model, and a fuzzy logic

DOI: 10.1201/9781003643166-6

Table 6.1 Study of trustworthiness and several factors

Behavioral Patterns	Frequency of communication: Individuals whohave mutual trust prefer to communicate more frequently. Mutual trust and comfort in communication support each other's friendship. Consistency: Trust is built over time through consistent behavior, but inconsistency can diminish trust. Reciprocity: Mutual trust is crucial as trust operates in both directions. If you expect someone to have faith in you, it is equally important for you to have trust in them. This means giving up control and giving others access to data, even if it makes you feel uncomfortable [6].
Network Structure	Connections: Connections can reveal a lot about a person. If an individual is associated with numerous esteemed and reliable individuals, it can have a favorable impact on their own standing. Centrality: Members or users who play a central role in a network, by having numerous connections and serving as links between various groups, typically wield greater influence. These members are regarded as trustworthy sources of information. This central position can foster trust. Influence: The impact of an individual's actions or beliefs on others can serve as an indicator of trustworthiness [7]. However, it's important to note some important points – Negative connections: While an individual may have numerous connections, it is essential to note that if those connections are not reliable, theycould not lead to trustworthiness. Centrality does not imply trust: Just because someone holds a central position in a network does not necessarily mean they are trustworthy; they could wield influence for negative purposes. Context is significant: The type of network also plays a role in determining trust dynamics. Trust within a professional network may differ from that within a social network.
Reputation and Feedback	Credibility of the source: Recommendations from trusted individuals carry more weight than those from strangers. Similarly, feedback from a reputable community source is highly valuable [8]. Content of the feedback: When sharing feedback, don't just rely on stars or basic labels. Look closely at the reviews. Volume of feedback: A single glowing review might not mean much, but a consistent pattern of positive feedback from various sources strengthens the case for trustworthiness. However, there are also limitations to consider: False reviews: There's always the possibility of fake reviews or endorsements, so a healthy dose of skepticism is important. Bias: Reviews can be biased, either positively or negatively. People might be more likely to leave a review if they had a particularly strong (positive or negative) experience. Limited scope: Reviews and feedback often focus on specific transactions or interactions. They might not give a complete picture of someone's overall trustworthiness.

(*Continued*)

Table 6.1 (*Continued*) Study of trustworthiness and several factors

Privacy and Security Measures	Security breach: If a data breach occurs on a platform, users Personal information will be exposed, and identity theft, fraud, and other security risks may occur. This can significantly undermine trust as users feel that their data is not secure. Data Protection Practices: It's really important how the platform deals with your data. If they have clear privacy rules and let you control your info, it builds trust. But if they hide how they use data or have a history of misusing it, that can really make trust go down. Privacy Settings: How much control you get over your privacy settings matters a lot. Platforms that let you decide exactly who sees your info make you feel more powerful and trusting [9]. Here are some additional points to consider: User perception: Even if a platform has strong security, if users don't get it or feel it's not enough, trust can still be shaken. Understanding and perception matter. Regulation: Government regulations concerning data privacy can also influence trust. Platforms that comply with stricter regulations might be seen as more trustworthy. In general, strict data protection and security rules are necessary to buildtrust on social media platforms. When users feel that their information isprotected and their privacy is respected, they are more likely to trust the platform and interact with it authentically.
Shared Interests or Objectives	Common Ground and Connection: Having common interests helps people connect and understand each other better. Knowing someone's likes and dislikes can make interactions more predictable and comfortable, reducing uncertainty and making a trustworthy environment. Also, working together toward common goals strengthen trust, as it shows mutual reliance and commitment. But it's important to note that just having similar interests might not be enough for a deep connection – facing challenges [41] together and sharing core values, like honesty and fairness, play a big role in building strong trust. In conclusion, having shared interests [8] or goals is a strong way to build trust. It forms a connection, makes things more predictable, and encourages working together. To keep trust strong over time, it's not enough to just have similar interests; it's important to share values and respect each other, too.

approach, aiming to offer insights into the intricate workings of trust in online social networks.

In this paper, the following **summarized key points are discussed:**

Limitations of Binary Trust

Relationships among two users in SNA may use a basic "trusted" or "untrusted" label. Believing in people isn't just a simple "yes" or "no." It is not very obvious, and it depends on different situations that have happened before and what someone personally likes.

Fuzzy approach for Trust among users

When employing a fuzzy approach for trust modeling, it involves applying fuzzy logic. This perspective considers trust as something that can vary in different degrees, rather than a simple yes or no. Fuzzy models represent trust in a broader range of ways, offering a more flexible and improved understanding of trust.

In this paper, it is aimed to explore and study following important ideas that help to understand how trust works in social networks

Factors considered:

Direct factors for Trust: These are based on your direct interactions with someone, like how long you've known them, how often you interact, and the types of interactions you have (e.g., casual chats vs. deep conversations).

Referenced factors for Trust: These are more public metrics that can indicate someone's influence and engagement on the platform, such as follower count, posting habits, likes received, and engagement rate.

Calculation methods:

Machine learning: Techniques like Logistic Regression, Decision trees, and k-Nearest Neighbors (k-NN) can analyze data on these factors and predict a user's trustworthiness.

Linear Combination Model: This method assigns weights to different factors (e.g., higher weight to direct interaction duration) and combines them with normalization for a Trust Score. Accuracy is then evaluated using a Confusion Matrix, which shows how well the model classifies users as trustworthy or not.

Fuzzy logic approach: This method defines fuzzy sets (ranges) for factors like influence score and uses fuzzy rules [10] based on linguistic variables ("high," "medium," "low") to evaluate trust. The final trust level is determined by combining the outcomes of these rules and converting them into a single numerical value.

6.2 LITERATURE SURVEY

Hatamleh et al. [11] in their paper Trust in Social Media: Enhancing Social Relationships, have discussed that trust impacts social connections formed through social media engagement. The study finds that trust acts as a moderator, strengthening the positive relationship between social media engagement and the development of social relationships [11].

Almogbel et al. [12] in their paper User Behavior in Social Networks toward Privacy and Trust: Literature Review study explores the link between privacy, trust, and user behavior on social networks. It examines methods for protecting user privacy and building trust within these platforms, highlighting limitations and suggesting future research directions [12].

Alkhamees et al. [13] in their paper User trustworthiness in online social networks: A systematic review highlights a research review focusing on studies between 2012 and 2020 that addressed this issue using various methods like bot detection, spam protection, and fake news recognition. The review aims to improve understanding of online user trustworthiness by examining recent research [13].

Ansheng et al. [14] in their paper Trusted Network Evaluation Model Based on Comprehensive Trust have discussed the method of calculating trust utilizing the three types of models: node behavior model, multiple-attribute decision-making model, and reputation model [14].

Li et al. [15] in their paper Trust in Social Networks: A Computational Model Based on the Trust Features of Language Use propose a trust calculation model based on threefeatures: node characteristics, relationships, and post content. This model assesses a node's trustworthiness, relationships with neighbors, and the quality of their posts [15].

Liu et al. [16] in their paper Three-Valued Subjective Logic: A Model for Trust Assessment in Online Social Networks propose a novel three-valued subjective logic (3VSL) model to assess trust in complex online social networks (OSNs). It leverages the Dirichlet-Categorical distribution for trust computation and achieves high accuracy on real-world datasets [16].

Du et al. [17] in their paper A new trust model for online social networks explore a trust model for social media networks. It considers user interactions and reputations to define trust values between users. The model leverages Gaussian kernel density estimation for reputation and uses interaction data to improve trust prediction accuracy [17].

Nepal et al. [18] in their paper Behavior-based propagation of trust in social networks with restricted and anonymous participation have discussed about a personalized recommender could provide an incentive for members to interact with other members in the community and also describe a trust propagation model based on three key factors: the density of interactions, the degree of separation, and the decay of friendship effect [18].

Kawser Wazed Nafi et al. [19] in their paper A Fuzzy Logic Based Certain Trust Model for E-commerce a new fuzzy logic-based Certain Trust model which considers the ambiguity and vagueness of different domains. Fuzzy-Based Certain Trust Model depends on certain values given by experts and developers that can be applied in a system like cloud computing, internet, website, e-commerce, etc. to ensure the trustworthiness of these platforms [19].

Thomas Dubois et al. [20] in their paper Predicting Trust and Distrust in Social Networks have used an inference algorithm that relies on a probabilistic interpretation of trust based on random graphs with a modified spring-embedding algorithm. This algorithm correctly classifies hidden trust edges as positive or negative with high accuracy [20].

Guha et al. [21] in their paper Propagation of trust and distrust proposes a framework for trust propagation schemes in social networks, including distrust. It analyzes how trust scores spread through a network and achieves accurate trust prediction even with limited user ratings [21].

6.3 METHODOLOGY

6.3.1 Preliminiers

- Social Structure:Social networks represent a simplified view of how people, groups, and even organizations connect with each other.
- Actors and Ties: The core of a social network consists of actors (individuals or groups) and the ties that link them. These ties can represent various relationships, from close friendships to business partnerships.
- Network Theory: Social networks can be visualized using graphs from network theory. In these graphs, actors are represented by nodes, and the ties between them are shown by links or edges.
- Graph: A graph is a pair of sets (V, E), where V is a set of vertices and E is the multi-set of edges that connect pairs of vertices. Typically, the graph is written in the form G (V, E) or G=(V, E).
- Degree: The degree of a node v, denoted deg(v), in an undirected and unweighted network G, is the number of other nodes of the network for which v has an edge. Degree is frequently described as an undirected, unweighted network; A node's degree is simply the number of edges that reach it [5].
- Centrality: Centrality measures how connected a node is. It quantifies the direct influence of a node on its local neighborhood [6].

6.3.2 Types of trust

- Reflexive Trust: A reflexive trust relationship exists when an individual trusts themselves. In social networks, this might represent self-reliance or self-confidence.
- Symmetric Trust: Symmetric trust occurs when trust between two individuals works in both directions. If person A trusts person B, then person B trusts person A to a similar degree. Symmetric trust fosters balanced relationships and mutual understanding in social networks [38].
- Transitive Trust: Transitive trust refers to the property that if person A trusts person B and person B trusts person C, then person A can potentially trust person C. This property allows trust to propagate through indirect connections in a network [22].

- Asymmetric Trust: Asymmetric trust exists when trust between individuals is not mutual. For instance, if person A trusts person B, it doesn't necessarily mean that person B trusts person A equally. Asymmetrical trust can create power dynamics or unequal relationships within a social network.
- Directional Trust: Developing and maintaining trust [32] relationships might have a specific direction. For example, person A might trust person B, but person B might not reciprocate that trust. This directional aspect impacts the dynamics and stability of relationships within the network [40].
- Partial Trust: Trust can be partial or context-specific [34]. Individuals might trust others in certain domains or for specific tasks while being cautious or distrustful in other aspects. This property allows for nuanced relationships in social networks.
- Dynamic Trust: Trust in social networks can change over time due to evolving interactions, experiences, or external factors. It's not static and can be influenced by various factors [39].

6.3.3 Machine learning approach for trust with respect to direct factor

The concept of trust in friendships can be influenced by various factors, and in this case, "Years of Knowing" and "Interaction Duration" are considered. Trust in friendship develops through a combination of time spent together and the quality of those interactions.

Let's analyze how the existence of friendship (Yes/No) correlates with these two factors:

Years of Knowing

- Positive Influence: A longer period of knowing someone may contribute to building trust. Over time, individuals can become more familiar with each other's values, behaviors, and reliability.
- Negative Influence: However, a long history does not guarantee trust. Negative experiences or changes in behavior over the years might erode trust.

Interaction Duration

- Positive Influence: Spending more time together can foster a deeper understanding and connection, contributing to trust. Regular positive interactions can strengthen the bond between friends.
- Negative Influence: On the contrary, short interaction duration might make it challenging to establish trust. Limited time spent together may result in a lack of understanding or familiarity (Table 6.2).

Table 6.2 Case studies of existence of friendship and trust

Case 1	*Friendship exist*	*Trust exist*
Case 2	Friendship not exist	Trust exist
Case3	Friendship exist	Trust not exist
Case 4	No friendship	No trust

This research focuses on a specific situation (case 1) to predict trust levels using a machine learning approach.

In mathematical terms, the representation of direct proportionality between the existence of friendship (F) and the existence of trust (T) using the proportionality constant (k):

$$T = k.F$$

This equation states that as the variable F (existence of friendship) increases, the variable T (existence of trust) will also increase, and vice versa. The constant k represents the proportionality factor between friendship and trust.

In the specific case where F=0, T=0, it is logical that without friendship, there is no basis for trust. It is unlikely for someone to confide in or rely on someone they are not friends with. Similarly, when F=1, k=1, and T=1, it indicates a direct proportionality between friendship and trust. In a strong friendship (F=1), there is a high level of trust (T=1) with a proportionality constant (k) of 1. This implies that the strength of trust is directly proportional to the strength of the friendship. In real-world scenarios, the relationship between friendship and trust may not always be as linear as described in this model. Trust can develop gradually within a friendship, and there may be setbacks that temporarily weaken trust. However, the fundamental idea remains true: trust thrives in strong friendships

This research explores how friendship and trust go hand-in-hand. They're like the building blocks of strong relationships, affecting how users interact with others. Our model explains how trust grows from friendship. Here's how it works in the model:

- No Friendship, No Trust (F = 0, T = 0): This makes sense! If you're not friends with someone, you wouldn't trust them with secrets or rely on them for help.
- Strong Friends, Strong Trust (F = 1, k = 1, T = 1): Here, close friends (F = 1) have a high level of trust (T = 1). The number k=1 shows that there is a linear relationship betweenfriendship and trust.

This "k" is like a special number showing how much friendship affects trust. This model helps to understand how friendship and trust work together.

Strengths of the Model

- Simplicity: The F=0, T=0 and F=1, T=1 conditions clearly establish the existence of trust stemming from friendship.
- Proportionality: The concept of a constant k captures the idea that stronger friendships lead to greater trust.

Here are some sentences explaining why the concept of using k=1 in a specific context can be a valuable approach:

- **Simplicity and Clarity**: "Setting k=1 offers a clear and straightforward model for understanding the core relationship between friendship and trust. It establishes a direct proportionality, making initial analysis and interpretation easier."
- **Valid for Initial Stages:** "In the early stages of friendship development, a k=1 model might be particularly relevant. As trust is initially built, the proportional relationship between friendship and trust can be quite strong."
- **Foundation for Further Exploration**: "While k=1 simplifies the model; it serves as a strong foundation for further research. Understanding this core relationship allows for the introduction of additional variables (represented by modifying k) to explore the complexities of trust-building in real-world friendships."

6.3.4 Mathematical algorithm of this model based on the given conditions; it can be formulated below step by step

Step 1: Input Parameters

- F (Friendship level): A binary variable where 0 indicates no friendship and 1 indicates the presence of friendship.
- k (Proportionality constant): A constant reflecting the strength of the relationship between friendship and trust.

Step 2: Algorithm

- Begin by defining the function to calculate trust (T) based on friendship (F) and the proportionality constant (k).
- Use the given conditions:
  - If F = 0, then T = 0.
  - If F = 1 and k = 1, then T = 1.
  - For other cases, calculate T based on the relationship between F, k, and T.

Step 3: Output

- The output will be the calculated value of trust (T) based on the input parameters (F and k).

Table 6.3 Friendship dataset [36]

ID	*Person A*	*Person B*	*Years of Knowing*	*Interaction Duration*	*Interaction Type*	*Moon Phase During Interaction*	*Friends*
1	A1	B1	3.87656	13.03511	Class	Waning_ Gibbous	1
2	A2	B2	2.836218	5.811429	At Work	Waxing_ Cresent	0
3	A3	B3	3.006119	4.882863	Over a Meal	New_Moon	0
4	A4	B4	2.960067	9.274924	Social_ Media	Waxing_ Gibbous	0
5	A5	B5	7.640688	8.843167	Class	First_ Quarter	1
6	A6	B6	10.12505	23.03403	Social_ Media	Third_ Quarter	1
7	A7	B7	8.916969	8.030543	Party	New_Moon	1
8	A8	B8	3.020859	4.206463	Party	Third_ Quarter	0
9	A9	B9	6.036803	29.03715	Over a Meal	Waning_ Gibbous	1
10	A10	B10	6.162918	29.28239	Social_ Media	Waning_ Gibbous	1

The example usage demonstrates how to utilize the algorithm by taking user input for F and k, and then outputs the corresponding trust level (Table 6.3).

While this dataset includes "years of knowing" and "interaction duration," it lacks a direct "trust" measurement. However, **this model suggests a connection between friendship strength (derived from these factors) and trust.** This analysis explores the connection between friendship and trust, even though the original dataset lacks a dedicated "trust" column. Two key conditions are considered:

- No Friendship, No Trust (F = 0, T = 0): This implies the absence of friendship (F = 0) signifies a lack of trust (T = 0). It highlights the fundamental link between friendship and trust development.
- Friendship and Trust Relationship: While the dataset doesn't provide a specific value for trust (T), a potential proportional relationship between friendship and trust can be inferred. Stronger friendships (higher values in the "friendship" column) might indicate a greater likelihood of trust existing.

While the data lacks a direct trust measure, this approach offers a valuable starting point. It explores the potential link between friendship strength and trust. This is based on the idea that no friendship (F=0) implies no trust (T=0), and strong friendships (F=1) suggest high trust (T=1).

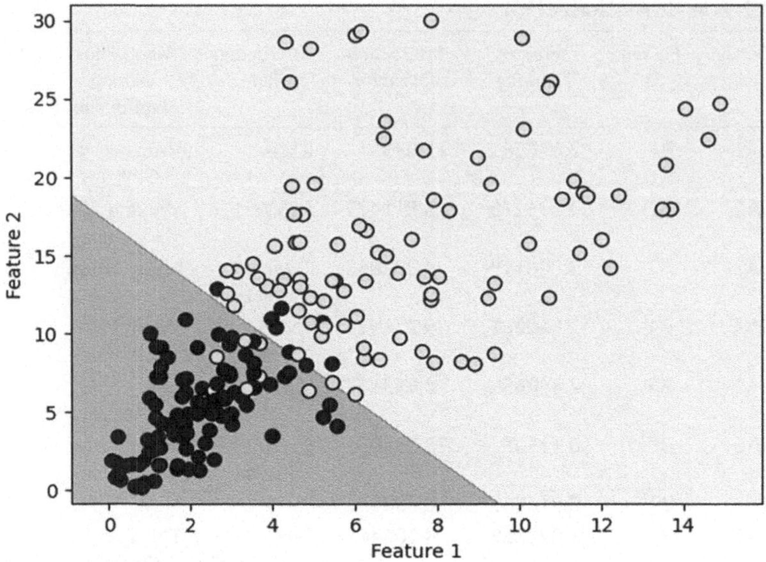

Figure 6.1 Logistic regression decision boundary for all data points.

6.3.5 Logistic Regression

This is a better choice for trust assessment because it predicts the probability of an event (trust being built) based on the input data (interaction data) (Figures 6.1 and 6.2).

$$\mathbf{P(y = 1|x)} = \frac{\mathbf{1}}{1 + e^{(-\beta 0 - \beta 1x1 - \beta 2x2 - \ldots - \beta kxk)}} \tag{6.1}$$

Accuracy: 93.03%
Intercept ($\boldsymbol{\beta}_0$): -9.409767210410827
Coefficient for Feature 1 ($\boldsymbol{\beta}_1$): 1.039134589206226
Coefficient for Feature 2 ($\boldsymbol{\beta}_2$): 0.5511454622104489

6.3.6 k-NN algorithm

The k-Nearest Neighbors (k-NN) algorithm is a widely used technique in machine learning for both classification and regression tasks. k-NN assumes that similar data points tend to have similar labels or values. It classifies a new data point based on the labels of its closest neighbors in the training data.In this paper, classification isdone, and accuracy is found.

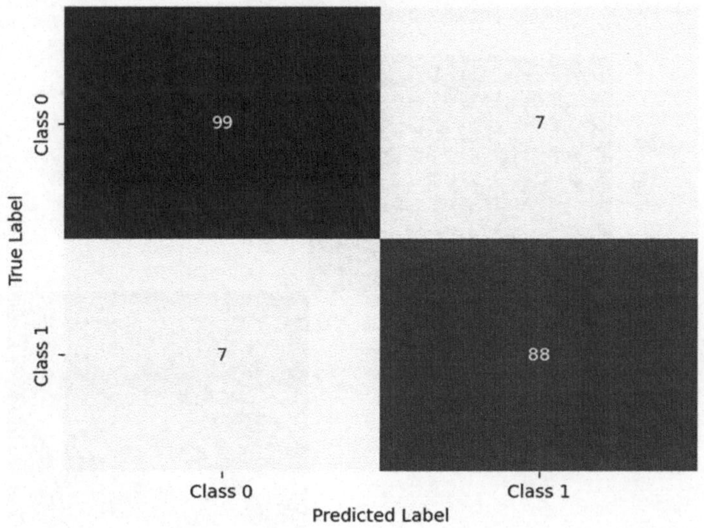

Figure 6.2 Confusion matrix for all data point (during logistic regression).

Steps Involved:
Training Phase:

- Stores the entire training dataset. This dataset consists of data points with known labels or values.

Classification Phase:

When presented with a new data point (who's label or value is unknown):

- Calculates the distance between the new data point and all the data points in the training set using a distance metric (like Euclidean distance).
- Identifies the k closest neighbors (nearest data points) to the new data point.

For classification: Determines the most frequent class label among the k neighbors. This becomes the predicted class label for the new data point (Figure 6.3).

6.3.7 Decision tree visualization

- The decision tree is built from a training dataset.
- The training dataset consists of data points with known labels or values.
- The algorithm iteratively splits the training dataset into smaller subsets based on the features of the data.
- The goal of the splitting process is to create subsets that are as pure as possible, meaning that all or most of the data points in a subset belong to the same class.

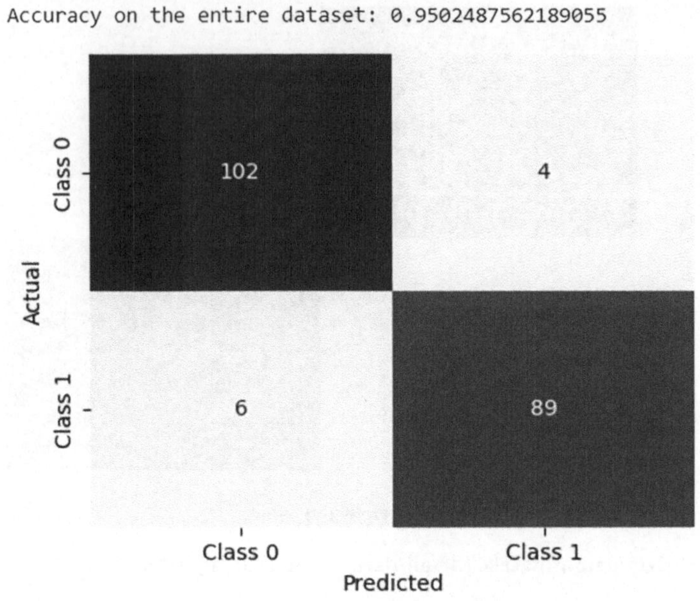

Figure 6.3 Confusion matrix for entire dataset (k-NN approach).

- The splitting process continues until a stopping criterion is met, such as a maximum depth of the tree or a minimum number of data points in a subset.
- Decision trees are a popular machine learning algorithm used for **classification** and regression tasks. They are easy to understand and interpret, making them a valuable tool for data exploration and model explanation. Visualizing decision trees can help us understand how the model makes decisions and identify potential areas for improvement (Figures 6.4 and 6.5).

Accuracy: 0.926829268292683

6.3.8 Linearcombination model for trust score calculation

Trust Score Calculation using Linear Combination Model with Normalization and accuracy Evaluation using Confusion Matrix (Table 6.4)

Input:

- Factors: X_1, X_2, X_3............, Xn.
- Weights: (w_1, w_2, w_3, w_4............,w_n) for each factor (X_1, X_2, X_3............, Xn)

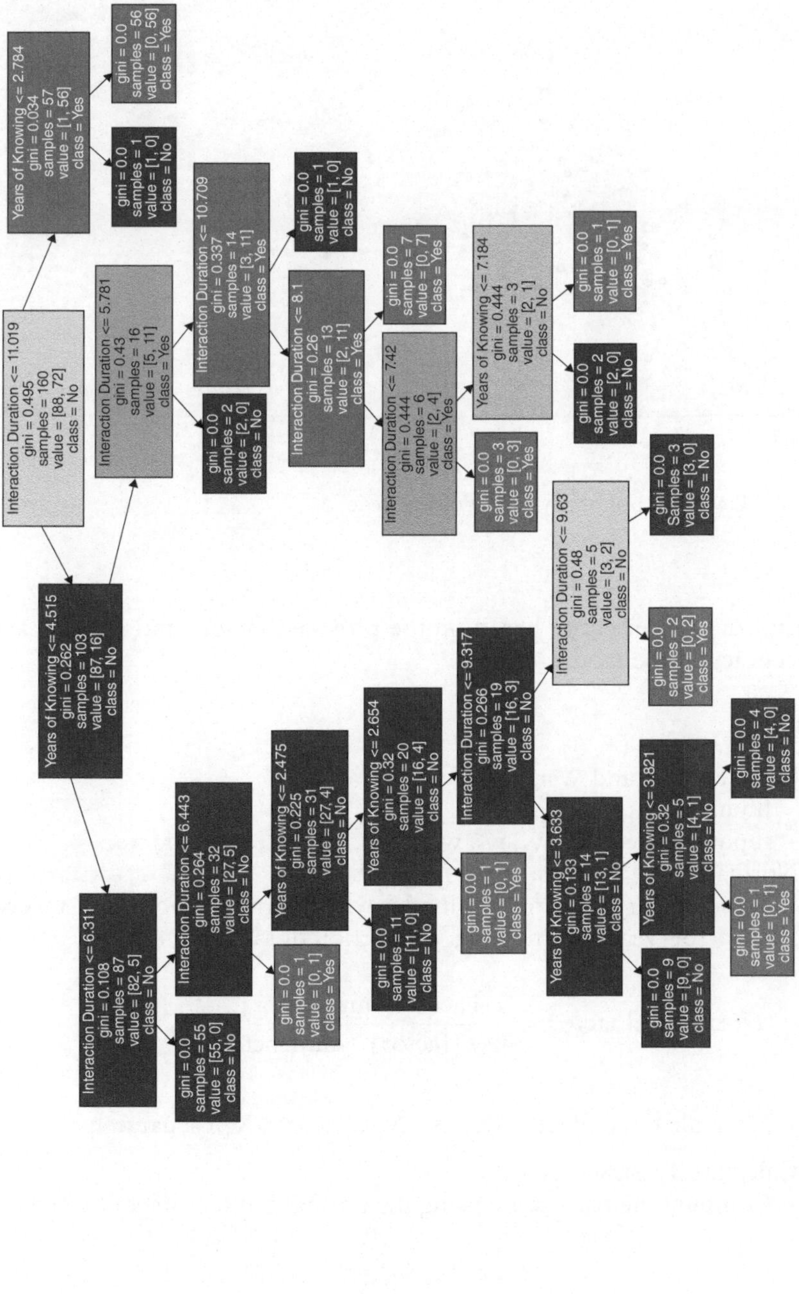

Figure 6.4 Decision tree visualization.

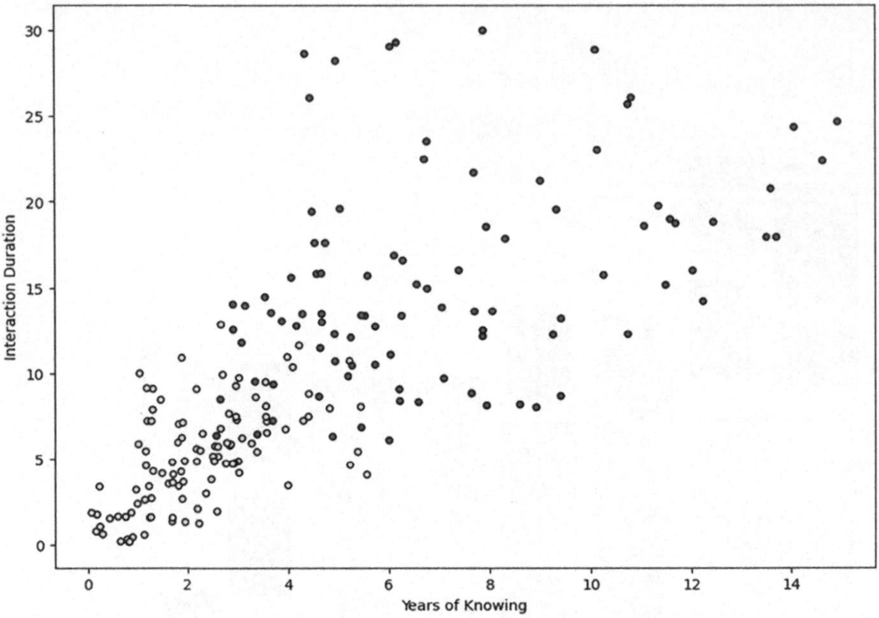

Figure 6.5 Decision boundary for decision tree.

Output:

Trust score for an entity based on the provided factors and weights and also accuracy of the model

Steps:

I. Input Factors and Weights:
   - Input the X_1, X_2, X_3…………, Xn.
   - Input the weights (w_1, w_2, w_3, ……………, w_n) for each factor.

II. Normalize the Factors:
   - Trust-Linear combination model with Normalization with respect to factor–years of knowing, interaction duration…. etc.

$$\text{Normalized factor} = \frac{\text{Factor} - \min(\text{factor})}{\max(\text{factor}) - \min(\text{factor})}$$

   - Normalize each factor (X_1, X_2, X_3…………, X_n.) separately.

III. Calculate Trust Score:
   - Compute the trust score using the normalized factors and weights:

Table 6.4 Calculation of trust score

X_1	X_2	$T = 0.5\ X_1 + 0.5\ X_2$
0.2563	0.4311	0.3437
0.1863	0.1887	0.1875
0.1977	0.1575	0.1776
0.1946	0.3049	0.24975
0.5098	0.2904	0.4001
0.6771	0.7668	0.72195
0.5958	0.2631	0.42945
0.1987	0.1348	0.16675
0.4018	0.9683	0.68505
0.4103	0.9765	0.6934
0.2851	0.237	0.26105
0.3032	0.5244	0.4138
0.1747	0.4247	0.2997
0.1924	0.1539	0.17315
0.0826	0.2991	0.19085
0.3343	0.6508	0.49255
0.5245	0.4015	0.463
0.0351	0.049	0.04205
0.473	0.3194	0.3962
0.3098	0.4458	0.3778
0.5297	0.2667	0.3982
0.363	0.2242	0.2936
...	...	...
0.6746	0.9622	0.8184
0.6289	0.4367	0.5328
0.7819	0.6223	0.7021
0.0709	0.0134	0.04215
0.2761	0.4219	0.349
0.3277	0.3528	0.34025
0.0732	0.1488	0.111
0.1741	0.2216	0.19785
0.0646	0.3291	0.19685
0.3489	0.3992	0.37405
0.1851	0.2509	0.218
0.2638	0.3612	0.3125
0.1902	0.1526	0.1714
0.3242	0.2061	0.26515

The mathematical structure of calculating Trust score is as follows:

$$T = w_1x_1 + w_2x_2 + w_3x_3 + \ldots\ldots\ldots\ldots\ldots + w_nx_n$$

where $x_1, x_2, \ldots\ldots\ldots x_n$ are the n basic factors (like years of knowing, interaction duration.... etc) in contribution to make trust score and $w_1, w_2\ldots\ldots$ and w_n (>0) are weight factors for respective factors. Also,

$$w_1 + w_2 + \ldots\ldots + w_n = 1$$

IV. Set some threshold value for trust like If T > 0.6 friendship exist otherwise friendship does not exist. Now, use the original value and predicted value to get accuracy using the confusion matrix.

In the confusion matrix (Figure 6.6):

$$\text{True Positive}(\text{TP}): 87$$
$$\text{True Negative}(\text{TN}): 77$$
$$\text{False Positive}(\text{FP}): 23$$
$$\text{False Negative}(\text{FN}): 14$$

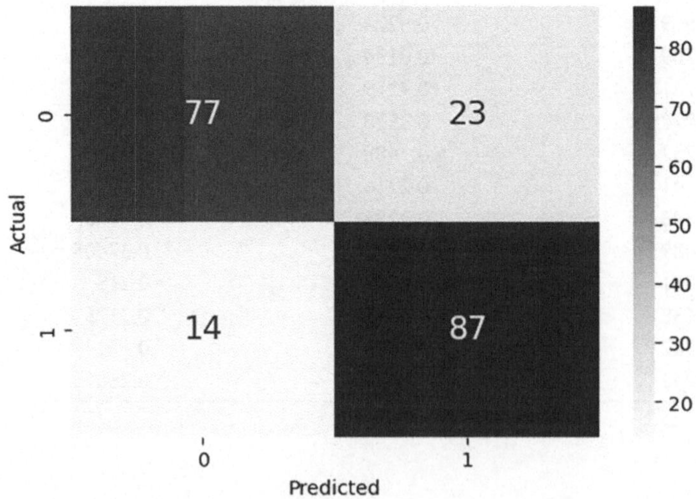

Figure 6.6 Confusion matrix (trust score).

Now, these values can be useful to calculate various performance metrics such as accuracy, precision, recall, and F1-score:

$$\text{Accuracy} = (\text{TP} + \text{TN}) / (\text{TP} + \text{TN} + \text{FP} + \text{FN})$$
$$\text{Precision} = \text{TP} / (\text{TP} + \text{FP})$$
$$\text{Recall} = \text{TP} / (\text{TP} + \text{FN})$$
$$\text{F1-score} = 2(\text{Precision}^{*}\,\text{Recall}) / (\text{Precision} + \text{Recall})$$

To find the accuracy, use the following formula:

$$\text{Accuracy} = (\text{True Positives} + \text{True Negatives}) / \text{Total Observations}$$

In this case:

$$\text{Accuracy} = (87 + 77) / (77 + 23 + 14 + 87)$$
$$\text{Accuracy} = 164 / 201\text{Accuracy} = 201164$$
$$\text{Accuracy} \approx 0.8169\text{Accuracy} \approx 0.8169$$

So, the accuracy of your model is approximately 81.69%81.69%.

6.3.9 Fuzzy logic approach

I. **Fuzzification:**
   a. Assign membership values to linguistic variables for each input variable.
   b. Determine the degree to which each data point belongs to the fuzzy sets associated with linguistic variables.

II. **Rule Evaluation:**
   a. Define fuzzy rules based on linguistic variables and their membership values.
   b. Evaluate each rule for each data point by applying the antecedents (if part) and determining the contribution of each rule.

III. **Aggregation:**
   a. Combine the contributions of each rule using a fuzzy logic operator (e.g., MAX or SUM) to obtain an aggregated fuzzy output for each data point.

IV. **Defuzzification:**
   a. Convert the aggregated fuzzy output into a crisp value [23, 28].

Fuzzy approach for trust with respect to referenced factor (Table 6.5)

Table 6.5 Influence score, followers, engagement rate [37]

rank	*influence_ score*	*posts*	*followers*	*avg_ likes*	*60_day_ eng_rate*	*new_post_avg_ like*	*total_likes*
1	92	3.3k	475.8m	8.7m	1.39%	6.5m	29.0b
2	91	6.9k	366.2m	8.3m	1.62%	5.9m	57.4b
3	90	0.89k	357.3m	6.8m	1.24%	4.4m	6.0b
4	93	1.8k	342.7m	6.2m	0.97%	3.3m	11.5b
5	91	6.8k	334.1m	1.9m	0.20%	665.3k	12.5b
6	91	5.6k	329.2m	3.5m	0.88%	2.9m	19.9b
7	92	5.0k	327.7m	3.7m	1.20%	3.9m	18.4b
8	92	2.0k	272.8m	3.6m	0.76%	2.0m	7.4b
9	89	4.1k	268.3m	2.4m	0.35%	926.9k	9.8b
10	91	7.4k	254.5m	1.9m	0.59%	1.5m	13.9b

I. Fuzzification:

Membership Function Influence Score

(a) *Very small influence*
Membership function

$$\mu_{VSI}(x) = \begin{cases} 1 & \text{if } x \le 22 \\ \dfrac{40-x}{18} & \text{if } 22 < x \le 40 \\ 0 & \text{otherwise if } x > 40 \end{cases}$$

(b) *Small Influence*
Membership function

$$\mu_{SI}(x) = \begin{cases} 0 & \text{if } x \le 22 \text{ or } x > 58 \\ \dfrac{x-22}{18} & \text{if } 22 < x \le 40 \\ \dfrac{58-x}{18} & \text{if } 40 < x \le 58 \\ 0 & \text{otherwise if } x > 58 \end{cases}$$

(c) *Medium influence*
Membership function

$$\mu_{MI}(x) = \begin{cases} 0 & \text{if } x \le 40 \text{ or } x > 76 \\ \dfrac{x-40}{18} & \text{if } 40 < x \le 58 \\ 1 & \text{if } 58 < x \le 76 \\ \dfrac{76-x}{18} & \text{if } 76 < x \le 94 \\ 0 & \text{otherwise if } x > 94 \end{cases}$$

(d) *High (H) Influence*
Membership function

$$\mu_{HI}(x) = \begin{cases} 0 & \text{if } x \le 58 \text{ or } x > 94 \\ \dfrac{x-50}{18} & \text{if } 58 < x \le 76 \\ 1 & \text{if } x > 76 \end{cases}$$

(e) *Very High Influence*
Membership function (Figure 6.7; Table 6.6)

$$\mu_{VHI}(x) = \begin{cases} 1 & \text{if } x \le 76 \\ \dfrac{x-76}{18} & \text{if } 76 < x \le 94 \\ 1 & \text{if } x > 94 \end{cases}$$

[200 rows × 6 columns]

Tools and Technologies Used [24, 25, 26, 28]

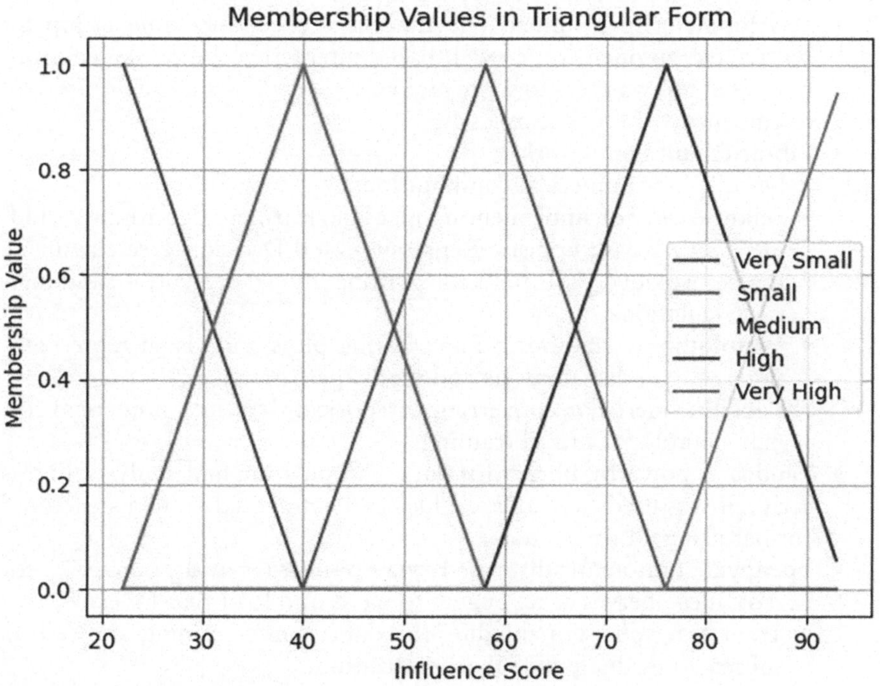

Figure 6.7 Influence Score and membership value.

Table 6.6 Output of different membership value of influence score

Serial number	*Influence score*	*Influence Score_ VeryLow_I*	*Influence Score_Low_I*	*Influence Score_ Medium_I*	*Influence Score_ High_I*	*Influence Score_ VeryHigh_I*
0	92	0	0.0	0.0	0.11	0.89
1	91	0	0.0	0.0	0.17	0.83
2	90	0	0.0	0.0	0.22	0.78
3	93	0	0.0	0.0	0.06	0.94
4	91	0	0.0	0.0	0.17	0.83
...	...	...	...	...	...	...
...	...	...	...	...	...	...
195	71	0	0.0	0.28	0.72	0.00
196	81	0	0.0	0.0	0.72	0.28
197	79	0	0.0	0.0	0.83	0.17
198	78	0	0.0	0.0	0.89	0.11
199	80	0	0.0	0.0	0.78	0.22

The following tools and technologies were employed to perform the comparative analysis and implement the predictive models:

- **Programming Language**: Python was used due to its flexibility, ease of use, and availability of powerful data science libraries.
- **Development Environment**:The analysis was conducted using **Jupyter Notebook**, an open-source web-based interactive computing environment that allows users to write and execute code, visualize results, and document workflows seamlessly.
- **Libraries and Frameworks:**
  - **NumPy:** For numerical computations.
  - **Scikit-learn:** For implementing machine learning algorithms including Naïve Bayes, Logistic Regression, and Decision Tree classifiers. It also provided tools for data splitting, model evaluation, and metric calculations.
  - **Matplotlib** and **Seaborn:** For creating plots and visual representations of data distributions and model performance.
  - **LabelEncoder:** For converting categorical data into numerical format suitable for model training.
- **Pandas:** A powerful library for data manipulation and analysis. In this script, it was used to read Excel files and extract data into a structured format using a `DataFrame`.
- **openpyxl** (automatically used via `pandas.read_excel()` for `.xlsx` files): Serves as an engine to parse and load Excel files.
  - **Train_test_split:** For dividing the dataset into training and testing subsets to evaluate model generalization.

Membership Function Followers

(a) *Very small Followers*
Membership function

$$\mu_{VSF}(x) = \begin{cases} 1 & \text{if } x \le 30000000 \\ \dfrac{40000000 - x}{10000000} & \text{if } 30000000 < x \le 40000000 \\ 0 & \text{otherwise if } x > 40000000 \end{cases}$$

(b) *Small Followers*
Membership function

$$\mu_{SF}(x) = \begin{cases} 0 & \text{if } x \le 30000000 \text{ or } x > 60000000 \\ \dfrac{x - 30000000}{15000000} & \text{if } 30000000 < x \le 45000000 \\ \dfrac{60000000 - x}{15000000} & \text{if } 45000000 < x \le 60000000 \\ 0 & \text{otherwise if } x > 58 \end{cases}$$

(c) *Medium Followers*
Membership function

$$\mu_{MF}(x) = \begin{cases} 0 & \text{if } x \le 50000000 \text{ or } x > 80000000 \\ \dfrac{x - 50000000}{15000000} & \text{if } 50000000 < x \le 60000000 \\ 1 & \text{if } 65000000 < x \le 75000000 \\ \dfrac{80000000 - x}{5000000} & \text{if } 7500000 < x \le 80000000 \end{cases}$$

(d) *High (H) Followers*
Membership function

$$\mu_{HF}(x) = \begin{cases} 0 & \text{if } x \le 70{,}000{,}000 \\ \dfrac{x - 70{,}000{,}000}{130{,}000{,}00018} & \text{if } 70{,}000{,}000 < x \le 200{,}000{,}000 \\ 1 & \text{if } x > 200{,}000{,}000 \end{cases}$$

(e) *Very High Followers*
Membership function (Figure 6.8; Table 6.7)

$$\mu_{VHF}(x) = \begin{cases} 1 & \text{if } x \le 150{,}000{,}000 \\ \dfrac{x - 150{,}000{,}000}{325{,}800{,}000} & \text{if } 150{,}000{,}000 < x \le 475{,}800{,}00 \\ 1 & \text{if } x > 475{,}800{,}00 \end{cases}$$

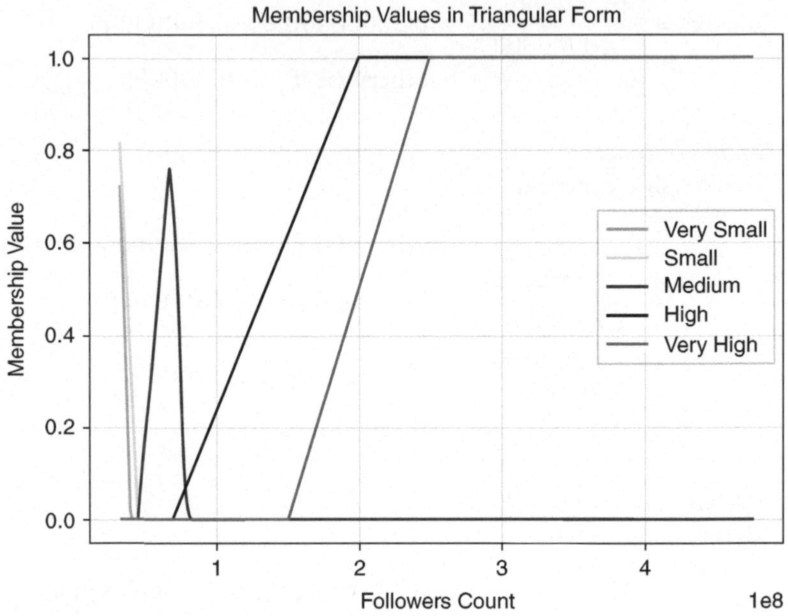

Figure 6.8 Followers count with memebership value.

Table 6.7 Output of different membership value of followers

Serial number	*followers*	*Followers_ Verysmall_F*	*Followers_ small_F*	*Followers_ medium_F*	*Followers_ high_F*	*Followers_ Veryhigh_F*
0	475800000	0	0	0	1	1
1	366200000	0	0	0	1	1
2	357300000	0	0	0	1	1
3	342700000	0	0	0	1	1
4	334100000	0	0	0	1	1
...	...	...	...	...	...	...
...	...	...	...	...	...	...
195	33200000	0.68	0.79	0	0	0
196	33200000	0.68	0.79	0	0	0
197	33200000	0.68	0.79	0	0	0
198	33000000	0.7	0.8	0	0	0
199	32800000	0.72	0.81	0	0	0

Creating membership functions for the "60-days Engagement rate" variable involves defining the linguistic categories (e.g., very small, small, medium, high, very high) based on the provided data. Here's an example:

Membership FunctionEngagement Rate

(a) *Very small Engagement rate*
Membership function

$$\mu_{VSE}(x) = \begin{cases} 1 & \text{if } x \le 0.75 \\ \dfrac{1.25 - x}{0.5} & \text{if } 0.75 < x \le 1.25 \\ 0 & \text{otherwise} \end{cases}$$

(b) *Small Engagement rate*
Membership function

$$\mu_{SE}(x) = \begin{cases} 0 & \text{if } x \le 0.75 \text{ or } x > 2 \\ \dfrac{x - 0.75}{0.75} & \text{if } 0.75 < x \le 1.5 \\ \dfrac{2 - x}{0.5} & \text{if } 1.5 < x \le 2 \end{cases}$$

(c) *Medium Engagement rate*
Membership function

$$\mu_{ME}(x) = \begin{cases} 0 & \text{if } x \le 1.25 \text{ or } x > 3 \\ \dfrac{x - 1.25}{1.25} & \text{if } 1.25 < x \le 2.5 \\ \dfrac{3 - x}{0.5} & \text{if } 2.5 < x \le 3 \end{cases}$$

(d) *High (H) Engagement rate*
Membership function

$$\mu_{HE}(x) = \begin{cases} 0 & \text{if } x \le 2 \text{ or } x > 4 \\ \dfrac{x - 2}{2} & \text{if } 2 < x \le 3 \\ \dfrac{4 - x}{01} & \text{if } 3 < x \le 4 \end{cases}$$

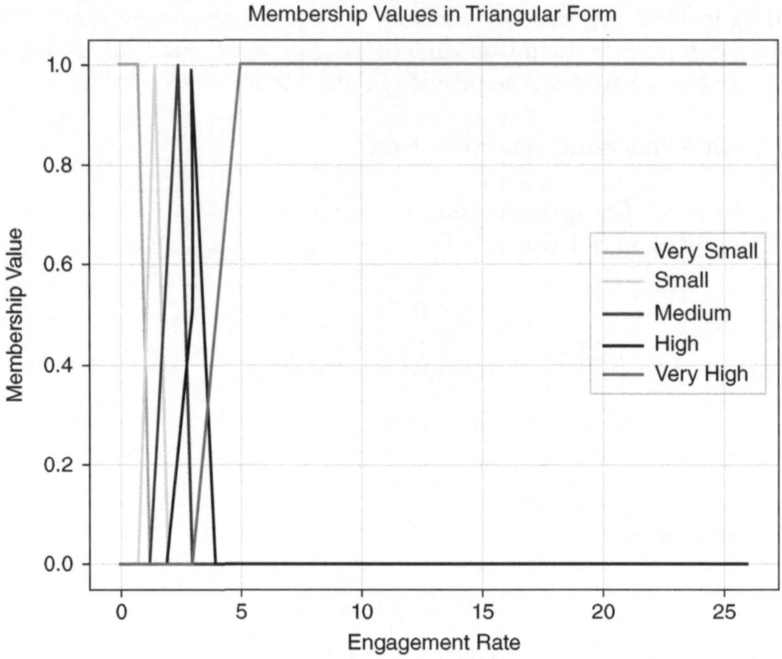

Figure 6.9 Engagement rate with memebership value.

(e) *Very High Engagement rate*
Membership function (Figure 6.9; Table 6.8)

$$\mu_{\text{VHE}}(\mathrm{x}) = \begin{cases} 0 & \text{if } \mathrm{x} \leq 3 \\ \dfrac{x-3}{2} & \text{if } 3 < \mathrm{x} \leq 5 \\ 1 & \text{if } \mathrm{x} > 5 \end{cases}$$

II. **Fuzzy Rule Evaluation**

Here are the 125 fuzzy rules (Table 6.9):

Rule 1: If Influence Score is Very Low_I AND Engagement Rate is extremely low_E AND Followers are Very small_F, THEN Trust Level is Extremely Low.

Rule 2: If Influence Score is Very Low_I AND Engagement Rate is extremely low_E AND Followers are small_F, THEN Trust Level is Extremely Low.

Rule 3: If Influence Score is Very Low_I AND Engagement Rate is extremely low_E AND Followers are medium_F, THEN Trust Level is Extremely Low.

Table 6.8 Output of different membership value of Engagement rate

Serial number	*60- days Engagement Rate*	*Engagement Rate_extremely low_E*	*Engagement Rate_low_E*	*Engagement Rate_ average_E*	*Engagement Rate_high_E*	*Engagement Rate_ veryhigh_E*
0	1.39	0	0.85	0.11	0	0
1	1.62	0	0.76	0.3	0	0
2	1.24	0.02	0.65	0	0	0
3	0.97	0.56	0.29	0	0	0
4	0.2	1	0	0	0	0
5	0.88	0.74	0.17	0	0	0
6	1.2	0.1	0.6	0	0	0
…	….	……	……	…	…	…
…	….	……	…..	…	…	…
195	1.4	0	0.87	0.12	0	0
196	0.64	1	0	0	0	0
197	0.26	1	0	0	0	0
198	1.42	0	0.89	0.14	0	0
199	0.3	1	0	0	0	0

Rule 4: If Influence Score is Very Low_I AND Engagement Rate is extremely low_E AND Followers are high_F, THEN Trust Level is Extremely Low.

Rule 5: If Influence Score is Very Low_I AND Engagement Rate is extremely low_E AND Followers are Very high_F, THEN Trust Level is Extremely Low.

Rule 6: If Influence Score is Very Low_I AND Engagement Rate is low_E AND Followers are Very small_F, THEN Trust Level is Very Low.

Rule 7: If Influence Score is Very Low_I AND Engagement Rate is low_E AND Followers are small_F, THEN Trust Level is Very Low.

Rule 8: If Influence Score is Very Low_I AND Engagement Rate is low_E AND Followers are medium_F, THEN Trust Level is Very Low.

Rule 9: If Influence Score is Very Low_I AND Engagement Rate is low_E AND Followers are high_F, THEN Trust Level is Low.

Rule 10: If Influence Score is Very Low_I AND Engagement Rate is low_E AND Followers are Very high_F, THEN Trust Level is Low.

Rule 11: If Influence Score is Very Low_I AND Engagement Rate is average_E AND Followers are Very small_F, THEN Trust Level is Very Low.

Rule 12: If Influence Score is Very Low_I AND Engagement Rate is average_E AND Followers are small_F, THEN Trust Level is Very Low.

Rule 13: If Influence Score is Very Low_I AND Engagement Rate is average_E AND Followers are medium_F, THEN Trust Level is Very Low.

Table 6.9 Calculation of all rule value

Serial no	*Influence score*	*Followers*	*Rule1*	*Rule2*	*Rule3*	*Rule4*	…	*Rule122*	*Rule123*	*Rule124*	*Rule 125*
0	92	475800000	0	0	0	0	…	0	0	0	0
1	91	366200000	0	0	0	0	…	0	0	0	0
2	90	357300000	0	0	0	0	…	0	0	0	0
3	93	342700000	0	0	0	0	…	0	0	0	0
4	91	334100000	0	0	0	0	…	0	0	0	0
….	…	…	…	…	…	…	…	…	…	…	…
195	71	33200000	0	0	0	0	…	0	0	0	0
196	81	33200000	0	0	0	0	…	0	0	0	0
197	79	33200000	0	0	0	0	…	0	0	0	0
198	78	33000000	0	0	0	0	…	0	0	0	0
199	80	32800000	0	0	0	0	…	0	0	0	0

Rule 14: If Influence Score is Very Low_I AND Engagement Rate is average_E AND Followers are high_F, THEN Trust Level is Low.
Rule 15: If Influence Score is Very Low_I AND Engagement Rate is average_E AND Followers are Very high_F, THEN Trust Level is Low.
Rule 16: If Influence Score is Very Low_I AND Engagement Rate is high_E AND Followers are Very small_F, THEN Trust Level is Low.
Rule 17: If Influence Score is Very Low_I AND Engagement Rate is high_E AND Followers are small_F, THEN Trust Level is Medium.
Rule 18: If Influence Score is Very Low_I AND Engagement Rate is high_E AND Followers are medium_F, THEN Trust Level is High.
Rule 19: If Influence Score is Very Low_I AND Engagement Rate is high_E AND Followers are high_F, THEN Trust Level is High.
Rule 20: If Influence Score is Very Low_I AND Engagement Rate is high_E AND Followers are Very high_F, THEN Trust Level is Very High.
Rule 21: If Influence Score is Very Low_I AND Engagement Rate is very high_E AND Followers are Very small_F, THEN Trust Level is Low.
Rule 22: If Influence Score is Very Low_I AND Engagement Rate is very high_E AND Followers are small_F, THEN Trust Level is High.
Rule 23: If Influence Score is Very Low_I AND Engagement Rate is very high_E AND Followers are medium_F, THEN Trust Level is Very High.
Rule 24: If Influence Score is Very Low_I AND Engagement Rate is very high_E AND Followers are high_F, THEN Trust Level is Very High.
Rule 25: If Influence Score is Very Low_I AND Engagement Rate is very high_E AND Followers are Very high_F, THEN Trust Level is Very High.
Rule 26: If Influence Score is Low_I AND Engagement Rate is extremely low_E AND Followers are Very small_F, THEN Trust Level is Extremely Low.
Rule 27: If Influence Score is Low_I AND Engagement Rate is extremely low_E AND Followers are small_F, THEN Trust Level is Extremely Low.
Rule 28: If Influence Score is Low_I AND Engagement Rate is extremely low_E AND Followers are medium_F, THEN Trust Level is Extremely Low.
Rule 29: If Influence Score is Low_I AND Engagement Rate is extremely low_E AND Followers are high_F, THEN Trust Level is Extremely Low.
Rule 30: If Influence Score is Low_I AND Engagement Rate is extremely low_E AND Followers are Very high_F, THEN Trust Level is Extremely Low.
Rule 31: If Influence Score is Low_I AND Engagement Rate is low_E AND Followers are Very small_F, THEN Trust Level is Very Low.
Rule 32: If Influence Score is Low_I AND Engagement Rate is low_E AND Followers are small_F, THEN Trust Level is Very Low.
Rule 33: If Influence Score is Low_I AND Engagement Rate is low_E AND Followers are medium_F, THEN Trust Level is Very Low.

Rule 34: If Influence Score is Low_I AND Engagement Rate is low_E AND Followers are high_F, THEN Trust Level is Low.
Rule 35: If Influence Score is Low_I AND Engagement Rate is low_E AND Followers are Very high_F, THEN Trust Level is Low.
Rule 36: If Influence Score is Low_I AND Engagement Rate is average_E AND Followers are Very small_F, THEN Trust Level is Very Low.
Rule 37: If Influence Score is Low_I AND Engagement Rate is average_E AND Followers are small_F, THEN Trust Level is Very Low.
Rule 38: If Influence Score is Low_I AND Engagement Rate is average_E AND Followers are medium_F, THEN Trust Level is Very Low.
Rule 39: If Influence Score is Low_I AND Engagement Rate is average_E AND Followers are high_F, THEN Trust Level is Low.
Rule 40: If Influence Score is Low_I AND Engagement Rate is average_E AND Followers are Very high_F, THEN Trust Level is Low.
Rule 41: If Influence Score is Low_I AND Engagement Rate is high_E AND Followers are Very small_F, THEN Trust Level is Low.
Rule 42: If Influence Score is Low_I AND Engagement Rate is high_E AND Followers are small_F, THEN Trust Level is Medium.
Rule 43: If Influence Score is Low_I AND Engagement Rate is high_E AND Followers are medium_F, THEN Trust Level is High.
Rule 44: If Influence Score is Low_I AND Engagement Rate is high_E AND Followers are high_F, THEN Trust Level is High.
Rule 45: If Influence Score is Low_I AND Engagement Rate is high_E AND Followers are Very high_F, THEN Trust Level is Very High.
Rule 46: If Influence Score is Low_I AND Engagement Rate is very high_E AND Followers are Very small_F, THEN Trust Level is Low.
Rule 47: If Influence Score is Low_I AND Engagement Rate is very high_E AND Followers are small_F, THEN Trust Level is High.
Rule 48: If Influence Score is Low_I AND Engagement Rate is very high_E AND Followers are medium_F, THEN Trust Level is Very High.
Rule 49: If Influence Score is Low_I AND Engagement Rate is very high_E AND Followers are high_F, THEN Trust Level is Very High.
Rule 50: If Influence Score is Low_I AND Engagement Rate is very high_E AND Followers are Very high_F, THEN Trust Level is Very High.
Rule 51: If Influence Score is Medium_I AND Engagement Rate is extremely low_E AND Followers are Very small_F, THEN Trust Level is Very Low.
Rule 52: If Influence Score is Medium_I AND Engagement Rate is extremely low_E AND Followers are small_F, THEN Trust Level is Medium.
Rule 53: If Influence Score is Medium_I AND Engagement Rate is extremely low_E AND Followers are medium_F, THEN Trust Level is High.
Rule 54: If Influence Score is Medium_I AND Engagement Rate is extremely low_E AND Followers are high_F, THEN Trust Level is High.
Rule 55: If Influence Score is Medium_I AND Engagement Rate is extremely low_E AND Followers are Very high_F, THEN Trust Level is High.

Rule 56: If Influence Score is Medium_I AND Engagement Rate is low_E AND Followers are Very small_F, THEN Trust Level is Medium.
Rule 57: If Influence Score is Medium_I AND Engagement Rate is low_E AND Followers are small_F, THEN Trust Level is Medium.
Rule 58: If Influence Score is Medium_I AND Engagement Rate is low_E AND Followers are medium_F, THEN Trust Level is Medium.
Rule 59: If Influence Score is Medium_I AND Engagement Rate is low_E AND Followers are high_F, THEN Trust Level is High.
Rule 60: If Influence Score is Medium_I AND Engagement Rate is low_E AND Followers are Very high_F, THEN Trust Level is High.
Rule 61: If Influence Score is Medium_I AND Engagement Rate is average_E AND Followers are Very small_F, THEN Trust Level is Medium.
Rule 62: If Influence Score is Medium_I AND Engagement Rate is average_E AND Followers are small_F, THEN Trust Level is Medium.
Rule 63: If Influence Score is Medium_I AND Engagement Rate is average_E AND Followers are medium_F, THEN Trust Level is Medium.
Rule 64: If Influence Score is Medium_I AND Engagement Rate is average_E AND Followers are high_F, THEN Trust Level is High.
Rule 65: If Influence Score is Medium_I AND Engagement Rate is average_E AND Followers are Very high_F, THEN Trust Level is High.
Rule 66: If Influence Score is Medium_I AND Engagement Rate is high_E AND Followers are Very small_F, THEN Trust Level is High.
Rule 67: If Influence Score is Medium_I AND Engagement Rate is high_E AND Followers are small_F, THEN Trust Level is High.
Rule 68: If Influence Score is Medium_I AND Engagement Rate is high_E AND Followers are medium_F, THEN Trust Level is High.
Rule 69: If Influence Score is Medium_I AND Engagement Rate is high_E AND Followers are high_F, THEN Trust Level is High.
Rule 70: If Influence Score is Medium_I AND Engagement Rate is high_E AND Followers are Very high_F, THEN Trust Level is High.
Rule 71: If Influence Score is Medium_I AND Engagement Rate is very high_E AND Followers are Very small_F, THEN Trust Level is High.
Rule 72: If Influence Score is Medium_I AND Engagement Rate is very high_E AND Followers are small_F, THEN Trust Level is High.
Rule 73: If Influence Score is Medium_I AND Engagement Rate is very high_E AND Followers are medium_F, THEN Trust Level is High.
Rule 74: If Influence Score is Medium_I AND Engagement Rate is very high_E AND Followers are high_F, THEN Trust Level is High.
Rule 75: If Influence Score is Medium_I AND Engagement Rate is very high_E AND Followers are Very high_F, THEN Trust Level is Very High.
Rule 76: If Influence Score is High_I AND Engagement Rate is extremely low_E AND Followers are Very small_F, THEN Trust Level is Medium.
Rule 77: If Influence Score is High_I AND Engagement Rate is extremely low_E AND Followers are small_F, THEN Trust Level is Medium.
Rule 78: If Influence Score is High_I AND Engagement Rate is extremely low_E AND Followers are medium_F, THEN Trust Level is High.

Rule 79: If Influence Score is High_I AND Engagement Rate is extremely low_E AND Followers are high_F, THEN Trust Level is High.
Rule 80: If Influence Score is High_I AND Engagement Rate is extremely low_EAND Followers are Very high_F, THEN Trust Level is High.
Rule 81: If Influence Score is High_I AND Engagement Rate is low_E AND Followers are Very small_F, THEN Trust Level is Medium.
Rule 82: If Influence Score is High_I AND Engagement Rate is low_E AND Followers are small_F, THEN Trust Level is High.
Rule 83: If Influence Score is High_I AND Engagement Rate is low_E AND Followers are medium_F, THEN Trust Level is High.
Rule 84: If Influence Score is High_I AND Engagement Rate is low_E AND Followers are high_F, THEN Trust Level is High.
Rule 85: If Influence Score is High_I AND Engagement Rate is low_E AND Followers are Very high_F, THEN Trust Level is High.
Rule 86: If Influence Score is High_I AND Engagement Rate is average_E AND Followers are Very small_F, THEN Trust Level is High.
Rule 87: If Influence Score is High_I AND Engagement Rate is average_E AND Followers are small_F, THEN Trust Level is High.
Rule 88: If Influence Score is High_I AND Engagement Rate is average_E AND Followers are medium_F, THEN Trust Level is High.
Rule 89: If Influence Score is High_I AND Engagement Rate is average_E AND Followers are high_F, THEN Trust Level is High.
Rule 90: If Influence Score is High_I AND Engagement Rate is average_E AND Followers are Very high_F, THEN Trust Level is Very High.
Rule 91: If Influence Score is High_I AND Engagement Rate is high_E AND Followers are Very small_F, THEN Trust Level is High.
Rule 92: If Influence Score is High_I AND Engagement Rate is high_E AND Followers are small_F, THEN Trust Level is High.
Rule 93: If Influence Score is High_I AND Engagement Rate is high_E AND Followers are medium_F, THEN Trust Level is High.
Rule 94: If Influence Score is High_I AND Engagement Rate is high_E AND Followers are high_F, THEN Trust Level is Very High.
Rule 95: If Influence Score is High_I AND Engagement Rate is high_E AND Followers are Very high_F, THEN Trust Level is Very High.
Rule 96: If Influence Score is High_I AND Engagement Rate is very high_E AND Followers are Very small_F, THEN Trust Level is High.
Rule 97: If Influence Score is High_I AND Engagement Rate is very high_E AND Followers are small_F, THEN Trust Level is High.
Rule 98: If Influence Score is High_I AND Engagement Rate is very high_E AND Followers are medium_F, THEN Trust Level is Very High.
Rule 99: If Influence Score is High_I AND Engagement Rate is very high_E AND Followers are high_F, THEN Trust Level is Very High.
Rule 100: If Influence Score is High_I AND Engagement Rate is very high_E AND Followers are Very high_F, THEN Trust Level is Very High.

Rule 101: If Influence Score is Very High_I AND Engagement Rate is extremely low_E AND Followers are Very small_F, THEN Trust Level is High.
Rule 102: If Influence Score is Very High_I AND Engagement Rate is extremely low_E AND Followers are small_F, THEN Trust Level is High.
Rule 103: If Influence Score is Very High_I AND Engagement Rate is extremely low_E AND Followers are medium_F, THEN Trust Level is High.
Rule 104: If Influence Score is Very High_I AND Engagement Rate is extremely low_E AND Followers are high_F, THEN Trust Level is High.
Rule 105: If Influence Score is Very High_I AND Engagement Rate is extremely low_E AND Followers are Very high_F, THEN Trust Level is High.
Rule 106: If Influence Score is Very High_I AND Engagement Rate is low_E AND Followers are Very small_F, THEN Trust Level is High.
Rule 107: If Influence Score is Very High_I AND Engagement Rate is low_E AND Followers are small_F, THEN Trust Level is High.
Rule 108: If Influence Score is Very High_I AND Engagement Rate is low_E AND Followers are medium_F, THEN Trust Level is High.
Rule 109: If Influence Score is Very High_I AND Engagement Rate is low_E AND Followers are high_F, THEN Trust Level is High.
Rule 110: If Influence Score is Very High_I AND Engagement Rate is low_E AND Followers are Very high_F, THEN Trust Level is High.
Rule 111: If Influence Score is Very High_I AND Engagement Rate is average_E AND Followers are Very small_F, THEN Trust Level is High.
Rule 112: If Influence Score is Very High_I AND Engagement Rate is average_E AND Followers are small_F, THEN Trust Level is High.
Rule 113: If Influence Score is Very High_I AND Engagement Rate is average_E AND Followers are medium_F, THEN Trust Level is High.
Rule 114: If Influence Score is Very High_I AND Engagement Rate is average_E AND Followers are high_F, THEN Trust Level is Very High.
Rule 115: If Influence Score is Very High_I AND Engagement Rate is average_E AND Followers are Very high_F, THEN Trust Level is Very High.
Rule 116: If Influence Score is Very High_I AND Engagement Rate is high_E AND Followers are Very small_F, THEN Trust Level is Very High.
Rule 117: If Influence Score is Very High_I AND Engagement Rate is high_E AND Followers are small_F, THEN Trust Level is Very High
Rule 118: If Influence Score is Very High_I AND Engagement Rate is high_E AND Followers are medium_F, THEN Trust Level is Very High.

Rule 119: If Influence Score is Very High_I AND Engagement Rate is high_E AND Followers are high_F, THEN Trust Level is Very High.
Rule 120: If Influence Score is Very High_I AND Engagement Rate is high_E AND Followers are Very high_F, THEN Trust Level is Very High.
Rule 121: If Influence Score is Very High_I AND Engagement Rate is very high_E AND Followers are Very small_F, THEN Trust Level is Very High.
Rule 122: If Influence Score is Very High_I AND Engagement Rate is very high_E AND Followers are small_F, THEN Trust Level is Very High.
Rule 123: If Influence Score is Very High_I AND Engagement Rate is very high_E AND Followers are medium_F, THEN Trust Level is Very High.
Rule 124: If Influence Score is Very High_I AND Engagement Rate is very high_E AND Followers are high_F, THEN Trust Level is Very High.
Rule 125: If Influence Score is Very High_I AND Engagement Rate is very high_E AND Followers are Very high_F, THEN Trust Level is Very High.

III. **Aggregation**
Results: Aggregation of maxiumum rule value (Figure 6.10)

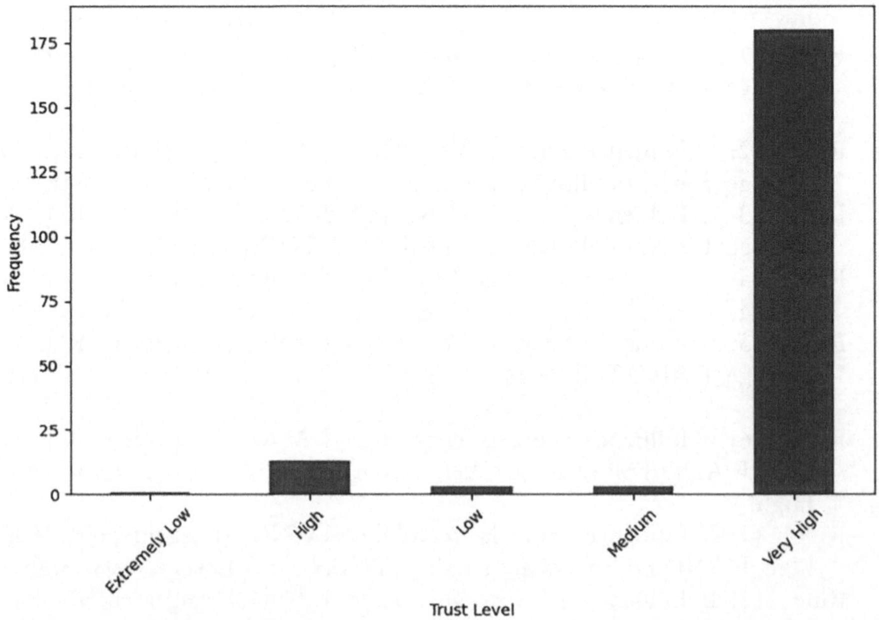

Figure 6.10 Trust level distribution.

Maximum rule values		Rule31	0.65	Rule63	0.11	Rule95	0.02
		Rule32	0.77	Rule64	0.06	Rule96	0.45
Rule1	0.00	Rule33	0.40	Rule65	0.00	Rule97	0.63
Rule2	0.00	Rule34	0.00	Rule66	0.44	Rule98	0.76
Rule3	0.00	Rule35	0.00	Rule67	0.44	Rule99	0.27
Rule4	0.00	Rule36	0.29	Rule68	0.00	Rule100	0.89
Rule5	0.00	Rule37	0.29	Rule69	0.00	Rule101	0.44
Rule6	0.00	Rule38	0.30	Rule70	0.00	Rule102	0.50
Rule7	0.00	Rule39	0.00	Rule71	0.25	Rule103	0.61
Rule8	0.00	Rule40	0.00	Rule72	0.25	Rule104	0.89
Rule9	0.00	Rule41	0.00	Rule73	0.06	Rule105	0.89
Rule10	0.00	Rule42	0.00	Rule74	0.17	Rule106	0.39
Rule11	0.00	Rule43	0.00	Rule75	0.00	Rule107	0.45
Rule12	0.00	Rule44	0.00	Rule76	0.72	Rule108	0.50
Rule13	0.25	Rule45	0.00	Rule77	0.79	Rule109	0.85
Rule14	0.00	Rule46	0.00	Rule78	0.76	Rule110	0.85
Rule15	0.00	Rule47	0.00	Rule79	0.44	Rule111	0.36
Rule16	0.00	Rule48	0.06	Rule80	0.39	Rule112	0.44
Rule17	0.00	Rule49	0.00	Rule81	0.70	Rule113	0.50
Rule18	0.09	Rule50	0.00	Rule82	0.80	Rule114	0.63
Rule19	0.00	Rule51	0.65	Rule83	0.50	Rule115	0.63
Rule20	0.00	Rule52	0.72	Rule84	0.38	Rule116	0.42
Rule21	0.00	Rule53	0.00	Rule85	0.28	Rule117	0.42
Rule22	0.00	Rule54	0.00	Rule86	0.47	Rule118	0.50
Rule23	0.00	Rule55	0.00	Rule87	0.63	Rule119	0.61
Rule24	0.00	Rule56	0.28	Rule88	0.67	Rule120	0.02
Rule25	0.00	Rule57	0.28	Rule89	0.47	Rule121	0.29
Rule26	0.41	Rule58	0.11	Rule90	0.22	Rule122	0.29
Rule27	0.61	Rule59	0.11	Rule91	0.51	Rule123	0.35
Rule28	0.00	Rule60	0.00	Rule92	0.51	Rule124	0.08
Rule29	0.00	Rule61	0.12	Rule93	0.67	Rule125	0.11
Rule30	0.00	Rule62	0.17	Rule94	0.39		

IV. **Defuzzification**

Trust Categorization in Social Networks: To understand and categorize trust levels in social networks, it is utilized for defuzzified values derived from various metrics and interactions. The process involves categorizing these defuzzified values into different levels of trust, which can then be analyzed for various insights. Here's a detailed explanation of the methodology:

Defuzzified Values: Defuzzified values are numerical representations of trust derived from various interactions and metrics in the social network. These values provide a quantifiable measure of trust that can be systematically analyzed.

Trust Level Categorization: A function "categorize_trust" is defined to map defuzzified values into distinct trust categories (Table 6.10):

- "Very High": for values >= 80
- "High": for values >= 60
- "Medium": for values >= 40
- "Low": for values >= 20
- "Extremely Low": for values < 20

6.4 CONCLUSION AND FUTURE SCOPE

Social network analysis (SNA) allows us to understand the dynamics of online interactions among individuals [30]. Trust is the key for building relationship [31, 43] in making friendships and working together online. By utilizing direct and referenced variables such as the duration of interaction, the score of influence, and the rate of engagement, these methodologies provide valuable understanding into the intricate process of trust development within online communities. Machine learning techniques like Logistic Regression and k-NN algorithms provide predictive capabilities, while alternative approaches like linear combination models offer a systematic way to calculate trust scores. Additionally, fuzzy logic-based [29] approaches introduce linguistic variables and membership functions to capture the uncertainty and subjectivity inherent in trust assessment. Through the use of fuzzification, fuzzy rule formulation, and aggregation techniques, these methods allow for a detailed understanding of trust dynamics in social networks.

In real-world friendships, trust might not always build proportionally with friendship. There could be periods where trust weakens despite a strong friendship (due to misunderstandings or arguments). The model doesn't account for external factors that can influence trust, such as shared experiences, betrayals from others (outside the friendship), or cultural norms. So, in the future, there is scope to explore a model where trust (T) is influenced by both the strength of friendship (F) and additional variables representing external factors.In the future, the development of more sophisticated algorithms and the use of machine learning approaches will be helpful to analyze

Table 6.10 Defuzzified Value and Trust Level

	Followers	*Defuzzified*		*Trust Level*
0	475800000	105.771186	0	Very High
1	366200000	105.107143	1	Very High
2	357300000	102.686813	2	Very High
3	342700000	103.211340	3	Very High
4	334100000	100.250000	4	Very High
5	329200000	99.060000	5	Very High
6	327700000	102.631868	6	Very High
7	272800000	101.656863	7	Very High
8	268300000	97.500000	8	Very High
9	254500000	100.250000	9	Very High
10	254000000	107.983146	10	Very High
11	237000000	100.250000	11	Very High
12	234100000	99.000000	12	Very High
13	222200000	99.457265	13	Very High
14	220400000	97.424242	14	Very High
15	211800000	95.382353	15	Very High
16	201600000	98.103448	16	Very High
17	195200000	96.329480	17	Very High
18	181500000	95.625000	18	Very High
19	177100000	97.050228	19	Very High
20	170300000	99.638655	20	Very High
21	152000000	95.290000	21	Very High
22	150700000	109.575000	22	Very High
23	140500000	83.606742	23	Very High
24	139100000	94.406977	24	Very High
25	135300000	94.060241	25	Very High
26	130900000	93.452555	26	Very High
27	125100000	91.962963	27	Very High
28	123400000	95.269841	28	Very High
29	119600000	94.294118	29	Very High
30	118500000	91.500000	30	Very High
31	111400000	93.814815	31	Very High
32	105200000	89.340909	32	Very High
33	85900000	72.400000	33	High
34	85600000	91.500000	34	Very High
35	82300000	91.500000	35	Very High
36	81300000	99.000000	36	Very High
37	81100000	94.000000	37	Very High
38	80900000	86.500000	38	Very High
39	76100000	90.661290	39	Very High
40	75300000	90.614286	40	Very High

(Continued)

Table 6.10 (Continued) Defuzzified Value and Trust Level

	Followers	*Defuzzified*		*Trust Level*
41	75300000	103.007463	41	Very High
42	74900000	90.605263	42	Very High
43	73900000	92.486726	43	Very High
44	73200000	92.378049	44	Very High
45	72700000	101.957265	45	Very High
46	70400000	89.000000	46	Very High
47	70100000	93.250000	47	Very High
48	69900000	103.111111	48	Very High
49	68900000	98.000000	49	Very High
50	68900000	105.500000	50	Very High
51	68700000	102.385965	51	Very High
52	68400000	99.147541	52	Very High
53	67700000	99.829268	53	Very High
54	67700000	81.160920	54	Very High
55	67300000	93.250000	55	Very High
56	66900000	101.273810	56	Very High
57	66300000	101.467742	57	Very High
58	66200000	86.250000	58	Very High
59	65900000	93.250000	59	Very High
60	65000000	92.261364	60	Very High
61	63900000	86.593750	61	Very High
62	63500000	87.750000	62	Very High
63	63100000	86.870968	63	Very High
64	62900000	106.870968	64	Very High
65	62800000	93.275862	65	Very High
66	62800000	92.608434	66	Very High
67	62700000	94.071429	67	Very High
68	61900000	92.000000	68	Very High
69	61800000	107.269663	69	Very High
70	59600000	90.500000	70	Very High
71	59500000	101.807692	71	Very High
72	58700000	84.746032	72	Very High
73	58100000	90.500000	73	Very High
74	58100000	90.916667	74	Very High
75	57600000	106.593750	75	Very High
76	56900000	41.206522	76	Medium
77	55900000	108.115385	77	Very High
78	55600000	110.500000	78	Very High
79	55200000	90.256098	79	Very High
80	55100000	97.464286	80	Very High
81	54600000	100.500000	81	Very High

(Continued)

Table 6.10 (Continued) Defuzzified Value and Trust Level

	Followers	*Defuzzified*		*Trust Level*
82	54500000	90.500000	82	Very High
83	54500000	110.500000	83	Very High
84	54100000	98.000000	84	Very High
85	53900000	90.500000	85	Very High
86	53500000	83.882353	86	Very High
87	53400000	100.833333	87	Very High
88	53400000	90.500000	88	Very High
89	53200000	93.000000	89	Very High
90	53000000	78.263158	90	High
91	52800000	93.000000	91	Very High
92	52800000	90.500000	92	Very High
93	52500000	14.323529	93	Extremely Low
94	52400000	90.500000	94	Very High
95	51200000	96.684211	95	Very High
96	50800000	90.500000	96	Very High
97	50700000	108.067568	97	Very High
98	50700000	94.300000	98	Very High
99	50200000	81.400000	99	Very High
100	49900000	90.500000	100	Very High
101	49700000	93.000000	101	Very High
102	49300000	110.500000	102	Very High
103	49200000	108.000000	103	Very High
104	49100000	108.000000	104	Very High
105	49000000	90.500000	105	Very High
106	48900000	90.500000	106	Very High
107	48700000	90.500000	107	Very High
108	48300000	103.000000	108	Very High
109	48200000	103.000000	109	Very High
110	48200000	98.000000	110	Very High
111	48100000	98.000000	111	Very High
112	47700000	90.500000	112	Very High
113	47300000	90.500000	113	Very High
114	46900000	60.500000	114	High
115	46900000	105.500000	115	Very High
116	46800000	90.500000	116	Very High
117	46500000	90.500000	117	Very High
118	46500000	110.500000	118	Very High
119	46500000	90.500000	119	Very High
120	46200000	110.500000	120	Very High

(*Continued*)

Table 6.10 (Continued) Defuzzified Value and Trust Level

	Followers	*Defuzzified*		*Trust Level*
121	45900000	90.500000	121	Very High
122	45900000	93.000000	122	Very High
123	45800000	90.500000	123	Very High
124	45400000	90.500000	124	Very High
125	45400000	103.000000	125	Very High
126	44800000	102.000000	126	Very High
127	44500000	39.500000	127	Very High
128	44200000	92.000000	128	Very High
129	43900000	77.000000	129	Low
130	43800000	102.000000	130	Very High
131	43800000	97.000000	131	High
132	43700000	107.000000	132	Very High
133	43400000	89.500000	133	Very High
134	43200000	89.500000	134	Very High
135	43100000	92.000000	135	Very High
136	42900000	89.500000	136	Very High
137	42200000	97.000000	137	Very High
138	42100000	109.500000	138	Very High
139	42100000	89.500000	139	Very High
140	41900000	109.500000	140	Very High
141	41800000	74.717391	141	Very High
142	41600000	97.000000	142	Very High
143	41500000	108.923077	143	High
144	41500000	89.500000	144	Very High
145	40800000	89.672414	145	Very High
146	40700000	89.280702	146	Very High
147	40700000	97.235849	147	Very High
148	40100000	101.655172	148	Very High
149	40000000	89.500000	149	Very High
150	40000000	88.475410	150	Very High
151	39900000	72.785714	151	Very High
152	39900000	86.879310	152	Very High
153	39200000	83.954545	153	High
154	39200000	89.329787	154	Very High
155	39100000	51.812500	155	Very High
156	39100000	89.382353	156	Very High
157	39000000	89.232704	157	Medium
158	38900000	85.475000	158	Very High
159	38800000	72.107692	159	Very High
160	38700000	86.433333	160	Very High

(Continued)

Table 6.10 (Continued) Defuzzified Value and Trust Level

	Followers	*Defuzzified*		*Trust Level*
161	38500000	88.785714	161	High
162	38300000	96.609244	162	Very High
163	38300000	91.381166	163	Very High
164	38100000	95.845070	164	Very High
165	37400000	85.694215	165	Very High
166	37400000	88.551020	166	Very High
167	37000000	87.953947	167	Very High
168	36900000	76.635294	168	Very High
169	36500000	87.722892	169	Very High
170	36500000	89.088235	170	High
171	36400000	100.766234	171	Very High
172	36000000	79.267857	172	Very High
173	35900000	98.004425	173	Very High
174	35900000	38.278846	174	High
175	35600000	99.131673	175	Very High
176	35600000	91.671171	176	Low
177	35500000	101.647826	177	Very High
178	35400000	89.952153	178	Very High
179	35300000	86.657658	179	Very High
180	35000000	69.720497	180	Very High
181	34800000	87.734694	181	Very High
182	34800000	86.732984	182	High
183	34700000	94.713415	183	Very High
184	34700000	81.327586	184	Very High
185	34700000	87.675127	185	Very High
186	34600000	83.131737	186	Very High
187	34200000	86.406091	187	Very High
188	34200000	78.661972	188	Very High
189	34100000	86.353535	189	Very High
190	33700000	81.476744	190	High
191	33600000	106.042254	191	Very High
192	33500000	58.772021	192	Very High
193	33500000	35.767857	193	Very High
194	33300000	74.624204	194	Medium
195	33200000	74.295082	195	Low
196	33200000	83.653061	196	High
197	33200000	81.226519	197	High
198	33000000	87.603604	198	Very High
199	32800000	82.185567	199	Very High
			Average Trust Level: 91.60837942499998	

trust more accurately by considering a wider range of factorsbeyond basic metrics like follower count or engagement rates. Future research can develop methods to analyze trust in social networks across different languages and cultures. Social networks can implement trust scores as part of a user's reputation, allowing others to measure their reliability and trustworthiness. Also, in the future, trust analysis can play a key role in identifying and combating the spread of false information online. Platforms can use this information to highlight trusted sources and reduce content from unreliable sources.

REFERENCES

1. Srivastav, M. K., & Gupta, S. (2023). An approach for exploring practical phenomena in social network analysis. *Journal of Mines, Metals and Fuels*, *71*(5), 583–587. https://doi.org/10.18311/jmmf/2023/34154
2. Srivastav, M. K., & Gupta, S. (2022). Analysis and prediction of friendship model in social networks like Facebook using binomial distribution and supervised learning. *SSRN*. https://ssrn.com/abstract=4090505 or http://dx.doi.org/10.2139/ssrn.4090505
3. Scott, J. (2023). *Social network analysis*.Bloomsbury Publishing India Pvt. Ltd.
4. Connolly, B. (2020). *Digital trust: Social media strategies to increase trust and engage customers*.Bloomsbury Publishing.
5. Deo, N. (1974). *Graph theory with applications to engineering and computer science*.Prentice Hall.
6. Shabariram, C. P. (2017). *Social network analysis*. Charulatha Publications.
7. Tsvetovat, M., & Kouznetsov, A. (2012). *Social network analysis for startups*. O'Reilly.
8. Chakraborty, T. (2021). *Social network analysis*. Wiley.
9. Wasserman, S., &Fast, K. (1994). *Social network analysis: Methods and applications*. Cambridge University Press.
10. Shinghal, R. (2013). *Introduction to fuzzy logic*. PHI Learning Private Limited.
11. Hatamleh, I. H. M., Almomani, A., AlZubi, H., & Hammad, M. (2023). Trust in social media: Enhancing social relationships. *Social Sciences*, *12*(7), 416. https://doi.org/10.3390/socsci12070416
12. Almogbel, R. S., & Alkhalifah, A. A. (2022). User behavior in social networks toward privacy and trust: Literature review. *International Journal of Interactive Mobile Technologies (IJIM)*, *16*(1), 38–51. https://doi.org/10.3991/ijim.v16i01.27763
13. Alkhamees, M., Alsaleem, S., Al-Qurishi, M., Alrubaian, M., & Hussain, A. (2021). User trustworthiness in online social networks: A systematic review. *Applied Soft Computing*, *103*, 107159. https://doi.org/10.1016/j.asoc.2021.107159
14. Ansheng, Y., & Haiping, H. (2021). Trusted network evaluation model based on comprehensive trust. *Chinese Journal of Electronics*, *30*(6), 1178–1188. https://doi.org/10.1049/cje.2021.07.028
15. Li, X., & Hu, H. (2020). Trust in social networks: A computational model based on the trust features of language use. In *2020 Asia Conference on Computers and Communications (ACCC)* (pp. 109–113). IEEE.

16. Liu, G., Yang, Q., Wang, H., & Liu, A. X. (2021). Trust assessment in online social networks. *IEEE Transactions on Dependable and Secure Computing*, *18*(2), 994–1007. https://doi.org/10.1109/TDSC.2019.2916366
17. Du, W., Lin, H., Sun, J., Yu, B., & Yang, H. (2016). A new trust model for online social networks. In *2016 First IEEE International Conference on Computer Communication and the Internet (ICCCI)* (pp. 300–304). IEEE. https://doi.org/10.1109/CCI.2016.7778930
18. Nepal, S., Bista, S. K., & Paris, C. (2014). Behavior-based propagation of trust in social networks with restricted and anonymous participation. *Computational Intelligence*, *31*(4), 642–668. https://doi.org/10.1111/coin.12041
19. Nafi, K. W., Kar, T. S., Hossain, M. A., & Hashem, M. M. A. (2013). A fuzzy logic-based certain trust model for e-commerce. In *2013 International Conference on Informatics, Electronics and Vision (ICIEV)* (pp. 1–6). IEEE. https://doi.org/10.1109/ICIEV.2013.6572693
20. DuBois, T., Golbeck, J., & Srinivasan, A. (2011). Predicting trust and distrust in social networks. In *2011 IEEE Third International Conference on Privacy, Security, Risk and Trust and 2011 IEEE Third International Conference on Social Computing* (pp. 418–424). IEEE.
21. Guha, R. V., Kumar, R., Raghavan, P., & Tomkins, A. (2004). Propagation of trust and distrust. In *Proceedings of the 13th International Conference on World Wide Web (WWW '04)* (pp. 403–412). ACM. https://doi.org/10.1145/988672.988727
22. Mohammadi, A., & Golpayegani, S. A. H. (2021). SenseTrust: A sentiment-based trust model in social network. *Journal of Theoretical and Applied Electronic Commerce Research*, *16*(6), 2031–2050. https://doi.org/10.3390/jtaer16060114
23. Ganesh, M. (2015). *Introduction to fuzzy set and fuzzy logic*. PHI Learning Private Limited.
24. Shankar, T. N. (2009). *Neural networks*. University Science Press.
25. Srinivasraghavan, A., & Joseph, V. (2020). *Machine learning*. Wiley.
26. Salaria, R. S. (2019). *Programming in Python*. Khanna Publishing.
27. Kosorukoff, A. (2011). *Social network analysis: Theory and applications*.D. L. Passmore.
28. Sridhar, S., & Vijayalakshmi, M. (2021). *Machine learning*. Oxford University Press.
29. Alavala, C. R. (2013). *Fuzzy logic and neural networks*. New Age International.
30. Meenakshi, A., & Armstrong, A. N. (2017). *Graph theory and application*. Sri Krishna Hitech Publishing.
31. Chervany, N. L. (1999). What is trust? A conceptual analysis and an interdisciplinary model. In *Proceedings of the 32nd Annual Hawaii International Conference on System Sciences*.
32. Lewicki, R. J., & Bunker, B. B. (1996). Developing and maintaining trust in work relationships. In: Kramer, R.M. and Tyler, T.R., Eds., *Trust in Organizations: Frontiers in Theory and Research*, Sage Publications, Thousand Oaks, 114–139. http://doi.org/10.4135/9781452243610.n7
33. Mayer, R. C., Davis, J. H., & Schoorman, F. D. (1995). An integrative model of organizational trust. *The Academy of Management Review*, *20*(3), 709–734. https://doi.org/10.2307/258792

34. Uslaner, E. M. (2002). Does political trust matter? A meta-analysis on the consequences of political trust. *Comparative Political Studies*, *35*(8), 1029–1050.
35. Krasnova, H., & Veltri, N. F. (2014). Measuring trust: A text analysis approach to compare existing trust questionnaires. In *Proceedings of the 22nd European Conference on Information Systems (ECIS)*.
36. Kaggle. (2024). *Who is a friend? [Data set]*. Retrieved June 11, 2024, from https://www.kaggle.com/c/whoisafriend/data?select=test.csv
37. Chitaxiang. (2024, May23). *Instagram influencer score analysis*.Kaggle. Retrieved June 11, 2024, from https://www.kaggle.com/code/chitaxiang/instagram-influencer-data-analysis/input
38. AgileStories, AgileBlog. (2020, July6). *Flowdays - Die agile Genossenschaft*. Retrieved June 11, 2024, from https://flowdays.net/en/blog-en
39. FasterCapital. (n.d.). *The influence of product quality and brand reputation on consumer purchase decisions*. Retrieved June 11, 2024, from https://fastercapital.com/topics/the-influence-of-product-quality-and-brand-reputation-on-consumer-purchase-decisions.html
40. Davis III, J. S., Moe, E., Stannard, D. J., &Moe, E. (2005). *WO2007076297A2 - Trust-based rating system [Patent]*. Google Patents. Retrieved June 11, 2024, from https://patents.google.com/patent/WO2007076297A2/en
41. Jones, A., &George, R. (2020). Trust and friendship in social networks. *Journal of Social Dynamics*, *12*(3), 45–62.
42. Srivastav, M. K., &Nath, A. (2015). A comprehensive study on complexity of relationships in social networks. In *Proceedings of the 2015 International Conference on Computational Intelligence and Communication Networks (CICN)* (pp. 718–723). Jabalpur, India. https://doi.org/10.1109/CICN.2015.146
43. Srivastav, M. K., Gupta, S., Priyadharshini, V. M., Som, S., Acharya, B., Gerogiannis, V. C., Kanavos, A., &Karamitsos, I. (2023). Utilizing degree centrality measures for product advertisement in social networks. In *Proceedings of the European, Mediterranean, and Middle Eastern Conference on Information Systems (EMCIS)*.

Chapter 7

Human-centred innovation in Society 5.0

Harnessing technology and Industry 4.0 for enhanced quality of life, social responsibility and sustainability

Anindita Saha

7.1 INTRODUCTION

The advanced innovations in science and technology, exemplified by the remarkable surge in computing power, play a pivotal role in advancing both business and society. This progress, however, unfolds against a backdrop of global-scale challenges such as the depletion of natural resources, climate change, widening economic disparities, and terrorism. Our contemporary era is characterized by heightened uncertainty and escalating complexity across all domains. Therefore, it becomes imperative to harness Information and Communication Technology (ICT) capabilities fully. This entails not only gaining new insights but also creating novel values by forging connections between "people and things" and bridging the realms of the "real and cyber" worlds. Such integration is crucial for effectively and efficiently addressing societal issues, enhancing the well-being of individuals, and sustaining robust economic growth. To navigate these challenges successfully, it is essential to foster a shared vision among diverse stakeholders at multiple levels, facilitating a collective journey towards a digitally transformed society [1]. The Cabinet Office of Japan, from which the concept of Society 5.0 arose, defines it as "a human-centered society that balances economic advancement with the resolution of social problems through a system that highly integrates cyberspace and physical space" [2, 3]. Put differently, Society 5.0 serves as a communicative framework through which the government conveys its vision for the future to both industry stakeholders and the general public. This model represents the culmination of extensive discussions involving experts from diverse fields and is grounded in thorough research on the historical trajectories of technology and social development. Society 5.0 hinges on seamlessly merging the real and digital worlds. Data collected from various real-world sources undergoes processing through technologies like AI and machine learning, aiming to extract meaningful insights. These insights are then applied back to the real world, informing decision-making and enhancing societal aspects. This closed-loop system, utilizing technologies such as big data analytics, fosters a more efficient, responsive, and adaptive society. Practical applications include optimizing transportation based

DOI: 10.1201/9781003643166-7

Figure 7.1 Society 5.0-essential elements.

on real-time data, personalizing healthcare through individual health data, and improving urban planning by analysing citizen behaviour. The integration of data and technology is a cornerstone in addressing complex societal challenges and elevating the overall quality of life in Society 5.0. [4]. A simple pictorial representation is given in Figure 7.1.

Examining the historical trajectory of human development reveals distinct societal stages. Society 1.0 was characterized by nomadic communities engaged in hunting and gathering, embodying a harmonious balance with nature. The advent of agriculture marked the onset of Society 2.0, introducing organized settlements and the formation of early nations. The industrial revolution propelled Society 3.0 into an era of mass production, transforming economies and societies. Society 4.0, the information society, harnessed interconnected information networks, prioritizing intangible assets to generate added value. This phase marked a shift towards a digital landscape, where information became a key driver of economic and social progress. Building upon these foundations, Society 5.0 emerges as the next evolutionary step. Positioned as an information society, it synthesizes the advancements of Society 4.0 while emphasizing a crucial shift towards a human-centred framework. In Society 5.0, the integration of cutting-edge technologies is geared not only towards economic prosperity but also towards the well-being of individuals and communities. It envisions a harmonious coexistence of technological innovation and human values, fostering sustainability and inclusivity. Society 5.0 represents a purposeful evolution, recognizing the importance of aligning technological progress with the fundamental needs and aspirations of humanity, thereby striving for a prosperous and human-centric future [5]. Industry 4.0, with its advanced technologies and automation, significantly influences Society 5.0 by fostering a seamless integration of digital systems into everyday life. This convergence aims to enhance societal well-being through innovations like smart cities, healthcare advancements, and sustainable practices. The synergy between Industry 4.0 and Society 5.0 promotes a human-centric approach, leveraging technology for the betterment of individuals and communities. This transformative connection envisions a harmonious coexistence where technological progress aligns with societal needs and values. Thus, Industry 4.0 acts as a catalyst in shaping Society 5.0 towards a more inclusive, intelligent, and sustainable future [6]. Society 5.0 bears the social responsibility of leveraging advanced technologies for inclusive and equitable development. It prioritizes addressing societal challenges, such as poverty, inequality, and environmental issues, through innovation and collaboration. Emphasizing human-centric solutions, Society 5.0 aims to ensure that technological advancements contribute to the well-being of all members of society. It fosters ethical practices, data privacy, and transparency to build trust in the digital era. Ultimately, the social responsibility of Society 5.0 is to create a sustainable and human-centred future by harnessing technology for the benefit of the entire community. Sustainability in Society 5.0 is foundational, aiming to harmonize technological progress with ecological balance [7]. This societal model promotes green technologies, circular economies, and responsible consumption to mitigate environmental impact. It emphasizes the integration of renewable energy, smart infrastructure, and eco-friendly practices to create a resilient and sustainable future. Society 5.0 envisions a circular and regenerative approach, where innovation aligns with environmental conservation for the well-being of present and future generations. By prioritizing sustainability, Society 5.0 strives to achieve a balanced coexistence between technological advancements and ecological harmony [8]. Implementing Society 5.0 presents challenges, notably in managing ethical implications of advanced technologies. Privacy concerns arise with the extensive use of AI, Big Data, and IoT, necessitating a delicate balance between societal benefits and individual data protection. Mitigating potential misuse, especially in AI and robotics, demands ethical guidelines to ensure responsible innovation. Additionally, fostering inclusivity and preventing technology-driven social disparities require proactive policies and diverse perspectives. Successfully navigating these ethical challenges is vital for Society 5.0 to fulfil its promise of a human-centred, technologically integrated future [9]. The digital divide poses a risk, creating disparities in access to technological benefits. Balancing innovation with societal values and ensuring inclusive development remains a complex task. Cybersecurity threats and the vulnerability of interconnected systems present ongoing challenges to the integrity of Society 5.0. Additionally, navigating the rapid pace of technological change and addressing resistance to adoption are key obstacles in realizing the full potential of this societal paradigm.

7.2 EVOLUTION OF SOCIETY 5.0 – A HUMAN-CENTRIC APPROACH

The evolution of Society 5.0 marks a profound shift in societal development, characterized by a human-centric approach and the seamless integration of advanced technologies. This visionary framework builds on the progression of human societies through earlier stages, from hunting and gathering to agrarian, industrial, and information societies. In Society 5.0, technology is leveraged to enhance the well-being and empowerment of individuals, placing human needs and aspirations at the forefront of development. This entails personalized experiences, the collaborative coexistence of humans and AI, and innovations in healthcare. Moreover, the integration of technologies like AI, the Internet of Things, robotics, and biotechnology creates a connected ecosystem that transcends sectors such as smart cities, education, and manufacturing. Sustainability, inclusivity, and ethical considerations are key pillars of Society 5.0, ensuring that the benefits of technological progress are accessible to all while maintaining a responsible and ethical approach to development [10]. Figure 7.2 depicts the concept in detail.

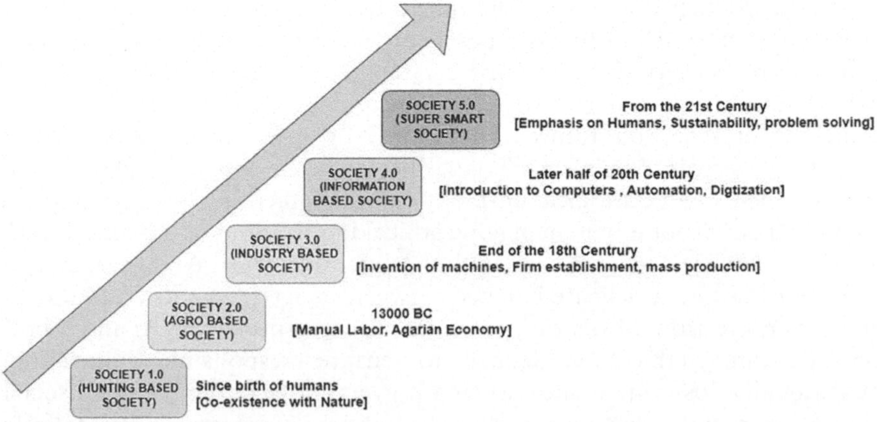

Figure 7.2 Evolution of Society 5.0.

7.2.1 Pre-industrial societies (Society 1.0 and 2.0)

Society 1.0 and Society 2.0 represent two distinct phases in the evolution of human civilization driven by technological advancements and social changes. The earliest human societies, often referred to as Society 1.0, were characterized by hunting and gathering. Society 1.0, often associated with the pre-industrial era, was characterized by agrarian economies, limited technological innovation, and predominantly rural communities. It revolved around subsistence farming and small, tightly-knit social structures. With the advent of agriculture, societies transitioned to an agrarian mode (Society 2.0). These societies were based on manual labour, and technological advancements were limited. Technically in sharp contrast with Society 1.0, Society 2.0 signifies the modern era, marked by the Industrial Revolution and the subsequent digital age. This phase witnessed rapid industrialization, urbanization, and the proliferation of technology. It led to mass urban centres, complex social structures, and a globalized economy [11]. Society 2.0 or Web 2.0, or Information Technology 2.0, encompasses websites that prioritize user-centric content, accessibility, and user-friendliness, emphasizing compatibility for multitasking and collaboration. This paradigm establishes a global virtual platform that fosters online interaction and communication, responding to client feedback and valuable input. Web 2.0 has proven to be a powerful means for expressing opinions, sharing ideas, and disseminating valuable content [12]. It has elevated digital communication by increasing blogging, research content, and facilitating creative domain naming. Notably, it boosts organizational workforce confidence, promoting efficient collective learning, a dynamic process. Web 2.0 allows the alteration of web page design and presentation for clients without changing technical specifications. Key advantages of Information Technology 2.0 include easy accessibility, content and media variety, dynamic learning communities, traceability features, and real-time communication and visualization [13].

The transition from Society 1.0 to 2.0 signifies a transformative shift in human existence, marking profound changes in lifestyles, economies, and interpersonal interactions. This evolution is propelled by advancements in industry, communication, and information technology. Society 2.0, marked by the advent of agriculture and organized settlements, brings about increased societal organization and the formation of early nations. The harnessing of these advances reshapes not only the way humans live but also how they engage with the world, setting the stage for subsequent societal stages that further intertwine human life with technological progress.

7.2.2 Industrial revolution (Society 3.0)

Popularly known as the Industrial Revolution, and starting in the late 18th century, Society 3.0, marked the transition to Society 3.0. It brought about mechanization, mass production, and significant technological advancements in manufacturing and transportation. Society 3.0 is a concept that represents the next phase in the evolution of human civilization, building upon the foundations of Society 2.0. While Society 2.0 was characterized by industrialization and digitalization, Society 3.0 envisions a society that is shaped by cutting-edge technologies such as AI, blockchain, nanotechnology, and biotechnology. This phase is marked by a high degree of interconnectivity, automation, and a focus on sustainable practices. In Society 3.0, we expect to see the emergence of smart cities, decentralized systems of governance and commerce, and an increased emphasis on environmental sustainability and resource efficiency. The boundaries between the physical and digital worlds become increasingly blurred, offering new opportunities for innovation, but also raising important ethical and privacy considerations. Society 3.0 represents a paradigm shift in how we organize and interact in our communities and the world at large, with technology playing a central role in shaping the future of human society.

7.2.3 Information age (Society 4.0)

The late 20th century saw the emergence of the Information Age, or Society 4.0, characterized by the widespread use of digital technology, the internet, and information systems. Information became a central asset, reshaping economies, communication, and daily life. Industry 4.0 represents a socio-technical system that redefines the interactions between individuals and organizations, the integration of technologies with production systems, and the dynamics of production and consumption. It introduces a fresh paradigm for the relationship between society and the industrial sector, centred around the ongoing process of digitalization [14]. Industry 4.0 merges traditional automation, following the established industrial model of the 20th century, with the distinctive elements of digital culture. This includes knowledge-driven systems, extensive utilization of sensor technology, ranging from network connectivity to comprehensive integration within IoT systems, enhanced process adaptability, and a shift from vertical specialization to horizontal process-orientated structures. As a result, this transition results in an escalation of complexity within industrial operations [15]. Industry 4.0 revolves around the comprehensive transformation of industrial elements through robust digital integration, closely tied to the ongoing fourth industrial revolution. This evolution strives for optimized and sustainable production methods. In this latest wave of industrialization, there is a notable surge in the efficient utilization of advanced data analytics, fog computing, edge computing, cloud computing, robotics, blockchain and the extensive interconnection of machines, enhancing the digital landscape [16].

7.2.4 Emergence of Society 5.0

The concept of Society 5.0 emerged in Japan as part of the country's vision for the future of society. It was first officially introduced by the Japanese government in the "5th Science and Technology Basic Plan" in 2016. Society 5.0 represents a response to the challenges and opportunities posed by rapid technological advancement and digital transformation [17]. At the core of Society 5.0 is the idea of prioritizing human needs, well-being, and empowerment. It envisions a society where technology serves as a means to enhance the lives of individuals, rather than a goal in itself. Society 5.0 would tackle complex global challenges such as climate change, healthcare, education, and resource management by harnessing advanced technology and innovative solutions. Ethical considerations would hold a central role, guaranteeing that technological advancements uphold human rights and privacy. Cutting-edge technologies, encompassing AI, biotechnology, quantum computing, and more, would play a pivotal part in advancing problem-solving and fostering sustainability. This advanced society would foster highly interconnected ecosystems, not just in terms of information but also in sustainable ecosystem management and resource utilization, marking a defining feature of Society 5.0 [18].

Society 5.0 represents a vision for a more interconnected, technologically advanced, and human-centred society that addresses contemporary challenges and leverages the potential of technology for the greater good. It is part of a global conversation on the future of societies in the digital age, and its principles resonate with discussions about the Fourth Industrial Revolution and the role of technology in shaping the world's future. While it has its roots in Japan, the concept of Society 5.0 is relevant and influential on a global scale.

7.3 TECHNOLOGY AND INDUSTRY 4.0: CATALYSTS FOR HUMAN-CENTRED PROGRESS

The term "Industry 4.0" was coined by Henning Kagermann and colleagues within the framework of the German federal government's high-tech strategy. This initiative emerged as a response to the ongoing digitization of manufacturing processes during the preceding decade [19]. Technology and Industry 4.0 represent pivotal catalysts for achieving human-centred progress by fundamentally reshaping the industrial landscape through advanced digitalization and automation. The integration of digital technologies, automation, and data-driven decision-making into industries revolutionizes the way businesses operate. The human-centric focus of Industry 4.0 emphasizes the well-being of the workforce and the delivery of products and services that better cater to individual needs. As a result, this technological transformation holds the potential to enhance the quality of life, promote economic growth, and create a more inclusive and adaptable society, marking a significant leap towards human-centred progress [20].

There are several technical facets that contribute to this transformation. To begin with, Industry 4.0 focuses on **Data-Driven Decision-Making**, thriving on the acquisition, analysis, and utilization of vast amounts of data. Advanced data analytics and Artificial Intelligence (AI) algorithms enable businesses to not only collect and process data from various sources but also make informed, real-time decisions based on this data. This data-driven decision-making optimizes processes, enhances efficiency, and allows companies to anticipate market trends and customer preferences, ultimately improving resource allocation and profitability. Further, the introduction of **IoT and Smart Manufacturing**, a core component of Industry 4.0, where a multitude of devices and sensors are interconnected, collecting and sharing data [21]. Smart manufacturing harnesses this connectivity to optimize production processes, increase operational efficiency, and reduce waste. Machines can communicate with each other to coordinate tasks, reducing downtime and improving overall production quality. **Predictive Maintenance** is another technical facet of Industry 4.0, which is empowered by the continuous monitoring of equipment and machinery through sensors. These sensors gather data on the condition of machinery and can predict when maintenance is required. This not only minimizes costly downtime and unexpected equipment failures but also extends the lifespan of assets, reducing maintenance costs [22]. Industry 4.0 also promotes **Efficient Resource Management**, through data-driven insights, such as Real-time monitoring of energy consumption, water usage, and raw material consumption, which allows for optimization and reduction of resource waste. This is not only environmentally responsible but also significantly reduces operational costs. The **Human-Machine Collaboration**, though Collaborative robotics or (cobots) exemplify the human-centric approach of Industry 4.0. These robots work alongside human employees, handling repetitive and physically demanding tasks, while humans focus on tasks requiring creativity, critical thinking, and decision-making. This symbiotic relationship improves productivity and job satisfaction while also enabling upskilling for the workforce to operate and maintain these technologies [23]. Industry 4.0 technologies, including AI and machine learning, enable mass **Customization and Personalization** of products and services. Businesses can use customer data to tailor products to individual preferences, resulting in higher customer satisfaction and loyalty. Figure 7.3 provides a holistic idea of the driving factors of Industry 4.0.

Quality Control is ensured by Industry 4.0, through advanced sensors and data analytics to emphasize rigorous quality control throughout the production process. This not only reduces defects and waste but also improves product reliability and safety. Companies can quickly detect and address quality issues, ensuring that customers receive high-quality products [24]. The implementation of Industry 4.0 necessitates a highly skilled workforce. Businesses and educational institutions need to collaborate to provide training and upskilling programs that equip individuals with the digital literacy and technical skills needed to operate and troubleshoot advanced technologies. This investment in **Workforce Development** benefits both employees and organizations [25]. **Innovation Acceleration** is an evident

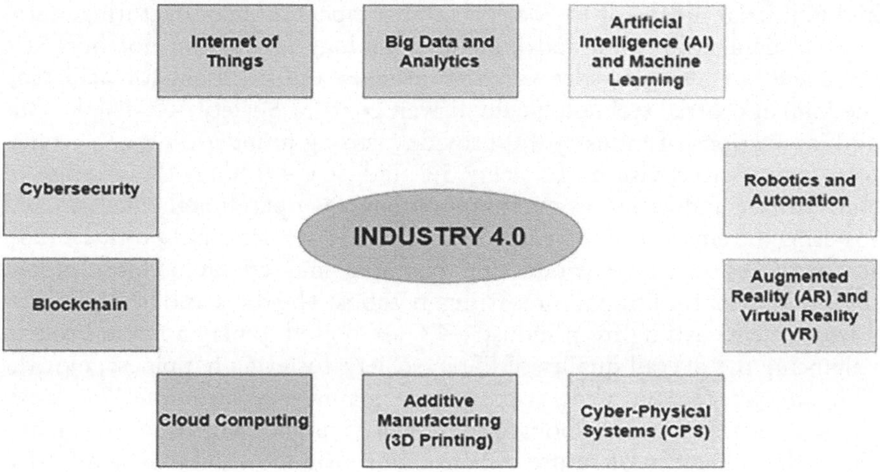

Figure 7.3 Industry 4.0 and the driving factors.

aspect of Industry 4.0 as it completely relies on digital technologies and data analytics. Companies can quickly prototype and test new ideas, bringing products and services to market faster. This fosters a competitive edge and stimulates economic growth. The **Optimization of Supply Chain** is ensured by Industry 4.0, as Digitalization of supply chains ensures better coordination, real-time insights, and more efficient management of inventory. This results in cost reductions, improved agility, and the ability to quickly adapt to market demands. Companies can also respond more effectively to disruptions, such as those caused by natural disasters or unforeseen events [26]. Industry 4.0 technologies play a significant role in advancing **Environmental Sustainability**. By optimizing resource consumption and improving operational efficiency, companies can reduce their environmental footprint. This aligns with broader ecological concerns and positions Industry 4.0 as a force for responsible and sustainable industrial progress [27].

Thus, Industry 4.0's technical facets drive a transformation in industries, promoting efficiency, innovation, and a human-centric approach. By leveraging data and technology, businesses can make informed decisions, optimize resource management, enhance product quality, and contribute to both economic growth and environmental sustainability. This technological evolution supports a more inclusive and adaptable society while equipping the workforce with the skills needed to thrive in the digital age.

7.4 INFLUENCE OF INDUSTRY 4.0 ON SOCIETY 5.0

Researches reveal that the influence of Industry 4.0 on Society 5.0 is profound, as Industry 4.0 acts as a foundational pillar for the realization of the goals and principles of Society 5.0. Industry 4.0 serves as a technological and industrial precursor to Society 5.0. Its impact on manufacturing, automation, data utilization, and human-technology interaction not only sets the stage but also provides valuable insights and practices for achieving the human-centric, technologically driven goals of Society 5.0. The lessons and innovations of Industry 4.0 provide a strong foundation for the transformative societal vision of Society 5.0. Industry 4.0 holds the promise of transforming industrial production by enhancing operational efficiency and fostering the emergence of new business models, services, and products [28]. It introduces real-time production planning and dynamic optimization, diverging from traditional forecasting methods. The tools and technological advancements arising from Industry 4.0 are poised to play a pivotal role in enhancing the overall quality of life in society, fostering happiness, motivation, and satisfaction while also allowing for more leisure time. This positive shift is anticipated to boost productivity, enabling humanity to actively shape the trajectory of future societies and promote equitable wealth distribution. As a natural progression from Industry 4.0, Society 5.0 emerges, initially championed in Japan in response to concerns about the ageing population. This societal paradigm aims to address the challenges posed by demographic shifts, leveraging the advancements of Industry 4.0 to create a future where technology is seamlessly integrated to enhance well-being and create a more balanced and inclusive society [29].

The most important influence of Industry 4.0 is the Digital Transformation of industries, that represents a profound shift in the basic operations of business. It involves the adoption of technologies like the Internet of Things (IoT), AI, and big data analytics to streamline processes, improve decision-making, and enhance efficiency. The digital transformation of industries is a critical precursor to the broader societal digital transformation in Society 5.0. This shift enables governments, organizations, and individuals to leverage technology to solve complex societal challenges and improve the overall quality of life. For example, digital health records and telemedicine are key components of this transformation, making healthcare services more accessible and efficient [30]. The introduction of resilience in supply chains by Industry 4.0, through data-driven insights and real-time monitoring, is essential for the stability and continuity of operations. These principles of supply chain resilience extend to Society 5.0, ensuring the efficient delivery of goods and services, especially during times of crisis. Society 5.0 leverages Industry 4.0's lessons to establish resilient supply chains that can adapt to disruptions, such as those caused by pandemics or natural disasters. This ensures that essential resources, including food, medical supplies, and energy, remain accessible to communities, promoting societal well-being and stability [31]. Cybersecurity, which is an essential component of technological growth, has been well adopted by Industry 4.0, thus necessitating robust cybersecurity measures to protect sensitive industrial data and systems. These security practices become even more critical in Society 5.0, where the protection of personal and societal data is paramount. As Industry 4.0

demonstrates, cybersecurity is not just about safeguarding industrial processes but also about protecting the integrity of data and digital infrastructure [32]. In Society 5.0, this extends to ensuring the confidentiality, availability, and integrity of personal data, financial transactions, and critical infrastructure, which are foundational to the well-being of individuals and society as a whole. One of the biggest influences of Industry 4.0 is Inclusivity and Accessibility, that encourages the development of user-friendly technology accessible to a broad audience. This focus on inclusivity extends to Society 5.0, where the aim is to make technology accessible to all, regardless of age, ability, or background. In the context of Industry 4.0, user interfaces and technologies are designed to be intuitive and easy to use [33]. In Society 5.0, this extends to addressing the needs of individuals with disabilities and the elderly, ensuring that technology serves as an enabler rather than a barrier. For example, voice recognition and assistive technologies play a vital role in making digital services accessible to those with limited mobility or vision. Core concepts of Industry 4.0, such as IoT and data analytics, are applied to create smart cities, which optimize urban services and resource management. These smart city principles align with the vision of Society 5.0, which seeks to create smarter, more efficient, and sustainable urban environments that enhance the well-being of urban populations. In Industry 4.0, smart city initiatives use data from IoT sensors to monitor traffic, energy consumption, waste management, and more [34]. This information is used to make real-time decisions that improve urban services, reduce energy consumption, and enhance quality of life. In Society 5.0, this concept is further expanded to create cities that are responsive to the needs of their inhabitants, utilizing data-driven insights to provide safer, cleaner, and more efficient urban environments. Industry 4.0 also promotes sustainable and eco-friendly manufacturing practices that minimize waste and energy consumption. These principles of sustainable manufacturing extend to Society 5.0, contributing to a more environmentally responsible and sustainable society [35]. In Industry 4.0, businesses adopt sustainable manufacturing practices that include the use of renewable energy, recycling, and energy-efficient processes. These practices not only reduce the environmental impact but also lead to cost savings. In Society 5.0, sustainability practices are applied to a broader spectrum of industries, from agriculture to construction, and are essential for achieving environmental goals. This shift towards sustainable practices is aligned with Society 5.0's vision of creating a balanced coexistence with the environment and a sustainable, resilient society. Industry 4.0 advances healthcare through telemedicine, wearable devices, and AI diagnostics. These innovations are integral to Society 5.0, which focuses on improving healthcare access and personalizing medical services to enhance the well-being of individuals. In Industry 4.0, telemedicine allows patients to consult with healthcare professionals remotely, reducing the need for physical visits to medical facilities. Wearable devices, such as smartwatches and fitness trackers, provide continuous health monitoring, enabling early disease detection and personalized treatment recommendations. AI diagnostics use machine learning algorithms to analyse medical data, improving the accuracy of diagnoses. These innovations in healthcare are foundational for Society 5.0, where healthcare is transformed into a more accessible, personalized, and data-driven system that prioritizes the well-being of individuals and communities [36]. Industry 4.0's emphasis on human-technology collaboration extends to addressing the needs of an ageing population. In Society 5.0, this support is expanded to create age-friendly societies where technology aids the elderly in daily life and healthcare. In Industry 4.0, human-robot collaboration is particularly valuable in assisting older individuals with tasks such as mobility, home maintenance, and healthcare. These concepts are integrated into Society 5.0, where technology serves as a supportive tool for the elderly, enabling them to maintain independence, stay connected, and access healthcare services more easily. The goal is to create a society where individuals can age with dignity and remain active, with technology playing a central role in enhancing their well-being and quality of life [37]. Industry 4.0 promotes resource efficiency and recycling, and these principles are applicable to Society 5.0. In Industry 4.0, businesses focus on minimizing waste and optimizing the use of raw materials. This concept is extended to Society 5.0, where a circular economy model is embraced to ensure that resources are used efficiently and waste is minimized. In a circular economy, products and materials are designed for reuse and recycling, reducing the consumption of virgin resources and reducing environmental impact. This transition to a circular economy aligns with Society 5.0's vision of sustainability and responsible resource management.

Industry 4.0's advancement in disaster preparedness and resilient manufacturing practices can be applied to build resilient communities in Society 5.0. In Industry 4.0, businesses implement disaster preparedness plans and resilient manufacturing processes to ensure continuity in the face of unexpected disruptions. These practices contribute to Society 5.0 by emphasizing the importance of building communities that can withstand and recover from environmental and societal challenges. This includes implementing early warning systems, community-based disaster management plans, and collaborative efforts to enhance the resilience of communities. The goal is to create communities that are not only capable of withstanding disasters but also capable of recovering and thriving in their aftermath [38]. Industry 4.0 often involves global supply chains and partnerships. These collaborative practices extend to Society 5.0, where international cooperation and information sharing are essential for addressing global challenges like climate change, cybersecurity, and pandemic preparedness. In Industry 4.0, businesses and industries collaborate across borders to source materials, components, and expertise. These principles are carried into Society 5.0, where global collaboration is pivotal in addressing complex global challenges [39]. This includes the sharing of data and knowledge between nations, cooperation on scientific research, and the development of international standards and regulations. In Society 5.0, global collaboration is critical for achieving sustainable and resilient solutions to the world's most pressing issues.

Industry 4.0 provides a strong technological and operational foundation for the realization of Society 5.0. The principles and practices developed in Industry 4.0, such as digital transformation, supply chain resilience, and sustainable manufacturing, serve as valuable lessons and frameworks for the broader societal transformation envisioned in Society 5.0. The alignment of these concepts promotes a more inclusive, efficient, and sustainable future while addressing complex societal challenges. A detailed comparison between the two has been shown in Table 7.1.

Table 7.1 Comparison Analysis of Industry 4.0 and Society 5.0

Aspect	*Industry 4.0*	*Society 5.0*
Focus	Primarily focuses on industrial processes, automation, and data exchange in manufacturing.	Primarily focuses on integrating technologies to address societal challenges and improve quality of life.
Technological Emphasis	Emphasizes technologies like IoT, AI, big data, robotics, and automation for optimizing production.	Emphasizes technologies like AI, IoT, robotics, biotechnology, and cyber-physical systems for societal well-being.
Integration	Integrates technologies to enhance efficiency, productivity, and resource utilization in industries.	Integrates technologies to create a harmonious and sustainable society, addressing various social issues.
Goal	Aims to revolutionize manufacturing and make it more efficient, flexible, and responsive.	Aims to create a human-centric society that leverages technology to address challenges and improve well-being.
Impact on Jobs	May lead to job displacement in certain industries as automation takes over routine tasks.	Emphasizes collaboration between humans and technology, potentially creating new job opportunities and enhancing existing ones.
Human-Centric Approach	Primarily driven by the optimization of industrial processes and production systems.	Prioritizes the well-being and empowerment of individuals, with technology serving human needs and enhancing the quality of life.
Application Areas	Manufacturing, supply chain, logistics, and industrial processes.	Healthcare, education, urban planning, environmental sustainability, and social services.

7.5 ENHANCING QUALITY OF LIFE THROUGH EQUITABLE TECHNOLOGICAL INTEGRATION IN SOCIETY 5.0

In Society 5.0, the equitable integration of technology is not just about advancing technology for its own sake; it is about harnessing the power of technology to improve the lives of individuals and communities in a fair and inclusive manner. This approach ensures that the benefits of technological progress are accessible to all, addressing inequalities and enhancing the overall quality of life while safeguarding against potential risks and ethical concerns. It represents a vision where technology serves as a tool to create more equitable, inclusive, and well-being-focused societies. Equitable technological integration is committed to providing **Universal Access** to technology. It means ensuring that not only the urban but also rural and underserved populations have reliable internet access, affordable devices, and digital literacy training. This equitable access empowers individuals to seek information, access education, apply for jobs, and connect with the world, leading to an improved quality of life. **Personalized Services, or the appropriate customization of services**, tailoring them to the specific needs of individuals, is another aspect of the enhanced social life in Society 5.0. In healthcare, personalized medicine uses genetic information to develop treatments unique to each patient, improving outcomes and reducing side effects. This personalized approach enhances the quality of life by providing more effective and less intrusive medical care [40]. Equitable technological integration exhibits **efficient public services**, reducing bureaucratic inefficiencies and providing citizens with faster and more accessible services. Digital government services, such as online tax filing and permit applications, save time and resources, improving the overall quality of life by simplifying interactions with government agencies. **Education and Lifelong Learning** become a reality in technology-driven education systems that not only provide access to quality education for people of all ages but also foster lifelong learning. Digital platforms offer a wide range of courses, allowing individuals to continually update their skills and pursue personal interests, ultimately enhancing their employability and overall well-being [41]. **Inclusive Employment** is fairly promoted as technology opens up new employment opportunities for a broader range of people. Remote work, for example, allows individuals with disabilities or those living in rural areas to participate in the job market. This inclusivity not only improves financial well-being but also provides a sense of purpose, contributing to a higher quality of life. Equitable technological integration also promotes **social inclusion** by providing tools and platforms for marginalized groups to connect and engage in society. Online communities, social networks, and digital advocacy groups empower individuals who may face geographical, physical, or social barriers, reducing isolation and improving social bonds, thus enhancing the overall quality of life [42]. Technology plays a pivotal role in enhancing

environmental sustainability, thereby contributing to an improved quality of life. Renewable energy technologies reduce pollution and lower energy costs, creating a healthier and more affordable living environment. Smart city solutions optimize resource usage and reduce congestion, improving the overall quality of urban life. The aspect of Mental Health and Well-being cannot be denied with Technology, that is increasingly being harnessed to support. Teletherapy and mental health apps provide convenient access to mental health services, reducing stigma and enhancing emotional well-being. Additionally, mindfulness and stress management apps offer tools for maintaining mental health and improving overall quality of life. Equitable technological integration aids in the **preservation and promotion of cultural heritage** [43]. Digital archives, virtual museums, and language preservation apps ensure that cultural identities remain intact. This enriches the lives of diverse communities, fostering a sense of cultural pride and continuity, which enhances overall well-being. Technology contributes to **community resilience** by providing early warning systems, disaster management apps, and community engagement platforms. These tools empower communities to withstand and recover from environmental and societal challenges. By being better prepared, communities can reduce the impact of disasters, minimize loss, and improve their quality of life in the face of adversity [44]. Global Connectivity is ensured through equitable technological integration, allowing individuals to connect with people and cultures from around the world. This cross-cultural exchange enriches lives by broadening perspectives, promoting global cooperation, and celebrating diversity. Global connectivity fosters understanding and collaboration, contributing to a more harmonious and interconnected world. Thus, observations reveal that equitable technological integration in Society 5.0 is a multifaceted approach that ensures that technology benefits everyone, regardless of their background or circumstances. It addresses disparities, promotes inclusivity, and leverages technology to improve various aspects of life, from education and healthcare to employment and community resilience. By focusing on equitable technological integration, Society 5.0 aims to create a more inclusive and well-being-centric society where technology is a powerful tool for improving the overall quality of life.

7.6 NURTURING SOCIAL RESPONSIBILITY IN SOCIETY 5.0

Social responsibility in Society 5.0 refers to the commitment of individuals, organizations, and governments to ethical and sustainable practices in the context of advanced technology and the digital society. The concept encompasses a vision of a society that prioritizes the well-being, equity, and inclusivity of all its members through the strategic application of advanced technologies and socio-economic principles. The most important key factor that guides social sustainability is the Human-Centric Approach, where the well-being and individual needs of its citizens are kept at the forefront of its agenda. This human-centric approach means that technology and societal structures are designed and orientated to improve the quality of life for all, addressing real-world challenges and making daily life more convenient and fulfilling [45]. Another essential feature of Society 5.0 revolves around Inclusivity, a central tenet where Technology is designed to be accessible and beneficial to everyone, regardless of their age, abilities, or background. This involves designing user-friendly interfaces, making digital services available in multiple languages, and providing support for those with disabilities. Education and Lifelong Learning or Continuous learning is emphasized as one of the major factors in Society 5.0, where Individuals are encouraged and equipped to adapt to the fast-paced changes in technology and the job market. Lifelong learning programs, online courses, and skill development initiatives ensure that people remain relevant in the digital society [46]. The healthcare system, which is another important aspect in Society 5.0, is underpinned by technology. Telemedicine enables remote consultations, and wearable devices monitor health parameters. Personalized healthcare solutions, guided by big data and AI, aim to prevent diseases and enhance overall well-being [47]. Social Inclusion is one of the key tenets in Society 5.0 that actively works to reduce inequalities and enhance social inclusion. This includes measures to ensure that everyone, regardless of socio-economic status, has access to the benefits of technological advancements, reducing the digital divide. Much importance is given to mental health in Society 5.0, as the digital society recognizes the importance of the same in today's world. Technology is employed to provide mental health support and counselling, ranging from AI chatbots for emotional well-being to online therapy platforms. Among other aspects, Society 5.0 leverages technology to enhance the quality of life for seniors as the demographic majority is comprised of them. Smart homes equipped with assistive technologies, wearable health devices, and transportation solutions cater to the needs of the elderly [48]. One of the most important features of Society 5.0 will be Disaster Response and Resilience where advanced technologies would be used for early warning systems and efficient disaster response. For example, sensors and AI can provide early alerts for earthquakes or climate-related disasters, enhancing community safety and resilience. Community Engagement would be another essential component of Society 5.0, where Technology would enable greater community participation and engagement. Digital platforms and e-participation tools empower citizens to contribute to decision-making processes, voice concerns, and collaborate on community projects. A vivid description of the responsibilities and actions necessary for nurturing social responsibilities of Society 5.0 (Figure 7.4).

The responsible use of technology is paramount in Society 5.0, but emphasis on ethical considerations, data privacy, and the responsible deployment of AI to prevent harm, discrimination, and misuse of technology is also

Inclusive Innovation	•Responsibility: Technology benefits all inclusively •Action : Foster inclusive research and development
Ethical Technology Development	•Responsibility: Ethical tech design, principled deployment. •Action:Promote tech inclusive ethical standards
Privacy Protection	•Responsibility: Protecting privacy in data age. •Action:Preserve privacy, and implement robust privacy measures
Environmental Sustainability	•Responsibility: Technology for green sustainable future •Action: Deveop eco-friendly technology by reducing carbon footprint
Community Engagement	•Responsibility: Involving communities for decision making •Action: Establish mechanism for public input and feedback
Education and Skill Development	•Responsibility: Focus on continuous learning •Action: Invest in education, provide training, empower individuals
Global Collaboration	•Responsibility: Global collaboration for shared challenges. •Action: Share knowledge, resources, and innovative solutions

Figure 7.4 Social responsibilities of Society 5.0.

essential for the implementation of the digital society for mankind. Ensuring that AI and automation are developed and used ethically and responsibly, with a focus on reducing biases and ensuring transparency in AI decision-making. Social 5.0's responsibility also entails using technology in ways that align with moral and ethical principles [49]. This involves considering the impact of technology on individuals, communities, and society as a whole and making decisions that prioritize the well-being of all. Society 5.0 would also be responsible for cultural preservation, where Digital initiatives are employed to preserve and promote cultural heritage, languages, and traditions. This includes digital archives, online language resources, and cultural exchange programs. Such a society would also promote gender equality in the workplace and society by using technology to identify and eliminate gender biases in recruitment, pay, and opportunities. In Society 5.0, technology is harnessed to provide support to vulnerable populations, such as refugees and displaced communities. This includes digital platforms for connecting displaced individuals with essential services, information, and educational resources. Responsible resource management is also an important asset of Society 5.0 that involves optimizing the utilization of natural resources, energy, and materials in an environmentally sustainable manner [50]. It requires minimizing waste, conserving resources, and adopting energy-efficient technologies. This practice not only reduces the environmental impact but also ensures long-term availability of resources, aligning with the social responsibility imperative of preserving the planet for future generations. Considering the environmental impact of technology and working towards sustainable practices is a responsibility in Society 5.0. This includes reducing carbon emissions, minimizing electronic waste, and promoting eco-friendly technology solutions. Digital government services and e-participation platforms enhance transparency, efficiency, and citizen engagement in governance. This includes digital voting systems and open data initiatives that promote accountability. Society 5.0 would also be responsible for the collaboration among stakeholders, including governments, businesses, academia, and civil society, which is crucial for achieving social responsibility goals. This enables the sharing of knowledge and resources to address societal challenges. There would also be Philanthropic Initiatives in Society 5.0 that would leverage digital platforms to support a wide range of social and environmental causes [51]. These efforts include online fundraising, crowdfunding, and digital campaigns that mobilize resources and engage a global community in addressing pressing issues, fostering a culture of giving and making a significant impact on societal and environmental challenges. Corporate Social Responsibility (CSR) in Society 5.0 extends beyond profit-driven objectives, requiring businesses to actively engage in ethical practices. It involves the implementation of sustainability initiatives, philanthropic contributions to societal and environmental causes, and ethical supply chain management, fostering a positive impact on communities and the environment while aligning with the values of a responsible and socially conscious organization. In short, social responsibility in Society 5.0 emphasizes ethical, inclusive, and sustainable technology use that prioritizes the well-being of individuals, communities, and the environment. It involves a commitment to responsible practices that harness the power of technology for the greater good while mitigating potential risks and ethical concerns.

7.7 A HOLISTIC APPROACH TOWARDS SUSTAINABILITY IN SOCIETY 5.0

Sustainability is a multifaceted concept encapsulating environmental, social, and economic dimensions. Environmental sustainability centres on responsible resource use and environmental preservation. Social sustainability seeks human well-being, equity, and diverse, harmonious communities, whereas economic sustainability involves resource-efficient economic growth. It's the intricate balance where environmental protection, social equity, and economic growth converge, aiming for a world where human activities harmonize with nature, individuals enjoy fulfilled lives, and equitable, sustainable economic practices cater to present and future needs. Sustainability is pivotal in addressing global challenges and shaping a resilient and equitable future [52, 53].

Sustainability within the context of Society 5.0 embodies a vision of an advanced society that is not only technologically interconnected but also environmentally and socially sustainable. **Environmental sustainability** within the context of Society 5.0 is characterized by the strategic integration of advanced technologies and data-driven solutions to address pressing environmental challenges. Key technical facets include the widespread adoption of green technologies, encompassing renewable energy sources, energy-efficient infrastructure, sustainable transportation systems, and eco-friendly manufacturing processes. Advanced technologies like the IoT and big data analytics are harnessed to optimize natural resource utilization, such as water and energy, minimizing waste and inefficiencies [54]. A core principle of Society 5.0 is the promotion of a circular economy, encouraging recycling, reuse, and repurposing of materials and products to reduce resource consumption and waste generation. Precision agriculture, employing sensor technology and data analysis, enhances crop management and reduces the environmental impact of farming. Smart transportation systems encompass electric vehicles, improved public transportation, and autonomous vehicles, which collectively mitigate emissions and congestion. Furthermore, technology-driven environmental monitoring and management provide real-time data on air and water quality, as well as early warning systems for natural disasters, facilitating rapid, data-informed responses to environmental threats. Society 5.0 also places importance on biodiversity conservation, leveraging technology to monitor and protect endangered species and their habitats. Waste reduction is achieved through advanced waste management solutions, including recycling, composting, and waste-to-energy technologies. Additionally, green infrastructure, such as green roofs and urban gardens, is developed to enhance urban sustainability [55]. Educational and awareness campaigns are vital components of Society 5.0, promoting environmental responsibility and sustainability practices among individuals and businesses. Regulatory frameworks and cross-sector collaborations are fundamental for ensuring adherence to environmental standards and advancing sustainability initiatives. Hence, Society 5.0 leverages technology and data for a holistic and technical approach to environmental sustainability, aiming to minimize the ecological footprint and foster a harmonious coexistence with the environment.

Economic sustainability in the context of Society 5.0 embodies a multidimensional strategy underpinned by the seamless integration of advanced technologies. Central to this approach is the pervasive use of digital technologies, including AI, big data analytics, and the IoT, which revolutionize business operations, driving efficiency, productivity, and innovation. Industry 4.0 principles, such as automation and smart manufacturing, optimize production processes, enhance supply chain logistics, and facilitate predictive maintenance. Advanced manufacturing technologies like 3D printing and nanotechnology bolster efficiency and minimize waste, while digital platforms, including e-commerce and financial technology (FinTech), enable global market access and financial inclusivity [56]. Data analytics and AI empower data-driven decision-making for resource allocation, market strategies, and financial management. Robust cybersecurity safeguards data and infrastructure, underpinning trust in digital transactions. Economic sustainability is also closely entwined with the adoption of sustainable energy sources and regulatory frameworks that foster innovation and ethical practices. In sum, the economic sustainability paradigm in Society 5.0 is intricately linked to technology-driven innovation, balanced growth, and environmental responsibility, fostering a resilient, inclusive, and forward-thinking economic landscape [57]. In short, economic sustainability in Society 5.0 is heavily reliant on digital transformation, advanced manufacturing, FinTech, and sustainable supply chains. It embraces innovation, workforce development, and global collaboration while adhering to ethical, environmental, and regulatory standards to create a resilient and inclusive economic environment. Technical advancements are at the forefront of achieving economic sustainability, fostering growth and prosperity for individuals and businesses [58] (Figure 7.5).

Social sustainability in Society 5.0 is characterized by the intricate fusion of advanced technologies and societal values, working in tandem to create a harmonious and inclusive human-centric environment. It places a strong emphasis on leveraging digital innovation to enhance the well-being of individuals and communities. This encompasses equitable access to education and healthcare, bridging the digital divide, and supporting continuous skills development to enable adaptation to rapid technological changes. Additionally, social sustainability in Society 5.0 prioritizes the preservation of cultural heritage and the promotion of ethical values, ensuring that identity and community cohesion remain intact in a rapidly evolving digital landscape [59]. Moreover, the concept of community resilience is embedded within the framework, where technology is utilized to build communities that are not only capable of withstanding environmental and societal challenges but also of recovering and thriving in their aftermath. Collaboration among diverse stakeholders, data-driven decision-making, and robust regulatory frameworks play pivotal roles in facilitating responsible technology deployment that upholds social sustainability [60].

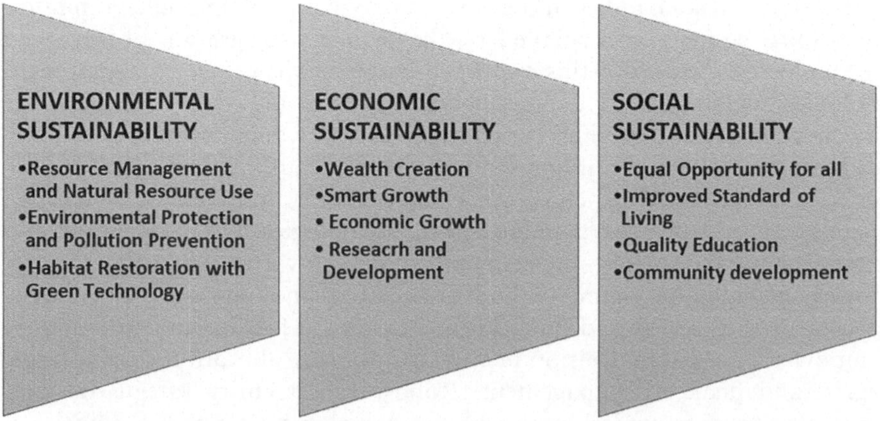

Figure 7.5 Sustainability in Society 5.0.

In the context of Society 5.0, social sustainability envisions a future where technology is a catalyst for building equitable, inclusive, and well-being-focused societies. The goal is to simultaneously address existing inequalities and ensure that the benefits of technological progress are accessible to all members of society. This involves fostering a sense of community, embracing diversity, and actively working to bridge digital divides. By leveraging technology to enhance social cohesion and prioritize individual well-being, Society 5.0 aims to create a harmonious and inclusive environment where the positive impacts of technological advancements contribute to the betterment of society as a whole [61].

7.8 CHALLENGES AND CONSIDERATIONS IN IMPLEMENTING SOCIETY 5.0

Implementing Society 5.0, which envisions an advanced and highly interconnected society leveraging cutting-edge technologies, an array of unresolved issues and complex technical challenges must be effectively tackled.

7.8.1 Technical challenges

Implementing Industry 5.0 introduces a host of technical challenges that require resolution for the effective integration of advanced technologies and human intelligence. These challenges encompass interoperability, data management, research and development, regulatory adherence, and workforce enhancement. To unlock the full advantages of Industry 5.0, manufacturers must demonstrate a commitment to invest in addressing these challenges.

Technology Integration: A key challenge lies in ensuring the interoperability of diverse technologies and systems. Given the integration of various elements like AI, machine learning, robotics, and the IoT, it's imperative to guarantee their seamless communication and effective collaboration. The primary challenge is to incorporate advanced technologies such as AI, IoT, biotechnology, and quantum computing, as it requires substantial investment and expertise [62]. Ensuring compatibility and seamless integration can be complex, as each such technology has its own specifications and standards. It demands careful planning, rigorous testing, and often the creation of custom interfaces. Continuous assessment of emerging technologies is necessary. Collaboration, clear communication, and adaptability are vital to overcome this complexity.

Privacy and Security: Data management is another hurdle to overcome. Given the immense data volumes within the Industry 5.0 ecosystem, manufacturers must implement robust data management systems capable of handling data's sheer volume, speed, and diversity. With increased data sharing and connectivity, there are heightened concerns about data privacy and cybersecurity [63]. Implementing robust measures to protect individuals' data and secure critical systems is paramount. Hence, appropriate data security measures are essential to safeguard against unauthorized access that could jeopardize the integrity of the manufacturing process. Embracing advanced technologies in Society 5.0 necessitates substantial commitments to research and development. Manufacturers must demonstrate a willingness to invest in the creation of novel technologies, procedures, and products to maintain competitiveness in the swiftly evolving environment [64]. Such investments are pivotal for staying abreast of emerging technologies and perpetuating innovation to address customer requirements. Developing effective data governance and analytics strategies is essential to harness the potential of this data while respecting privacy. Global cooperation is essential to craft effective solutions and foster unified standards for the successful implementation of Society 5.0. Society 5.0 challenges often surpass national borders, making international collaboration vital as far as data sharing, cybersecurity, and global technology standards are concerned.

Regulatory Frameworks: Beyond the technical hurdles, regulatory challenges also demand attention. As mentioned, the intricacies of data privacy and cybersecurity regulations pose significant challenges. As data generation and sharing span various technologies and systems, safeguarding data privacy and fortifying systems against cyber threats becomes of utmost importance. Existing regulations may not be equipped to handle the challenges posed by Society 5.0. Governments and institutions need to adapt and establish frameworks that govern technology use, data protection, and other key aspects. As technologies advance, ethical questions about their use arise. Society 5.0 must navigate issues like AI ethics, biotechnology regulations, and digital rights to ensure responsible deployment [65].

Workforce Adaptation: The transition to a highly advanced society may require reskilling and upskilling the workforce to adapt to new technologies and roles. A key human challenge lies in the demand for a highly skilled workforce. Incorporating advanced technologies into manufacturing necessitates employees with advanced technical proficiencies, such as programming and data analysis [66]. Regrettably, a shortage of such skilled workers is a growing concern, making it progressively more challenging to recruit and retain qualified talent. This skills gap poses a substantial obstacle for companies aiming to embrace Industry 5.0. Addressing these changes while minimizing job displacement is crucial. Manufacturers need to allocate resources to training and development initiatives to prepare their employees for a technology-driven landscape. This involves cultivating new skill sets like data analysis, machine learning, and robotics programming, which can prove demanding for current personnel.

7.8.2 Human challenges

Apart from the technical challenges, there exist few human-related challenges in Society 5.0 as this aspect is inevitable in a broader perspective [67].

Cultural shift in workforce: One of the most challenging factors in Society 5.0 deals with the imperative for a cultural transformation within organizations, that needs Industry to have a strong emphasis on enhanced collaboration and communication between human workers and machines. Consequently, employees must be open to change and receptive to new work methodologies. However, achieving this cultural shift can be arduous, as long-tenured employees accustomed to traditional work practices may exhibit resistance to change [68].

Communication between employees and management: A correlated challenge pertains to the necessity for proficient communication between employees and management. Society 5.0 demands substantial coordination and collaboration between human workers and machines. Hence, it is imperative for managers to effectively convey information to employees regarding the ongoing changes and their underlying significance. Additionally, employees should have a platform to voice their perspectives and offer feedback to management, ensuring the successful execution of Society 5.0.

Replacement of workforce: A major human challenge centres on the potential job displacement resulting from automation. As machines gain intelligence and capabilities, certain roles could become redundant, necessitating worker retraining for new positions. Addressing this challenge involves the implementation of reskilling and upskilling initiatives, equipping employees with the essential skills to adapt to the evolving manufacturing landscape. However, these programs can pose financial burdens, particularly for smaller companies. Additionally, it is vital for companies to uphold ethical and responsible practices in the adoption of Industry 5.0 [69].

Implementing Society 5.0 introduces environmental challenges, including heightened energy consumption and electronic waste proliferation. The increased use of advanced technologies raises concerns about ecological footprints and resource extraction. To address these issues, it is crucial to focus on developing energy-efficient technologies, promoting sustainable energy sources, and implementing circular economy principles for responsible electronic waste management. By integrating environmental considerations into the technological development process, Society 5.0 can strive for a balanced and sustainable future that not only enhances human well-being but also preserves the health of the planet [70]. The development of sustainable practices and technologies is of paramount importance in mitigating these issues. Among the others, in the context of Society 5.0, where healthcare takes centre stage, effectively meeting the requirements of an ageing population while ensuring accessible and high-quality healthcare services represents a multifaceted and intricate challenge. Resource management may be another serious challenge in Society 5.0, which becomes paramount in advanced societies, encompassing sustainable handling of essential resources like energy, water, and raw materials. The implementation of resource-efficient practices and technologies is a critical aspect to contemplate [71].

Ultimately, the implementation of Industry 5.0 introduces various human challenges that demand resolution for a seamless transition into this new manufacturing era. These challenges encompass the demand for a highly skilled workforce, the imperative cultural transformation within organizations, proficient communication between employees and management, potential job displacement due to automation, and the ethical and responsible integration of Industry 5.0. Tackling these challenges empowers companies to unlock the advantages of Industry 5.0 and maintain competitiveness in an ever-expanding global market.

7.9 FUTURE PROSPECTS

Society 5.0 embodies a transformative vision centred on the seamless integration of digital technologies, such as AI, the IoT, and big data, into the fabric of society. This concept strives to forge a human-centred society that adeptly balances economic progress with addressing pressing social challenges. At its core, Society 5.0 envisions a harmonious coexistence between technology and humanity, recognizing the potential of digital advancements to address complex issues. It seeks to tackle critical concerns like environmental sustainability, where technological integration is seen as a powerful tool for creating eco-friendly solutions and mitigating the impact of human activities on the planet. The concept is particularly attuned to the challenges posed by ageing populations and social inequalities. By strategically leveraging technology, Society 5.0 aims to enhance the quality of life for individuals, especially in an ageing society, where innovations can contribute to improved healthcare, personalized services, and greater inclusivity [72]. Positioning humans at the heart of innovation, Society 5.0 builds upon the outcomes of Industry 4.0, emphasizing the transformative power of technology. This involves not only technological integration but also a strategic approach aimed at fostering social responsibility and promoting sustainability. It envisions a future where technology is not just a driver of economic growth but a catalyst for positive societal change, actively contributing to the well-being of individuals and communities. In fact, Society 5.0 represents a forward-looking paradigm that seeks to harness the influence of technology to create a more inclusive, sustainable, and human-centric society [73]. Society 5.0 is experiencing remarkable advancements across a spectrum of sectors, revolutionizing industries such as healthcare, manufacturing, textiles, education, and food. These transformations are driven by a confluence of cutting-edge technologies that redefine the way we live and work. The integration of transformative technologies, including cloud computing, Blockchain, big data analytics, the IoT, and the upcoming 6G networks, is propelling Society 5.0 into a new era of unprecedented connectivity and innovation. Central to this evolution are the rapid advancements in key technologies such as AI, Big Data, Robotics, Deep Learning, and Machine Learning. These technologies collectively pave the way for a future characterized by automated living. AI, for instance, empowers systems to emulate human cognitive functions, enhancing decision-making processes across various sectors. Big Data analytics [74] extracts valuable insights from massive datasets, informing strategic decisions and optimizing resource allocation. Robotics and automation redefine manufacturing processes, making them more efficient and adaptable. As Society 5.0 continues to unfold, these technological advancements contribute to a landscape where cognitive abilities are harnessed to realize effective and efficient solutions. This not only streamlines processes in diverse industries but also sets the stage for a more interconnected, intelligent, and automated future. The synergy of these transformative technologies marks a significant leap toward a society that leverages innovation to address complex challenges and enhance the overall quality of life [75]. The ecosystem is envisioned to deliver sustainability across economic, environmental, social, and political dimensions. The central focus is on people and the generation of value. The upcoming workforce needs the expertise and understanding to discern various production systems, enabling them to make well-informed decisions among distinct approaches: relying solely on human effort, exclusively on technological efforts, or through collaborative efforts between humans and technology. In the future, industrial workers must shift to socio-technological systems, requiring continuous learning to enhance career prospects, work-life balance, and job development. As Society 5.0 transforms production, ongoing training becomes crucial for effective collaboration between humans and smart technologies, emphasizing versatile skills and digital education [76]. Other future trends may include Cognitive computing that seeks to emulate human thought processes through computerized models. It employs self-learning algorithms, data mining, pattern recognition, natural language processing, and other techniques to enable computers to interpret and understand information in a manner similar to the human brain. Human-machine interaction involves communication and interaction between humans and machines through the user interface. Natural user interfaces, such as gestures, are employed to capture attention, enabling intuitive control of machines through human-like behaviours. This direction is pivotal for Industry 5.0, for transitioning to operator-centric production in a digital environment that values both human and robotic characteristics by placing humans at the system's centre and integrating technologies accordingly [77].

7.10 CONCLUSION

In conclusion, Society 5.0 represents a visionary and transformative approach to societal evolution, transcending previous stages by seamlessly integrating advanced technologies with human-centric values. Envisioned as a harmonious ecosystem, Society 5.0 strives for sustainability across economic, environmental, social, and political dimensions, placing people at its core. At the heart of Society 5.0 is the recognition of the exponential advancements in technologies such as AI, Big Data, Robotics, and Machine Learning. These technologies are not viewed as mere tools but as integral components in the fabric of a new societal paradigm. The vision is to pave the way for an automated lifestyle that enhances the quality of life while fostering social responsibility and sustainability. The concept of Society 5.0 places a strong emphasis on creating value for individuals and communities, underscoring its human-centric nature. As industries evolve towards Industry 4.0 and beyond, the future workforce faces the imperative to navigate socio-technological production systems. Continuous learning becomes paramount in this scenario, ensuring that individuals can adeptly choose among purely human, purely technological, or collaborative approaches. In the context of this technological evolution, Industry 5.0 emerges as a transformative force, significantly impacting human-machine collaboration and the management of future production systems. The seamless integration of smart technologies with human capabilities takes centre stage, necessitating ongoing training to cultivate a skilled workforce capable of thriving in the dynamic landscape of the future. Crucially, Society 5.0 emphasizes the need for a multifaceted skill set and digital education. The collaborative partnership between humans and smart technologies demands a commitment to continuous adaptation and learning. This recognition of the importance of education extends beyond traditional academic boundaries, highlighting the necessity for a workforce that is not only technologically literate but also possesses a deep understanding of societal dynamics and ethical considerations. The collaborative partnership between humans and smart technologies is not without challenges. Society must address issues related to privacy, security, and the ethical use of technology. As Society 5.0 takes shape, it becomes imperative to establish robust frameworks and regulations that guide the responsible development and deployment of advanced technologies. This implies that Society 5.0 offers a compelling vision of a balanced, inclusive, and sustainable future. The seamless integration of technology into the fabric of society empowers individuals, fosters collaboration, and addresses complex societal challenges. It underscores the imperative for continuous adaptation, learning, and a collective commitment to shaping a prosperous and human-centric society. As we embark on the journey towards Society 5.0, it is crucial to recognize that the vision outlined is not a static destination but an ongoing process of evolution. The dynamic nature of technology and society requires a commitment to innovation, ethical considerations, and a collective effort to ensure that the benefits of technological advancement are accessible to all. Society 5.0, with its human-centric ethos and technological integration, provides a roadmap for a future that is not only advanced but also inclusive and sustainable.

REFERENCES

1. Shiroishi, Y., Uchiyama, K. and Suzuki, N., 2018. Society 5.0: For human security and well-being. *Computer*, *51*(7), pp. 91–95.
2. "The 5th Science and Technology Basic Plan," *Government of Japan*, 22 Jan. 2016; http://www8.cao.go.jp/cstp/english/basic/5thbasicplan.pdf
3. Onday, O., 2019. Japan's society 5.0: Going beyond Industry 4.0. *Business and Economics Journal*, *10*(2), pp. 1–6.
4. Fukuyama, M., 2018. Society 5.0: Aiming for a new human-centered society. *Japan Spotlight*, *27*(5), pp. 47–50.
5. Yin, Y., Stecke, K.E. and Li, D., 2018. The evolution of production systems from Industry 2.0 through Industry 4.0. *International Journal of Production Research*, *56*(1-2), pp. 848–861.
6. Mourtzis, D., Angelopoulos, J. and Panopoulos, N., 2022. A Literature Review of the Challenges and Opportunities of the Transition from Industry 4.0 to Society 5.0. *Energies*, *15*(17), p. 6276.
7. Kasinathan, P., Pugazhendhi, R., Elavarasan, R.M., Ramachandaramurthy, V.K., Ramanathan, V., Subramanian, S., Kumar, S., Nandhagopal, K., Raghavan, R.R.V., Rangasamy, S. and Devendiran, R., 2022. Realization of sustainable development goals with disruptive technologies by integrating Industry 5.0, Society 5.0, smart cities and villages. *Sustainability*, *14*(22), p. 15258.
8. Serpa, S. and Ferreira, C.M., 2019. Society 5.0 and sustainability digital innovations: A social process. *Journal of Organizational Culture, Communications and Conflict*, 23, pp. 1–14.
9. Tavares, M.C., Azevedo, G. and Marques, R.P., 2022. The challenges and opportunities of era 5.0 for a more humanistic and sustainable society—a literature review. *Societies*, *12*(6), p. 149.
10. Fukuda, K., 2020. Science, technology and innovation ecosystem transformation toward society 5.0. *International journal of production economics*, *220*, p. 107460.
11. Fuchs, C., 2011. Web 2.0, prosumption, and surveillance. *Surveillance & Society*, *8*(3), pp. 288–309.
12. Boateng, R., Mbarika, V. and Thomas, C., 2010. When Web 2.0 becomes an organizational learning tool: evaluating Web 2.0 tools. *Development and Learning in Organizations: An International Journal*, *24*(3), pp. 17–20.
13. Sharma, G. and Baoku, L., 2013. Customer satisfaction in Web 2.0 and information technology development. *Information technology & people*, *26*(4), pp. 347–367.
14. Mazali, T., 2018. From industry 4.0 to society 4.0, there and back. *AI & Society*, *33*(3), pp. 405–411.
15. Önday, Ö., 2020. Society 5.0-its historical logic and its structural development. *Journal of Scientific Reports*, *2*(1), pp. 32–42.

16. Nair, M.M., Tyagi, A.K. and Sreenath, N., 2021, January. The future with industry 4.0 at the core of society 5.0: Open issues, future opportunities and challenges. In *2021 international conference on computer communication and informatics (ICCCI)* (pp. 1–7). IEEE.
17. Gladden, M.E., 2019. Who will be the members of Society 5.0? Towards an anthropology of technologically posthumanized future societies. *Social Sciences*, *8*(5), p. 148.
18. Schoitsch, E., 2019. Beyond smart systems: Creating a society of the future (5.0) resolving disruptive changes and social challenges. *Innovation and Transformation in a Digital World: 27th Interdisciplinary Information Management Talks (IDIMT 2019)*, Kutna Hora, pp. 4–7.
19. Kagermann H. (2014). Industry 4.0. Behind Germany's new economic push. *Nikkei Asian Review*. December 16. http://asia.nikkei.com/Business/Companies/Behind-Germany-s-neweconomic-push. Accessed November 24, 2017.
20. Lemstra, M.A.M.S. and de Mesquita, M.A., 2023. Industry 4.0: A tertiary literature review. *Technological Forecasting and Social Change*, *186*, p. 122204.
21. Barari, A. and Tsuzuki, M.S.G., 2023. Smart manufacturing and Industry 4.0. *Applied Sciences*, *13*(3), p. 1545.
22. Compare, M., Baraldi, P. and Zio, E., 2019. Challenges to IoT-enabled predictive maintenance for industry 4.0. *IEEE Internet of Things Journal*, *7*(5), pp. 4585–4597.
23. Baratta, A., Cimino, A., Gnoni, M.G. and Longo, F., 2023. Human robot collaboration in Industry 4.0: A literature review. *Procedia Computer Science*, *217*, pp. 1887–1895.
24. Godina, R. and Matias, J.C., 2019. Quality control in the context of industry 4.0. In *Industrial Engineering and Operations Management II: XXIV IJCIEOM, Lisbon, Portugal, July 18–20 24* (pp. 177–187). Springer International Publishing.
25. Li, L., 2022. Reskilling and upskilling the future-ready workforce for industry 4.0 and beyond. *Information Systems Frontiers*, *26*, pp. 1697–1712.
26. Zhang, G., Yang, Y. and Yang, G., 2023. Smart supply chain management in Industry 4.0: the review, research agenda and strategies in North America. *Annals of Operations Research*, *322*(2), pp. 1075–1117.
27. Khan, S.A.R., Razzaq, A., Yu, Z. and Miller, S., 2021. Retracted: Industry 4.0 and circular economy practices: A new era business strategies for environmental sustainability. *Business Strategy and the Environment*, *30*(8), pp. 4001–4014.
28. Ellitan, L., 2020. Competing in the era of industrial revolution 4.0 and society 5.0. *Jurnal Maksipreneur: Manajemen, Koperasi, dan Entrepreneurship*, *10*(1), pp. 1–12.
29. Zengin, Y., Naktiyok, S., Kaygın, E., Kavak, O. and Topçuoğlu, E., 2021. An investigation upon industry 4.0 and society 5.0 within the context of sustainable development goals. *Sustainability*, *13*(5), p. 2682.
30. Menezes, B., Yaqot, M., Hassaan, S., Franzoi, R., AlQashouti, N. and Al-Banna, A., 2022, October. Digital Transformation in the Era of Industry 4.0 and Society 5.0: A perspective. In *2022 2nd International Conference on Emerging Smart Technologies and Applications (eSmarTA)* (pp. 1–6). IEEE.

31. Frederico, G.F., 2021. From supply chain 4.0 to supply chain 5.0: Findings from a systematic literature review and research directions. *Logistics*, *5*(3), p. 49.
32. Deshmukh, A., Patil, D.S., Soni, G. and Tyagi, A.K., 2023. Cyber security: New realities for Industry 4.0 and Society 5.0. In *Handbook of Research on Quantum Computing for Smart Environments* (pp. 299–325). IGI Global.
33. Dovleac, L., Chițu, I.B., Nichifor, E. and Brătucu, G., 2023. Shaping the inclusivity in the new society by enhancing the digitainability of sustainable development goals with education. *Sustainability*, *15*(4), p. 3782.
34. Gurjanov, A.V., Zakoldaev, D.A., Shukalov, A.V. and Zharinov, I.O., 2020, November. The smart city technology in the super-intellectual Society 5.0. *Journal of Physics: Conference Series*, *1679*(3), p. 032029).
35. Skobelev, P.O. and Borovik, S.Y., 2017. On the way from Industry 4.0 to Industry 5.0: From digital manufacturing to digital society. *Industry 4.0*, 2(6), pp. 307–311.
36. Kurniasih, D., Setyoko, P.I. and Saputra, A.S., 2022. Digital transformation of health quality services in the healthcare industry during disruption and society 5.0 era. *International Journal of Social and Management Studies*, *3*(5), pp. 139–143.
37. Garg, P.K., 2022. The future healthcare technologies: a roadmap to society 5.0. In *Geospatial Data Science in Healthcare for Society 5.0* (pp. 305–318). Singapore: Springer Singapore.
38. Mavrodieva, A.V. and Shaw, R., 2020. Disaster and climate change issues in Japan's Society 5.0—A discussion. *Sustainability*, *12*(5), p. 1893.
39. Aderibigbe, J.K., 2022. Accentuating Society 5.0 New Normal: The Strategic Role of Industry 4.0 Collaborative Partnership and Emotional Resilience. In *Agile Management and VUCA-RR: Opportunities and Threats in Industry 4.0 towards Society 5.0* (pp. 39–55). Emerald Publishing Limited.
40. Murphy, C., Carew, P.J. and Stapleton, L., 2022. Ethical personalisation and control systems for smart human-centred Industry 5.0 applications. *IFAC-PapersOnLine*, *55*(39), pp. 24–29.
41. Mintarti, S.U., Handayani, S. and Febrianto, I., 2023, January. The innovation of evaluation instrument based on cone of experience: A way for improving student competitiveness in Society 5.0. In *Unima International Conference on Social Sciences and Humanities (UNICSSH 2022)* (pp. 1183–1190). Atlantis Press.
42. Goede, M., 2020. *Society 5.0; we and I*. University of Governanace/Goede Consultants.
43. Murphy, C., Carew, P.J. and Stapleton, L., 2023. A human-centred systems manifesto for smart digital immersion in Industry 5.0: a case study of cultural heritage. *AI & Society*, *39*, pp. 2401–2416.
44. Barrett, B.F., DeWit, A. and Yarime, M., 2021. Japanese smart cities and communities: Integrating technological and institutional innovation for Society 5.0. In *Smart Cities for Technological and Social Innovation* (pp. 73–94). Academic Press.
45. Bölükbaşi, F., 2021. Revolution and Society 5.0: Japanese human centric approach and sectoral changes. *Toplumsal Politika Dergisi*, 2(2), pp. 1–28.
46. Mansur, H., Utama, A.H., Mohd Yasin, M.H., Sari, N.P., Jamaludin, K.A. and Pinandhita, F., 2023. Development of inclusive education learning design in the era of Society 5.0. *Social Sciences*, *12*(1), p. 35.

47. Ciasullo, M.V., Orciuoli, F., Douglas, A. and Palumbo, R., 2022. Putting Health 4.0 at the service of Society 5.0: Exploratory insights from a pilot study. *Socio-Economic Planning Sciences*, *80*, p. 101163.
48. Srinivasa, K.G., Siddesh, G.M. and Manisekhar, S.R. eds., 2022. *Society 5.0: Smart Future Towards Enhancing the Quality of Society*. Springer Nature Singapore.
49. Peyvandi, A., Majidi, B., Peyvandi, S. and Patra, J.C., 2022. Privacy-preserving federated learning for scalable and high data quality computational-intelligence-as-a-service in Society 5.0. *Multimedia Tools and Applications*, *81*(18), pp. 25029–25050.
50. Dewi, H.P., 2021, June. Determination of green marketing strategies through marketing communication in the business world in the society 5.0 era. In *18th International Symposium on Management (INSYMA 2021)* (pp. 181–187). Atlantis Press.
51. Siddesh, G.M. and Kulkarni, A.S., 2022. Introduction to Smart Big data analytics and smart real-time analytics in Society 5.0. In *Society 5.0: Smart Future Towards Enhancing the Quality of Society* (pp. 35–55). Singapore: Springer Nature Singapore.
52. Johri, P., Singh, J.N., Sharma, A. and Rastogi, D., 2021, December. Sustainability of coexistence of humans and machines: an evolution of industry 5.0 from industry 4.0. In *2021 10th International Conference on System Modeling & Advancement in Research Trends (SMART)* (pp. 410–414). IEEE.
53. Roblek, V., Meško, M., Bach, M.P., Thorpe, O. and Šprajc, P., 2020. The interaction between internet, sustainable development, and emergence of society 5.0. *Data*, *5*(3), p. 80.
54. Narvaez Rojas, C., Alomia Peñafiel, G.A., Loaiza Buitrago, D.F. and Tavera Romero, C.A., 2021. Society 5.0: A Japanese concept for a superintelligent society. *Sustainability*, *13*(12), p. 6567.
55. Sidek, S., Khadri, N.A.M., Hasbolah, H., Yaziz, M.F.A., Rosli, M.M. and Husain, N.M., 2021, August. Society 5.0: Green logistics consciousness in enlightening environmental and social sustainability. *IOP Conference Series: Earth and Environmental Science*, *842*(1), p. 012053).
56. Martins, V.W.B., Rampasso, I.S., Anholon, R., Quelhas, O.L.G. and Leal Filho, W., 2019. Knowledge management in the context of sustainability: Literature review and opportunities for future research. *Journal of Cleaner Production*, *229*, pp. 489–500.
57. Smuts, H. and Van der Merwe, A., 2022. Knowledge management in society 5.0: A sustainability perspective. *Sustainability*, *14*(11), p. 6878.
58. Potočan, V., Mulej, M. and Nedelko, Z., 2020. Society 5.0: balancing of Industry 4.0, economic advancement and social problems. *Kybernetes*, *50*(3), pp. 794–811.
59. Žižek, S.Š., Mulej, M. and Potočnik, A., 2021. The sustainable socially responsible society: Well-being society 6.0. *Sustainability*, *13*(16), p. 9186.
60. Chang, D.L., Sabatini-Marques, J., Da Costa, E.M., Selig, P.M. and Yigitcanlar, T., 2018. Knowledge-based, smart and sustainable cities: A provocation for a conceptual framework. *Journal of Open Innovation: Technology, Market, and Complexity*, *4*(1), p. 5.
61. Thakur, R., Borkar, P.S. and Agarwal, M., 2022. Smart Society 5.0 for Social and Technological Sustainability. In *Decision Analytics for Sustainable*

Development in Smart Society 5.0: Issues, Challenges and Opportunities (pp. 299–319). Singapore: Springer Nature Singapore.
62. Shiroishi, Y., Uchiyama, K. and Suzuki, N., 2019. Better actions for society 5.0: using AI for evidence-based policy making that keeps humans in the loop. *Computer*, *52*(11), pp. 73–78.
63. Dominic, N., Pratama, N.R., Cornelius, K., Senewe, S.H. and Pardamean, B., 2022, October. Society with trust: A scientometrics review of zero-knowledge proof advanced applications in preserving digital privacy for Society 5.0. In *Conference on Innovative Technologies in Intelligent Systems and Industrial Applications* (pp. 69–78). Cham: Springer Nature Switzerland.
64. Ghosh, T., Roy, A., Misra, S. and Raghuwanshi, N.S., 2021. CASE: A context-aware security scheme for preserving data privacy in IoT-enabled society 5.0. *IEEE Internet of Things Journal*, *9*(4), pp. 2497–2504.
65. Dhirani, L.L., Mukhtiar, N., Chowdhry, B.S. and Newe, T., 2023. Ethical dilemmas and privacy issues in emerging technologies: A review. *Sensors*, *23*(3), p. 1151.
66. Battini, D., Berti, N., Finco, S., Zennaro, I. and Das, A., 2022. Towards industry 5.0: A multi-objective job rotation model for an inclusive workforce. *International Journal of Production Economics*, *250*, p. 108619.
67. George, A.H., Fernando, M., George, A.S., Baskar, T. and Pandey, D., 2021. Metaverse: The next stage of human culture and the internet. *International Journal of Advanced Research Trends in Engineering and Technology (IJARTET)*, *8*(12), pp. 1–10.
68. Csiszer, A., 2022. Towards Society 5.0 in perspective of agile society. In *Agile Management and VUCA-RR: Opportunities and Threats in Industry 4.0 towards Society 5.0* (pp. 169–190). Emerald Publishing Limited.
69. Kolade, O. and Owoseni, A., 2022. Employment 5.0: The work of the future and the future of work. *Technology in Society*, *71*, p. 102086.
70. Almalki, F.A., Alsamhi, S.H., Sahal, R., Hassan, J., Hawbani, A., Rajput, N.S., Saif, A., Morgan, J. and Breslin, J., 2023. Green IoT for eco-friendly and sustainable smart cities: future directions and opportunities. *Mobile Networks and Applications*, *28*(1), pp. 178–202.
71. Mishra, P., Thakur, P. and Singh, G., 2022. Sustainable Smart City to Society 5.0: State-of-the-Art and Research Challenges. *SAIEE Africa Research Journal*, *113*(4), pp. 152–164.
72. Salgues, B., 2018. *Society 5.0: Industry of the Future, Technologies, Methods and Tools*. John Wiley & Sons.
73. Carayannis, E.G. and Morawska-Jancelewicz, J., 2022. The futures of Europe: Society 5.0 and Industry 5.0 as driving forces of future universities. *Journal of the Knowledge Economy*, *13*(4), pp. 3445–3471.
74. Rahmanto, F., Pribadi, U. and Priyanto, A., 2021, March. Big data: What are the implications for public sector Policy in society 5.0 era?. *IOP Conference Series: Earth and Environmental Science*, 717(1), p. 012009.
75. Tyagi, A.K., Dananjayan, S., Agarwal, D. and Thariq Ahmed, H.F., 2023. Blockchain—Internet of Things applications: Opportunities and challenges for Industry 4.0 and Society 5.0. *Sensors*, *23*(2), p. 947.
76. Alves, J., Lima, T.M. and Gaspar, P.D., 2023. Is Industry 5.0 a human-centred approach? A systematic review. *Processes*, *11*(1), p. 193.
77. Adel, A., 2022. Future of industry 5.0 in society: Human-centric solutions, challenges and prospective research areas. *Journal of Cloud Computing*, *11*(1), pp. 1–15.

Chapter 8

Medicine 4.0

Harnessing emerging technologies as catalysts for societal transformation

Nasir Vadia, Ramesh Parmar, Ashishkumar Kyada, and Vijaykumar Sutariya

8.1 INTRODUCTION

In the wake of the digital revolution, a paradigm shift has emerged, prompting a reevaluation of manufacturing, operational strategies, and health services within the pharmaceutical and health industry. This transformation is primarily driven by the growing challenges related to logistics efficiency and escalating energy costs, which are exerting a profound influence on the global production landscape and its corresponding distribution networks. Crucially, the rapid evolution of technology, characterized by groundbreaking advancements in Big data analytics, Artificial Intelligence (AI), the Internet of Things (IoT), Robotics, 3D printing, Augmented reality (AR)/ Virtual reality (VR), Blockchain, Telemedicine, and Cloud computing, has significantly altered the capabilities and intrinsic value of the global health care system. This transformative wave necessitates a comprehensive digital makeover for manufacturing and operations (Haleem et al. 2021a). Concurrently, the workforce must undergo a comprehensive retraining process to harness the potential of these emerging technologies effectively. At the core of this transformation lies a meticulous evaluation of the total delivered cost. This imperative exercise will determine the optimal locations for sourcing, manufacturing, and assembly activities on a global scale for total health management. In essence, what the pharma and health industry requires is a digital transformation—a comprehensive realignment with the digital era. This transformation is not merely a choice but an imperative for staying competitive, ensuring operational efficiency, and enhancing overall sustainability in a rapidly evolving global landscape. It is a call to action for the pharma and health sectors to embrace the opportunities presented by digital technologies and embark on a journey toward a more agile, efficient, and responsive future. To gain insight into the primary evolutionary phases in the field of medicine, we can succinctly delineate four distinctive epochs. Medicine 1.0 primarily revolves around the management of established maladies. Transitioning to medicine 2.0, we witness a shift toward patient-centric healthcare, underpinned by technology, outcome-oriented approaches, and an emphasis on preventive measures. This evolution confers

 DOI: 10.1201/9781003643166-8

substantial advantages upon individuals, healthcare providers, and organizations alike. Subsequently, the emergence of Medicine 3.0 marks a pivotal juncture characterized by the transcendence of operational barriers, necessitated by the mandatory infusion of technology into healthcare practices. The adoption of lean models, encompassing both production and managerial aspects, propels a drive for unparalleled adaptability. Innovations such as the utilization of "smart materials" usher in a digital revolution within the healthcare sector, fundamentally transforming its landscape. In the realm of Medicine 4.0, emerging technologies play a pivotal role, representing a central pillar. Addressing the pressing issue of health inequalities in our society demands a paradigm shift. This transformative era, referred to as Medicine 4.0, embraces novel technologies like 3D printing and the Internet of Things (IoT). Industry 4.0, which fosters the convergence of cyber-physical systems, serves as the bridge between the analog and digital worlds. Artificial Intelligence (AI), Robotics, the Internet of Medical Things (IoMT), and radio-frequency identification (RFID) are poised to usher in a revolutionary transformation in the field of medicine. Time-efficient and cost-effective wearables, medical devices, and robotic aids are poised to redefine the standards of care, exceeding prior expectations. These monumental strides in medical technology offer substantial savings in terms of time, resources, and patient anxiety, ultimately paving the way toward tailored healthcare solutions. While the digitalization of medicine underscores the importance of cutting-edge technology, it should be noted and emphasized that it shall not, and indeed must not, supplant the indispensable human connection that physicians bring to the realm of patient care (Li and Carayon 2021; Gupta and Singh 2023; Pace et al. 2018)

8.2 EMERGING TECHNOLOGIES IN HEALTHCARE

The healthcare sector is in a perpetual pursuit of innovation and advancement, driven by the imperatives of improved outcomes, heightened quality of care, enhanced accessibility, and cost containment. This evolution is characterized by a strategic shift toward leveraging technology not only for specialized treatment but also for bolstering preventive and primary care services (Thimbleby 2013). Figure 8.1 represents some important technologies associated with current healthcare systems.

A noteworthy manifestation of this transition is the burgeoning popularity of wearable devices designed to remotely monitor patient health data. As we gaze into the future, numerous emerging and established technologies hold the potential to revolutionize the healthcare landscape. Virtual reality (Tawseef et al. 2022), augmented reality (Campisi et al. 2020), wearable tech (Lu et al. 2020), machine learning (Siddique and Chow 2021), and artificial intelligence (Davenport and Kalakota 2019) are poised to assume pivotal roles, empowering healthcare entities to operate with greater

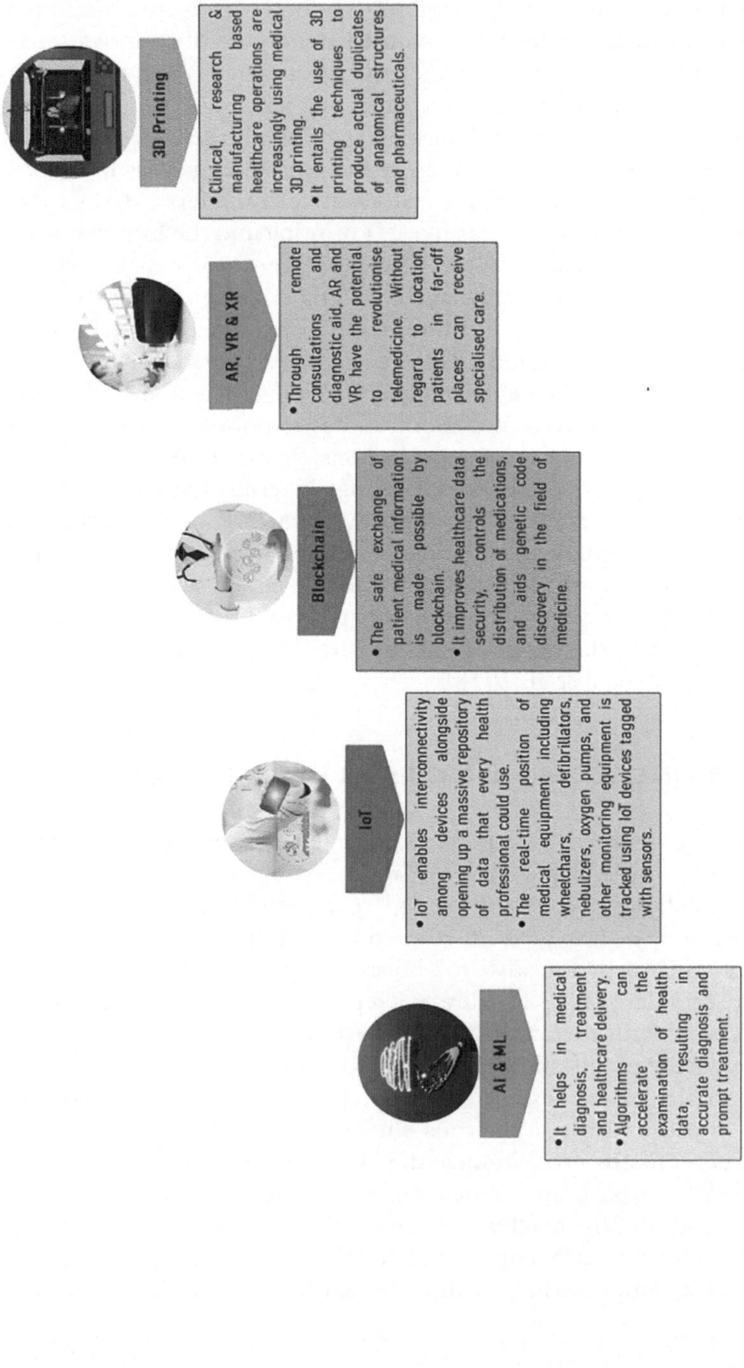

Figure 8.1 Advanced technologies associated with healthcare systems.

efficiency. The rapid pace of technological progress ensures that healthcare trends remain in a constant state of flux. Incorporating digital technology into healthcare holds the potential to establish sustainable healthcare systems, rebalance doctor–patient dynamics, and innovate disease management strategies, ultimately fostering improved health outcomes for individuals and communities (Rezaei Aghdam et al. 2020). The transformation begins with enhancing our approach to health, medicine, and healthcare using digital solutions (Bertalan et al. 2017). These technologies play a pivotal role in enhancing patient safety by identifying and mitigating risks in primary care settings. For instance, electronic sensors enable the continuous monitoring of vital signs and physical activity, aiding healthcare professionals in overseeing patients vulnerable to falls (Wenbin et al. 2022). Facilitating accurate and secure electronic sharing of prescription information reduces the occurrence of preventable adverse drug events (Osmani et al. 2023). Additionally, point-of-care diagnostic testing, encompassing conditions such as diabetes, HIV, COVID-19, and malaria, offers swift and crucial insights for treatment decisions (Maksut et al. 2020). The impact of digital technologies is evident in the evolution of medical and assistive devices in the 21st century, with innovations like 3-D printing revolutionizing the production of medical equipment, orthotics, and prosthetics (Ventola 2023).

8.2.1 Artificial intelligence and machine learning

Artificial Intelligence (AI) and Machine Learning (ML) have heralded a transformative era characterized by the emulation of human intelligence and cognitive capabilities. This paradigm shift is marked by the ever-expanding spectrum of conceivable applications spanning diverse industries. However, the most dynamic and promising arena for AI lies within the domain of healthcare. This section illuminates the revolutionary influence of AI and ML in the field of medicine, with a particular emphasis on their pivotal roles in the detection and treatment of ailments (Alowais et al. 2023). AI and ML also explore their extensive contributions to medical imaging, patient data management, medication development, remote consultation, medical statistics, surgical procedures, and personalized treatment plans (Zeeshan et al. 2020). Leading AI technologies are presented in Figure 8.2.

In addition to these areas, we explore the seamless integration of AI in primary healthcare, underscoring its indispensable role in elevating clinical decision-making, optimizing practice management, enhancing diagnostic capabilities, and advancing the training of primary care practitioners (Chris et al. 2021). The potential of AI and ML to metamorphose healthcare is boundless, promising heightened efficiency and tailored patient care while fostering continuous progress in the field of medical sciences (Zubair et al. 2021). AI holds the potential to revolutionize the healthcare sector in its entirety. AI-powered algorithms possess the capability to expedite tasks such as mining medical records, devising treatment strategies, and even

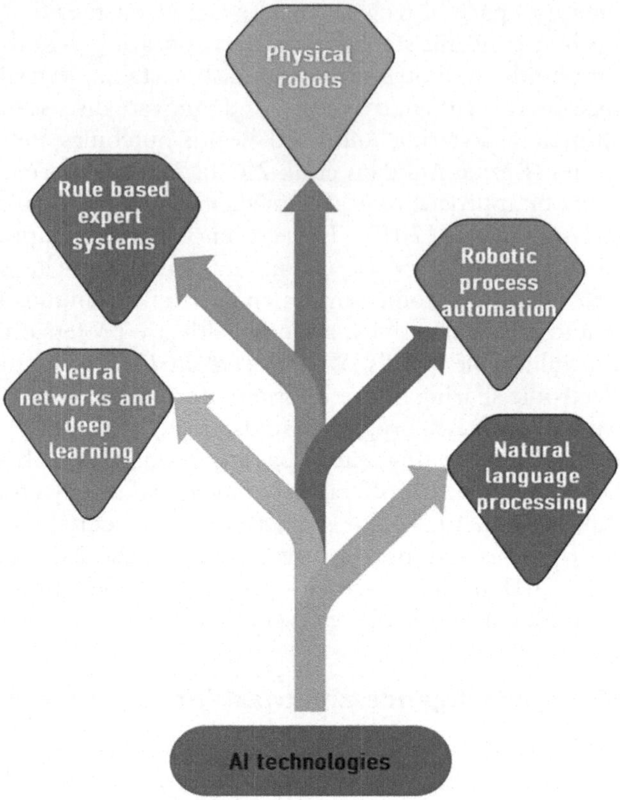

Figure 8.2 Types of AI technologies.

formulating pharmaceutical compounds at a pace that surpasses current capacities within the healthcare landscape, including those of medical professionals (François-Xavier et al. 2023; Junaid et al. 2021; Vora et al. 2023). Notably, the recent emergence of generative AI technologies has democratized access to these capabilities, making them accessible and valuable for both the general populace and healthcare practitioners in their daily routines. By harnessing the computational prowess of computers and their algorithms, this collaborative effort aims to efficiently sift through colossal volumes of data, vastly outperforming human scientists and medical experts in terms of speed and accuracy. The primary objective is to uncover intricate patterns and prognostic insights, thereby advancing the accuracy of disease diagnoses, refining treatment protocols, and bolstering public health and safety measures. According to Statista's data, the AI healthcare sector, initially valued at $11 billion in 2021, is anticipated to surge to a remarkable $187 billion by the year 2030. This substantial growth trajectory foreshadows substantial transformations within the healthcare ecosystem, encompassing medical practitioners, healthcare institutions, pharmaceutical and biotechnology enterprises, and other stakeholders. The heightened utilization of AI in healthcare can be attributed to several factors, including enhanced machine learning (ML) algorithms, heightened data accessibility, cost-effective hardware options, and the widespread deployment of 5G technology (Ghasan and Shah 2022). These factors collectively accelerate the pace of innovation in the healthcare domain. Notably, AI and ML technologies possess the capability to efficiently process vast volumes of health-related data, ranging from electronic health records and clinical research to genetic profiles, at speeds far surpassing human capabilities. Dr. Bertalan Mesko described the different AI-based technologies used in different fields of healthcare (Szigetvári and Bertalan 2023), illustrated in Figure 8.3.

Automation, machine learning, and artificial intelligence have had a significant impact on healthcare-related institutions, such as physicians, hospitals, insurance companies, and associated sectors. In numerous instances, these impacts have proven to be notably beneficial, often surpassing their effects on other sectors. The extensive adoption of AI and ML has led to their widespread utilization across various domains, encompassing tasks such as the management of medical records and diverse data sets, the formulation of treatment strategies, digital consultations, the emergence of virtual nursing, efficient medication management, accelerated drug development, the advancement of precision medicine, continuous health monitoring, in-depth analysis of healthcare systems, and a plethora of other applications.

8.2.2 Internet of Things (IoT) in wearable medical devices

In the pre-Internet of Things (IoT) era, patient–doctor interactions were primarily limited to in-person visits and conventional telecommunication methods. Continuous health monitoring and personalized recommendations were impractical for healthcare providers and hospitals. The advent of IoT-enabled devices has revolutionized the healthcare sector by enabling remote patient monitoring, enhancing patient safety and well-being, and empowering healthcare professionals to deliver exceptional care. IoT technology has also streamlined doctor–patient interactions, resulting in improved patient engagement and satisfaction (Albahri et al., 2021). Furthermore, remote health monitoring has proven effective in reducing hospitalization durations and preventing readmissions, ultimately contributing to substantial cost savings and enhanced treatment outcomes. Undoubtedly, IoT is reshaping the healthcare landscape, redefining the dynamics of device-human interactions in the delivery of healthcare solutions, benefiting patients, families, physicians, hospitals, and insurance companies alike. This technology facilitates real-time tracking of patients' health status, enabling healthcare providers to monitor treatment plan adherence and promptly respond to any urgent medical requirements. IoT integration empowers healthcare professionals to adopt a proactive approach to patient care, harnessing data from IoT devices to refine treatment strategies and achieve desired clinical outcomes. Beyond patient health monitoring, the healthcare sector benefits

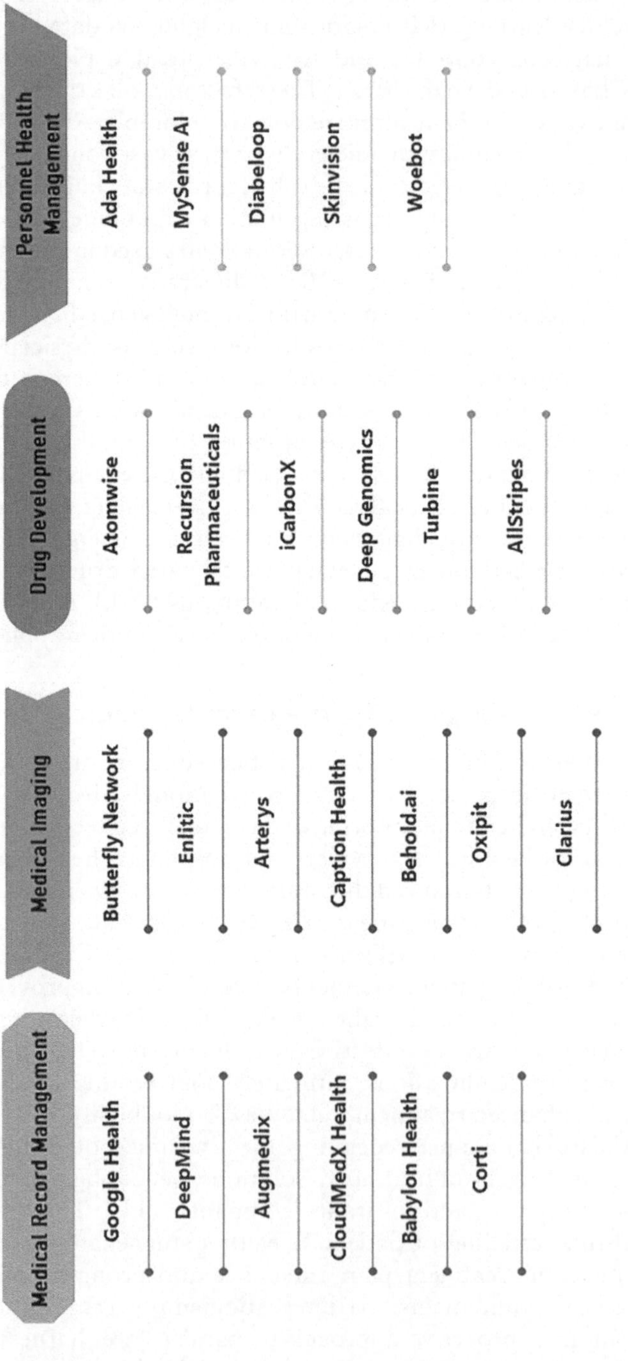

Figure 8.3 Diverse AI-based technologies used in healthcare.

from IoT in various facets. IoT devices equipped with sensors play a pivotal role in tracking the live locations of critical medical equipment, such as wheelchairs, defibrillators, nebulizers, oxygen pumps, and other monitoring apparatus (Mohapatra 2023). Real-time analysis of staff deployment across different locations enhances operational efficiency. Controlling the spread of infections within healthcare facilities is of paramount concern, and IoT-enabled hygiene monitoring devices serve as a formidable tool in preventing patient infections (Dheeraj et al. 2023). Additionally, IoT devices facilitate asset management tasks, encompassing pharmacy inventory control and environmental monitoring, including temperature and humidity regulation in refrigeration units (Ramasamy and Kadry 2021). The integration of IoT also presents an array of opportunities for health insurance providers. By harnessing the wealth of data generated by health monitoring devices, insurers can bolster their underwriting and claims operations. This data empowers them to identify fraudulent claims and prospects for underwriting, thereby enhancing transparency in underwriting, pricing, claims processing, and risk assessment. Customers, in turn, gain a clearer understanding of the rationale behind every decision and its subsequent outcomes, courtesy of data-driven decision-making processes facilitated by IoT. Furthermore, IoT devices offer insurance companies a reliable means of validating claims through the data captured by these devices, ensuring fair and accurate claim settlements. The IoT interface of patients and health professionals is presented in Figure 8.4.

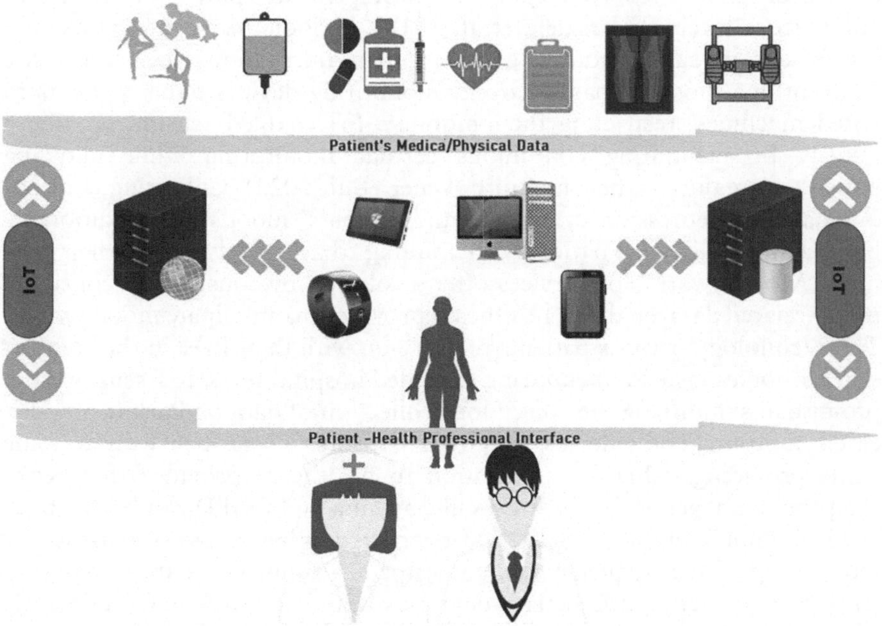

Figure 8.4 IoT interface of patients and health professionals.

IoT has emerged as a transformative force, allowing healthcare providers to transcend traditional clinical confines. Home-based monitoring systems have revolutionized patient care by enabling continuous health surveillance and eliminating the need for frequent and costly in-person consultations (Islam et al. 2023). IoT devices usher in a wealth of novel prospects for healthcare practitioners to oversee patient well-being while simultaneously empowering individuals to take charge of their health. In parallel, the diverse spectrum of wearable IoT gadgets presents a spectrum of advantages and intricacies, enriching the landscape for both healthcare providers and their patients. The utilization of Internet of Things (IoT) devices in healthcare has become increasingly prevalent, with remote patient monitoring emerging as its foremost application. Below is a quick summary of the many IoT application paths available at present. This innovative approach enables the automatic collection of critical health metrics, including heart rate, blood pressure, and temperature, from patients who are not physically present at healthcare facilities (Abdulmalek et al. 2022). This transformative technology eliminates the need for patients to travel to healthcare providers or manually record their health data. Traditionally, monitoring glucose levels has posed significant challenges. Routine checks and manual data recording can be inconvenient and only provide a snapshot of glucose levels at specific moments. Fluctuations in glucose levels may go unnoticed with periodic testing. IoT devices have emerged as a game-changer in this context, offering continuous and automatic glucose monitoring. These devices alleviate the burden of manual record-keeping and promptly alert patients to problematic glucose levels (Valenzuela et al. 2019). Periodic heart rate checks are insufficient in safeguarding against sudden fluctuations in heart rates. Conventional hospital-based cardiac monitoring devices tether patients to wired machines, restricting their mobility. IoT-enabled sensors provide a remedy by facilitating continuous cardiac monitoring while allowing patients to maintain their mobility (Umer et al. 2022). Collecting continuous data on depression symptoms and patients' moods has traditionally been challenging, with patients often failing to accurately report their feelings. "Mood-aware" IoT devices offer a solution by consistently collecting mood-related data, enabling healthcare providers to anticipate mood swings. This technology permits patients to carry on with their lives in the comfort of their homes instead of enduring extended hospital stays. IoT sensors hold promise in simplifying the continuous collection of data on Parkinson's disease symptoms. These devices empower patients to lead their lives at home while providing valuable information to healthcare providers for better symptom management. Conditions like asthma and COPD can bring about sudden, unpredictable attacks. IoT-connected inhalers assist patients by monitoring attack frequency and gathering environmental data to pinpoint triggers (Garin et al. 2023). This data aids healthcare providers in devising more effective treatment strategies. Beyond the mentioned applications, IoT's impact on healthcare extends to ingestible sensors, connected contact lenses, and even robotic surgery. These innovations hold immense potential to revolutionize health management and improve patient outcomes.

8.2.3 Blockchain applications in healthcare

Decentralized and publicly accessible digital ledger technology, commonly known as blockchain, has emerged as a transformative tool in various industries. Blockchain operates by recording transactions across multiple computers, ensuring that once a record is established, it remains unalterable without necessitating changes to subsequent blocks (Gad et al. 2022). This interconnected structure creates an immutable chain of data, enhancing security and trust. In the domain of healthcare, blockchain's distributed ledger technology has found applications in the secure transfer of patient medical records, bolstering the fortifications of healthcare data (Roberto et al. 2023), optimizing the management of pharmaceutical supply chains (Zakari et al. 2022), and empowering healthcare researchers to unlock the intricacies of genetic code (Singh et al. 2023). In blockchain technology, data is upheld across distributed networks instead of relying on a centralized database, thereby enhancing both stability and resistance to potential security breaches (Shi et al. 2020). Blockchain serves as a remarkable platform for the innovation and competition of contemporary, forward-thinking business models when juxtaposed with traditional enterprises. In the healthcare and pharmaceutical sectors, Blockchain emerges as a transformative solution to combat counterfeit medications, enabling comprehensive traceability of pharmaceuticals and facilitating the identification of the origins of falsification. Furthermore, Blockchain offers an unassailable guarantee of patient record confidentiality, acting as an immutable repository for medical histories that are impervious to alteration. This decentralized network seamlessly integrates with standard hospital hardware, thereby empowering researchers to harness computational capabilities for the estimation of therapies, medications, and remedies for a wide array of illnesses and disorders while capitalizing on the resources conserved by these devices. In light of the challenges posed by privacy and security concerns, traditional electronic health records and personal health record-based systems have faced limitations in facilitating the secure exchange of health information, leading to reluctance among stakeholders to engage in collaborative efforts. Consequently, healthcare costs have risen, placing substantial burdens on both patients and healthcare providers. In response to these trust-related issues, contemporary research and policymaking endeavors have turned their attention toward blockchain technology. According to insights from IBM, numerous prominent healthcare organizations anticipate that blockchain will play a pivotal role in transforming the healthcare landscape. This transformation is anticipated to occur through the enhancement of healthcare management systems and the establishment of a decentralized framework for the exchange of electronic healthcare data. Safeguarding patient information stands as a paramount concern. Unauthorized alterations to patient records can pose significant challenges for healthcare providers in accurately diagnosing and treating medical conditions. Alarmingly, from 2009 to 2017, over 176 million patient records fell victim to security breaches, wherein malicious actors not only gained access to sensitive medical data but also exploited financial information for unethical purposes. In this context, blockchain technology emerges as a formidable solution—a tamper-proof, decentralized, and incorruptible framework that promises to revolutionize the accessibility and security of patient data. Pharmaceuticals are not manufactured within hospital facilities; rather, they are produced within laboratories and pharmaceutical enterprises situated globally. Following production, these medicinal products are subsequently distributed to nations according to their respective requirements. Therefore, ensuring the integrity and transparency of the medical supply chain is of paramount importance to both importers and exporters. The utilization of blockchain technology offers an effective solution to this challenge, as it incorporates attributes such as transparency, decentralization, and tamper resistance. The significant advantages of integrating patient records into a blockchain ledger for enhanced healthcare management. Leveraging Machine Learning and Artificial Intelligence for disease prediction, alongside data analytics for cost-effective service offerings, represents a transformative step toward improved patient care. Furthermore, the incorporation of blockchain technology ensures meticulous record management, minimizing errors, and optimizing healthcare processes. This innovative approach has the potential to shape the future of healthcare, leading to more efficient and patient-centric healthcare systems. Electronic health record (EHR) systems represent a digital repository of health data generated and maintained by diverse healthcare institutions. Blockchain technology addresses the limitations associated with EHR systems, specifically focusing on issues of accessibility, interoperability, and authentication. By seamlessly linking EHRs and fostering shared ownership among stakeholders, blockchain emerges as a promising solution to revolutionize healthcare data management (Han et al. 2022). Innovators have pioneered an Ethereum-powered blockchain platform designed to emulate the intricacies of the recruitment procedure within the realm of scientific research. This groundbreaking blockchain system affords universal accessibility to trial data for all researchers, concurrently safeguarding the confidentiality of trial participants.

8.2.4 Virtual and augmented reality in medical training

The field of medicine has perennially held a paramount position within the spectrum of human endeavors. In this relentless pursuit of excellence, the synergy between scientific visionaries and technological trailblazers persists, tirelessly endeavoring to endow the healthcare sector with unparalleled precision and efficacy solutions. At the epicenter of this dynamic intersection lies revolutionary technology, as a beacon of hope for immediate life-saving interventions and as a harbinger of extended and enriched human existence. Virtual reality (VR) technologies have emerged as a highly promising domain with the potential to enhance medical training and elevate the precision of therapeutic interventions. Notably, recent studies have highlighted the revolutionary influence of virtual reality applications in the field of surgical training (Ntakakis et al. 2023). The integration of virtual reality (VR) technologies into surgical education has resulted in notable improvements, as supported by empirical data from an enormous amount of research. Furthermore, a striking 87% of investigations have demonstrated that healthcare specialists who underwent VR-based training exhibited marked enhancements in their procedural precision and accuracy. These findings collectively underscore the substantial potential of VR technologies in revolutionizing medical training and practice. In contemporary medical education, VR has emerged as a cutting-edge pedagogical tool, revolutionizing the training of healthcare professionals. VR technology affords users the unique opportunity to engage with diverse scenarios within immersive digital environments. For instance, individuals can fully immerse themselves in the intricacies of human anatomy, observing it from a multitude of angles and scales. This transformative capability empowers students and interns to undergo interactive training experiences of an elevated caliber, concurrently enabling practicing physicians to refine and enhance their professional competencies. Different possible modules of training through AR/VR are presented in Figure 8.5.

While abundant empirical evidence substantiates the efficacy of VR training across diverse sectors, research with a medical emphasis reinforces the compelling argument for the adoption of VR-based training within the

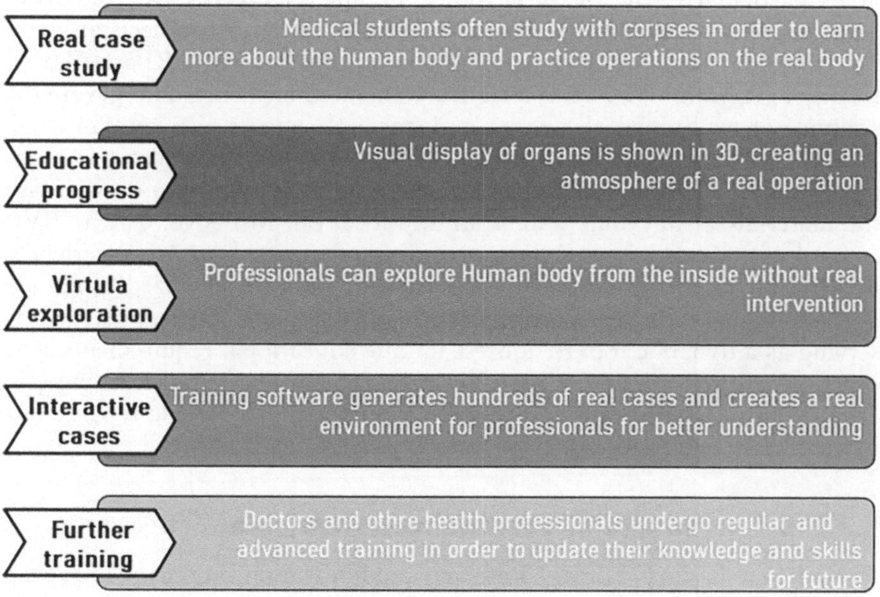

Figure 8.5 Training modules through AR/VR.

healthcare domain. These studies demonstrate that healthcare VR training accelerates clinicians' learning curves, mitigates errors, and enhances knowledge retention. Both Augmented Reality (AR) and VR offer an engaging and immersive educational environment for medical practitioners and students. These technologies furnish a three-dimensional perspective on medical scenarios, thereby enriching comprehension of intricate anatomical structures and medical processes. The interactivity inherent in AR and VR has the potential to elevate knowledge retention and facilitate more effective recall. Incorporating AR/VR technologies offers a versatile platform for simulating a diverse spectrum of medical conditions and scenarios, spanning routine procedures to exceptionally rare, intricate, and emergent cases. This immersive capability empowers healthcare practitioners to acquire proficiency in a broad spectrum of situations that may deviate from the norm of their customary training experiences. AR and VR represent formidable tools for patient education, facilitating the elucidation of intricate medical procedures and conditions through immersive, comprehensible visualizations (Aliwi et al. 2023). This innovative approach augments patients' grasp of their health status, thus promoting a proactive approach to managing their wellbeing. Moreover, the integration of AR and VR technologies extends beyond individual patient education to encompass collaborative learning among medical professionals dispersed across various geographical locations. This digital convergence fosters a sense of teamwork, facilitates seamless knowledge dissemination, and heightens the proficiency in decision-making abilities among healthcare practitioners. The cultivation of psychological readiness for high-pressure situations is of paramount importance. This holds particularly true for professions entailing emergency care, intensive care, and surgical disciplines, where swift and well-informed decision-making proves pivotal. The integration of Augmented Reality (AR) and Virtual Reality (VR) technologies has effectively liberated medical training from the confines of traditional classroom and hospital settings (Venkatesan et al. 2021). This transformative shift allows learners unrestricted access to training modules, affording them the flexibility and convenience to engage with the materials at any time and from any location. Moreover, AR and VR technologies have unlocked the potential for tailoring learning experiences to the individualized needs of each trainee. This personalized approach accommodates variances in learning pace and comprehension levels, thereby serving as a dynamic tool to address unique educational requirements. The outcome of such advancements is an enhancement in overall learning efficacy, promising significant improvements in the education and preparedness of healthcare professionals.

8.2.5 3D printing and personalized medicine

According to findings from the Research and Innovation Unit of the European Commission, personalized medicine represents a promising solution to confront the limitations posed by conventional pharmaceuticals, which often prove ineffective in treating a substantial portion of patients (Nimmesgern et al. 2017). Additionally, it addresses the escalating healthcare expenditures driven by the growing prevalence of chronic ailments and an increasingly aged demographic. Within this framework, personalized medicine offers tailored prevention and treatment approaches, meticulously fine-tuned for individual patients or specific patient cohorts, thereby eliminating financial resources squandered on trial-and-error interventions. Three-dimensional (3D) printing has emerged as a formidable tool in the domain of personalized medicine, offering healthcare professionals a diverse array of techniques to craft tailor-made pharmaceuticals and medical devices (Vanessa and Kumar, 2021). Initially, these technologies were devised for the production of tablets, evolving from basic formulations that solely encompassed a specific dose of a non-commercially available drug to intricate systems capable of accommodating multiple drugs with distinct release profiles within a single tablet, all customized to meet the unique needs of individual patients. Furthermore, 3D printing has paved the way for the creation of personalized metallic prostheses, parenteral implants, and various other categories of medical apparatus (Serrano et al. 2023). Some commonly employed 3D printing technologies are illustrated in Figure 8.6. Notably, in recent years, the utilization of 3D printing technologies for the fabrication of medicines containing biopharmaceuticals or drugs encapsulated within nanovehicles, commonly referred to as nanomedicines, has garnered increasing attention within the scientific community. Due to the compact and portable design of 3D printers, coupled with their user-friendly features, along with the capability to produce pharmaceuticals on an as-needed basis, this technology presents a seamless integration opportunity within various healthcare environments.

These settings include hospital wards, in-patient pharmacies, specialized clinics, and community pharmacies. The on-demand dispensing of medications in such contexts offers substantial advantages to clinical pharmacy practice. These advantages encompass enhancing medication accessibility and acceptance, streamlining the often laborious and time-consuming process of preparing extemporaneous doses, and expediting patient discharge times. Moreover, the swift production of customized medication doses becomes attainable, particularly in resource-constrained situations, such as hospital emergency departments, acute medical care units, disaster-stricken areas, ambulances, military facilities, and regions with limited economic resources. 3D printing or additive manufacturing, a progressive technique, unfolds the successive deposition of material in a stratified fashion to systematically craft a tangible three-dimensional model. This sophisticated process operates in tandem with computer-aided design (CAD) software, which serves as the conduit for transmitting imperative instructions to a 3D printer (Pravin and Sudhir 2018). The printer, in turn, translates the digital model into discrete two-dimensional (2D) cross-sections, subsequently orchestrating the assembly of cohesive, solid layers, thus fabricating the desired objects

Figure 8.6 3D printing technologies.

with precision and efficiency. Utilizing a computerized model, this innovative approach orchestrates the precise layer-by-layer assembly of three-dimensional structures. Its transformative influence on the pharmaceutical sector is extensive, given that 3D printing techniques have the potential to revolutionize product manufacturing by offering customized dosages, drug compositions, geometries, and release attributes, capabilities unattainable through traditional methods such as tablet production and encapsulation (Samiei 2020). A watershed moment occurred in July 2015 when the United States Food and Drug Administration (FDA) granted its approval for the first 3D-printed pharmaceutical, Spritam® (Lamichhane et al. 2023).

8.3 DIGITAL HEALTH AND REMOTE PATIENT MONITORING

8.3.1 Telemedicine and telehealth platforms

In the realm of healthcare, a well-rounded medical team, encompassing physicians, clinical nurses, dietitians, pharmacists, psychiatrists, and pathologists, assumes a pivotal role in delivering comprehensive patient care. Nonetheless, the geographical dispersion of team members can pose challenges to effective coordination, potentially resulting in suboptimal care delivery. Telemedicine, defined as the utilization of information and communication technologies to provide clinical healthcare to patients at a distance, holds the potential to address various systemic challenges within healthcare. These challenges encompass a scarcity of medical supplies, non-adherence of healthcare personnel to established guidelines, patient non-compliance with treatment regimens, limited access to critical healthcare information and data, and difficulties in ensuring patient follow-up (Haleem et al. 2021b). Integrating telemedicine with decision support systems offers a solution that streamlines healthcare provision through the use of protocol checklists, prompts, and alerts. This synergy enhances communication between healthcare providers and their patients while optimizing provider schedules. Furthermore, the promotion of electronic health records and health management information systems within the framework of telemedicine facilitates the seamless collection of routine healthcare data (Talal et al. 2020). These advancements collectively hold the promise of revolutionizing healthcare by fostering greater collaboration among care providers, mitigating geographic barriers, and optimizing data-driven healthcare delivery. The integration of hardware, software, and communication channels within the telemedicine system enables efficient information sharing and remote consultations across distant locations. The hardware comprises key components such as a computer, printer, scanner, and videoconferencing system. Through the software, patient data, encompassing images, reports, and videos, can be readily collected. The communication channel facilitates the connection between the two sites. Most telemedicine applications rely on two main technologies. The first, known as "store and forward," involves capturing and storing digital images to be sent to another location via a computer (Alenoghena et al. 2023). This method finds frequent use in telemedicine services like teledermatology, telepathology, and teleradiology, particularly for non-urgent cases, where medical diagnostic and consultation services are efficiently delivered within 24 to 48 hours (Goharinejad et al. 2021). Alternatively, for situations requiring in-person consultations, interactive television technology is commonly used. In this scenario, the patient, accompanied by their healthcare provider, typically a nurse practitioner or telemedicine coordinator, is based at the originating site, while the specialist is located at a referral site, often an urban hospital. Real-time interactions facilitated by videoconferencing technology allow consultations in various medical disciplines, including psychiatry, internal medicine, rehabilitation, cardiology, pediatrics, obstetrics and gynecology, and neurology (Dasgupta and Deb 2008). Additionally, several other innovative systems have emerged, including teleneuropsychology, telenursing, telepharmacy, and telerehabilitation. Teleneuropsychology enables patients with cognitive disorders or suspected cognitive impairments to receive remote neuropsychological consultations and assessments via telephone (Sarno et al. 2022). Various internet-driven frameworks for telemedicine are available to streamline the integration of

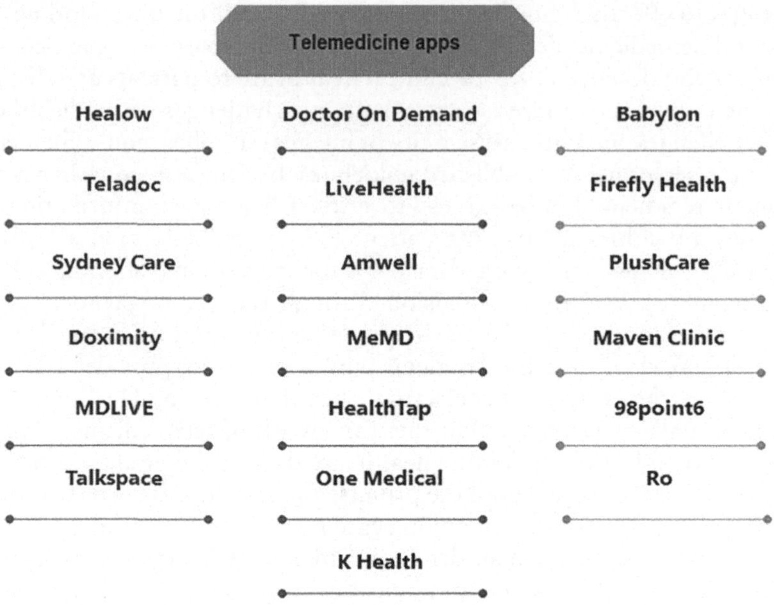

Figure 8.7 Diverse mobile apps of telemedicine.

patients with various hospitals and mobile medical specialists, some well-known mobile apps are presented in Figure 8.7.

Telenursing involves providing nursing care from a distance, utilizing communication technologies for consultations, diagnoses, and the monitoring of medical conditions and symptoms. Telepharmacy serves as a valuable platform for providing medical advice to patients when direct interaction with a pharmacist is not feasible. This approach allows patients to benefit from remote counseling and medication monitoring, subject to regulatory approval for refill authorization. Telerehabilitation leverages advanced communication technology, including video conferencing and webcams, to conduct clinical assessments and therapy for rehabilitation patients (Fiani et al. 2020). The incorporation of a robust visual component enhances the communication of symptoms and contributes to improved clinical outcomes. Amid the backdrop of the COVID-19 pandemic, the utilization of telehealth services has seen a remarkable and swift expansion, establishing itself as an efficient, cost-effective, and physically distant means of connecting patients with their healthcare providers (Tiwari et al. 2023). Telehealth, as defined by the Centers for Medicare and Medicaid Services, entails the utilization of telecommunications and information technology to facilitate health assessment, diagnosis, intervention, consultation, supervision, and information exchange across geographical distances (Dhruv et al. 2021). Telehealth services can be characterized by either synchronous interactions, in which clinicians and patients engage in real-time communication, or asynchronous approaches, where data trends or messages are periodically shared with clinicians, often serving as a valuable tool for remote patient monitoring. Considering the current healthcare scenario, there has been a growing dependence on the telehealth system for delivering superior healthcare services. Over the past year, the telehealth landscape has significantly transformed, aided by recent laws that facilitate the smooth assimilation of telehealth and surmount obstacles associated with payment schemes. As telehealth gradually becomes the standard clinical practice, it is crucial to comprehensively assess and gain a profound comprehension of both the potential benefits and risks it offers. The viewpoint of individuals undergoing medical care underscores the potential of telehealth services in generating cost savings by reducing travel time, minimizing extended work leaves, and curbing expenses linked with emergency department and clinic visits. Moreover, the cost-effectiveness of telehealth is not confined to patients alone but also extends to the broader healthcare system. Efficient sorting of cases through telehealth platforms leads to significant savings by preventing the improper utilization of healthcare resources. Telemedicine strategies have proven effective in addressing the issues of healthcare accessibility and promptness, not only in primary care but also in specialized healthcare settings. The implementation of telementoring initiatives, exemplified by programs like Project ECHO (Extension for Community Healthcare Outcomes), has empowered healthcare providers in remote regions to engage with specialists across various disciplines. These connections enable the exchange of knowledge and consultation on modern, top-notch healthcare protocols (Joanna et al. 2021). The enhancement of healthcare accessibility, a persistent challenge in public health, has seen progress through the adoption of telemedicine services. In times of public health crises, the need for increased healthcare resources and personnel can be effectively addressed by deploying healthcare providers from unaffected regions through telehealth systems. Incorporating a well-structured strategy for communication, collaboration, and compensation for telehealth interventions into local emergency preparedness guidelines can greatly enhance response capabilities during emergencies. The efficacy of telehealth interventions in such critical situations was exemplified through the deployment of "Virtual First Responders" during health emergencies. Implementing provisional measures to rapidly deploy telehealth resources has the potential to offer immeasurable support during periods of widespread public health crises. In fields with diminished workforce availability resulting from amplified demand or reduced supply, such as the attrition of clinicians due to retirement, telehealth services emerge as an invaluable asset. Leveraging extant healthcare professionals to extend their working hours from the convenience of their residences and encouraging semi-retired clinicians to engage in telehealth services presents a promising strategy for ameliorating the supply–demand disparity within specific sub-specialties.

8.3.2 Wearable health devices and sensors

In nearly all spheres of our daily routines, the overarching aim is to streamline and simplify tasks, seamlessly integrating them into our lives. This drive for convenience is particularly pronounced in the realm of healthcare, elucidating the surge in the adoption of wearable technology. Within the healthcare domain, wearable technology encompasses a diverse array of medical devices and accompanying adjuncts, leveraging sensors, actuators, software, and electronic patches affixed to the skin to monitor a patient's well-being, detect irregularities, and even provide therapeutic interventions (Majumder et al. 2017). These innovations span from wearable monitors for vital signs to intelligent wristwatch glucose monitors and wearable devices for pain alleviation. As of 2020, the global market for wearable technology reached a substantial $40.65 billion, with projections indicating a compound annual growth rate of 13.8% from 2021 to 2028 (Tan 2022).

In June 2023, Google revealed its most recent Feature Drop, introducing a host of enhancements for Fitbit gadgets. Notable improvements involve the integration of an array of exercise modes within the Exercise menu of the device. This new feature is accessible for Charge 5, Luxe, and Inspire 3. Additionally, Google integrated a menstrual health tile for logging periods and monitoring the cycle state, a feature available for Versa 4 and Sense 2. Moreover, users across all Fitbit devices can now benefit from the Daily Readiness Score, a valuable metric designed to gauge whether the body requires rest or exercise. In January 2023, a collaboration between Masimo and Royal Philips was extended, focusing on the integration of the advanced health-tracking capabilities of the Masimo W1 smartwatch to enhance patient monitoring in the realm of home telehealth. This partnership aims to leverage Philips's extensive patient monitoring infrastructure, merging it with the W1 watch to propel the telehealth and telemonitoring landscape forward. In April 2023, Garmin disclosed the expansion of regional availability for the Dexcom Connect IQ applications, catering to individuals with Type 1 or Type 2 diabetes utilizing the Dexcom G6 or Dexcom G7 Continuous Glucose Monitoring System. These innovative apps enable users to conveniently monitor their glucose levels and trends through a compatible smartwatch or cycling computer. Polar revealed the integration of the SleepWise function across its Ignite, Unite, Vantage M, and Vantage V models, enabling users to conveniently monitor their "Boost from sleep" data via the Polar Flow app (Miller 2019). This significant update supplements the existing provision of the feature on the Polar Ignite 2, Vantage M2, and Grit X devices, which were initially announced in February 2023. In May 2023, Medtronic, the entity that acquired Covidien in 2015, revealed its plans for the forthcoming takeover of EOFlow Co. Ltd. This strategic move is aimed at incorporating the innovative EOPatch device, renowned for its characteristics of being wearable, tubeless, and entirely disposable, thus enabling Medtronic to cater to the requirements of the diabetic population more effectively (Whooley 2023). At the beginning of 2022, OMRON launched its remote patient monitoring offerings at the 2022 Consumer Electronics Show (CES). Simultaneously, the company revealed a line of interconnected blood pressure monitors alongside a sophisticated mobile application designed to assist users in managing their cardiovascular well-being (Healthcare 2022). In May 2022, Withings disclosed the launch of its ScanWatch Horizon hybrid smartwatch in the United States. This innovative device is equipped to continuously track the wearer's ECG, heart rate, blood oxygen levels, breathing patterns, and physical and sleep activities, both on land and underwater (Eadicicco 2022). Notably, the ScanWatch Horizon boasts an impressive water resistance capability of up to 10 ATM while offering an extended battery life lasting up to 30 days. In September 2022, VitalConnect disclosed an infusion of funds from Health Insight Capital, the investment division of HCA Healthcare Inc. This financial boost is poised to expedite the expansion of VitalConnect's cardiac monitoring repertoire, facilitating the introduction of cutting-edge technologies designed to benefit both patients and healthcare providers (Kelly 2022).

8.3.3 Remote diagnostics and monitoring systems

Chronic illnesses encompass a range of physical and mental health conditions, including hypertension, diabetes, cardiovascular diseases, obesity, and stroke, among others. These conditions collectively constitute the predominant sources of health-related risks in the human population and are responsible for a substantial majority of global mortality(. The prevalence of chronic diseases has surged in tandem with population expansion, and the existing hospital infrastructure faces constraints in accommodating the growing number of patients. Additionally, the management of chronic diseases necessitates specialized home-based care to cater to the diverse needs of patients and administer tailored therapeutic interventions. Regrettably, a significant portion of caregivers and families find themselves lacking the requisite time and skills to provide such care. Consequently, the quality of life for individuals grappling with chronic diseases remains perpetually in jeopardy. The growing necessity of developing electronic health systems, including remote patient monitoring (RPM), electronic health record (EHR) systems, mobile health (m-health), telemedicine, e-visits, and e-consultations, among others, is increasingly recognized (El-Rashidy et al. 2021). These systems play a vital role in the continuous monitoring, diagnosis, prediction, and treatment of patients, leading to a reduction in healthcare costs and enabling patients to maintain their daily routines while their vital signs are continuously supervised. Furthermore, they enable physicians to provide consistent follow-up care for patients, transcending the limitations of in-person hospital visits. Patient monitoring systems equip patients with valuable insights into their symptoms and treatments and foster independent living, ultimately enhancing their overall quality of life (Vahdat et al. 2014). Simultaneously, PM systems serve as pivotal tools within hospital settings, aiding in patient prioritization based on their critical conditions, thereby optimizing the delivery of essential medical care. Devices for remote patient monitoring are widely accessible nowadays. Figure 8.8 presents the collection and management of data in a disease monitoring system.

Real-time patient monitoring systems (RPMS) offer the dual benefits of promptly identifying illnesses and ensuring continuous patient health status monitoring. Rapid responses are crucial in cases of detected untimely fatalities, prompting the immediate implementation of emergency interventions (Lakmini et al. 2017). Furthermore, these systems contribute to cost reduction in healthcare by effectively employing various communication technologies. Patients undergoing treatment can seamlessly engage in their regular daily activities. Moreover, RPMS enhancements have shown a positive impact on the provision of emergency care for traffic accidents and overall mobility. The utilization of wireless sensors and communication in the context of remote patient monitoring facilitates the acquisition and transmission of crucial data to healthcare facilities. This process involves the deployment of diverse sensor types, including wearable, implantable, and non-contact variants, to ensure comprehensive data capture. Subsequently, the collected data undergoes processing and conversion through an internal or external controller to enable seamless wireless transmission to the hospital. A combination of short- and long-range communication techniques is employed to achieve effective data relay, ensuring efficient data delivery to the end-user. RPMS aims to replicate the conventional patient supervision techniques typically employed by healthcare professionals such as physicians, nurses, and caregivers. Consequently, they typically perform the functions above outside the hospital environment. Vital signs tracking and disease identification represent two illustrative applications of such systems. Physiological vital signs, including body temperature, heart rate, blood oxygen saturation, blood pressure, and respiration rate, are routinely assessed through Vital Monitoring. The Remote Patient Monitoring System (RPMS) facilitates these assessments outside hospital settings, thereby fostering greater hospital efficiency and potentially reducing hospital admissions by enabling increased at-home patient monitoring. The remote monitoring process may involve the evaluation of one or more physiological parameters, the frequency of which depends on the patient's condition and can be either real-time or periodic (Miranda et al. 2023). Diagnostic techniques for diseases involve the analysis of physiological markers and health anomalies to pinpoint specific illnesses. Elevated blood pressure and vertigo, for instance, can indicate hypertension, while an elevated body temperature accompanied by headaches may suggest various cold-related ailments. A novel disease diagnostic approach has been developed by researchers, focusing exclusively on the identification of one or more particular maladies.

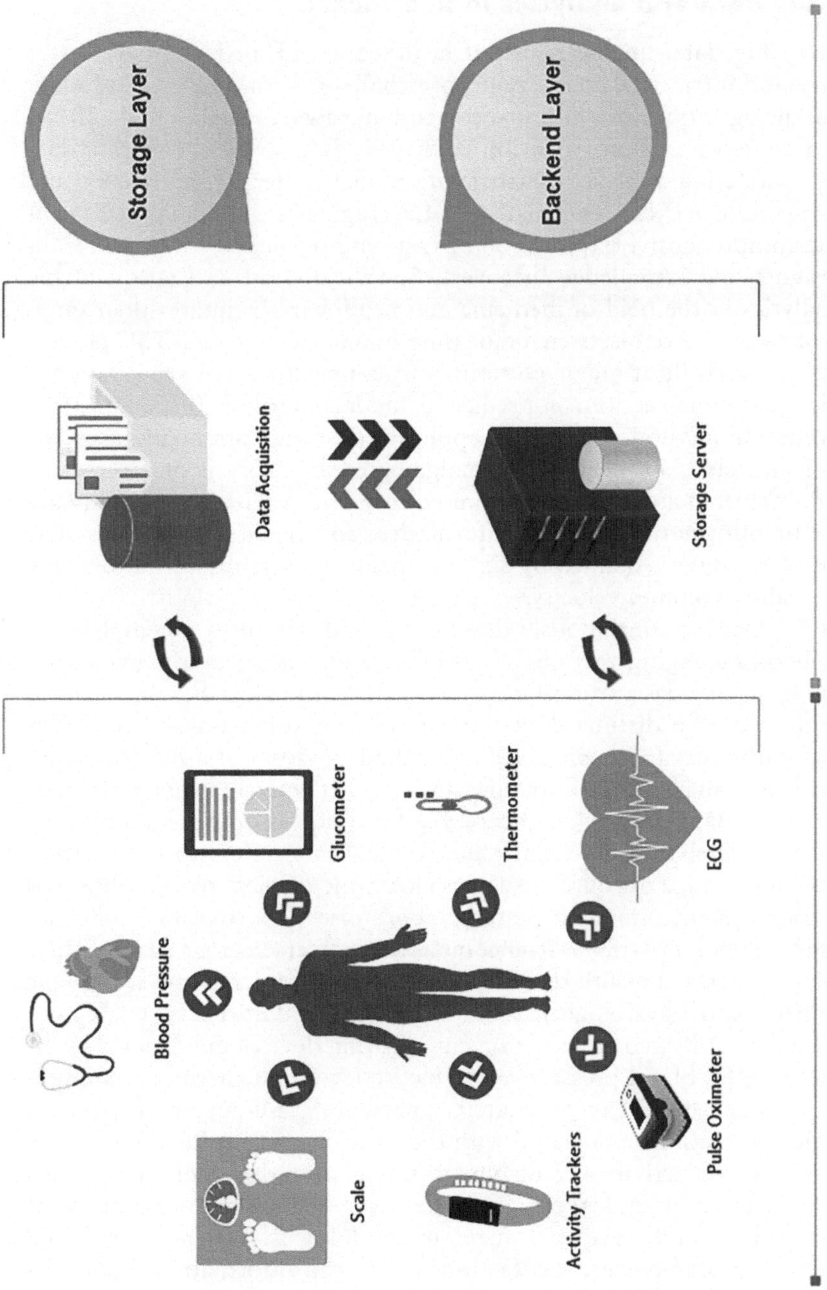

Figure 8.8 Integration of data collection and data management.

8.4 DATA-DRIVEN HEALTHCARE

8.4.1 Big data and analytics in healthcare

The term "big data" in the realm of healthcare and medicine pertains to extensive and intricate datasets that pose challenges for analysis and management using conventional software and hardware (Belle et al. 2015). Big data analytics includes a fusion of diverse data sources, ensuring data quality, conducting analysis, constructing models, interpreting results, and validating findings (Mikalef et al. 2019). Leveraging big data analytics facilitates the comprehensive exploration of vast data resources, leading to valuable insights and knowledge discovery. Specifically, the utilization of big data analytics in the field of medicine and healthcare facilitates the examination of extensive datasets encompassing numerous patients. This process involves the identification of patterns and connections between datasets, alongside the construction of predictive models through the application of data mining methodologies. The application of big data analytics in the realm of medicine and healthcare amalgamates the analysis of diverse scientific domains, including but not limited to bioinformatics, medical imaging, sensor informatics, medical informatics, and health informatics. The concept of big data is defined by six key attributes, often referred to as the 6 "Vs": value, volume, velocity, variety, veracity, and variability. The concept of Big Data is continuously developing, and presently, its emphasis is not solely on managing vast quantities of data but rather on the extraction of value from this data (Suter-Crazzolara 2018). Big Data is sourced from diverse origins with distinct data characteristics and is managed by different units within organizations, leading to the formation of a Big Data continuum. The primary goal of organizations is to effectively oversee, process, and scrutinize Big Data. In healthcare, Big Data flows comprise a multitude of data types, including different data. Clinical data refers to information derived from a range of sources, such as electronic medical records, hospital information systems, imaging centers, laboratories, pharmacies, and other healthcare service providers. It encompasses a diverse set of data, including patient-generated health records, unstructured physician notes, genomic information, and physiological monitoring data. Additionally, it includes biometric data collected from various monitoring devices that track parameters like weight, blood pressure, and glucose levels. Furthermore, comprehensive financial data is incorporated, representing a complete account of economic transactions associated with the provision of healthcare services. Scientific research activities contribute to this pool of data, comprising outcomes from research endeavors, including drug trials, the development of medical devices, and novel treatment methodologies. Patient-contributed data, detailing preferences, satisfaction levels, and information from self-monitoring systems for activities such as exercise, sleep, and dietary intake, are also included. Data derived from social media platforms further enriches this comprehensive dataset. The data in question are sourced not only from patients themselves but also from various organizations and institutions, as well as an array of monitoring devices, sensors, and instruments (Dash et al. 2019). Within the healthcare sector, data generated thus far is stored in both physical and digital formats. The healthcare industry has always been a prolific generator of data, largely driven by the necessity to maintain patient medical records. However, the challenge posed by Big Data in healthcare extends beyond sheer volume; it encompasses an unparalleled diversity in data types, formats, and the speed at which it necessitates analysis for continuous information provisioning. Given the ever-expanding diversity and volume of data sources, there is an imperative for advanced analytical tools and technologies, in conjunction with Big Data analysis methods, that can not only meet but also surpass the capabilities required for managing healthcare data effectively. Dealing with the vast expanse of Big Data presents a primary obstacle, necessitating effective strategies for processing this abundance of information and utilizing it to inform data-centric decisions across numerous domains. Within the domain of healthcare data, a notable challenge lies in tailoring the storage, analysis, and interpretation of extensive datasets to suit clinical environments (Batko and Ślęzak 2022). Specifically engineered for this purpose, data analytics systems in the healthcare sector aim to comprehensively elucidate, integrate, and visually represent intricate data. This approach serves to enhance the efficiency of gathering, retaining, scrutinizing, and illustrating large-scale healthcare data. The integration of different data analytics and data criteria is presented in Figure 8.9.

8.4.2 Predictive analytics for disease outbreaks

Big Data Analytics plays a pivotal role in the realm of patients, healthcare providers, and medical institutions. This revolutionary technology ensures that patients are equipped with comprehensive knowledge, empowering them to make informed decisions about their healthcare. Furthermore, it guarantees that patients receive tailored treatments that are optimally effective for their specific conditions. In the contemporary landscape, large-scale data has become ubiquitous, manifesting in various forms such as social network data, meteorological records, omic data, web-search data, disease reports, epidemic data, and statistics (Hassan et al. 2022). The accessibility of these vast datasets from diverse origins has facilitated their widespread generation and accumulation. Leveraging a data science approach, for instance, a predictive analytics framework, enables the extraction of valuable insights and knowledge from disparate big data sources. This extracted intelligence can subsequently be translated into actionable wisdom for informed decision-making. Healthcare predictive analytics entails the utilization of statistical algorithms and machine learning techniques to forecast forthcoming probabilities by analyzing past data (Calster et al. 2019). Essentially, it involves leveraging existing knowledge to anticipate future events. The utilization of

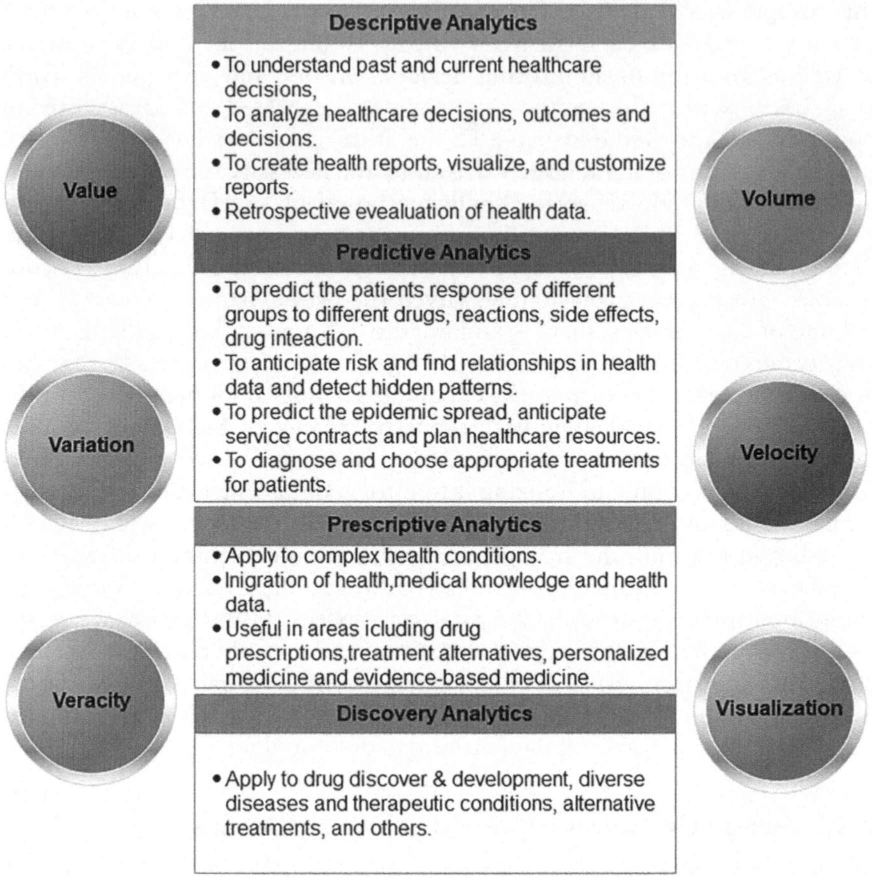

Figure 8.9 Integration of data analytics and data criteria.

predictive modeling involves harnessing extensive data on various global diseases and syndromes, incorporating it into a systematic classification system, and employing advanced machine learning techniques to pinpoint critical cases, thereby enhancing the precision of potential danger detection. The concept for chronological data management in disease diagnosis employing predictive analytics is shown in Figure 8.10.

Consequently, predictive analytics can expedite the comprehension of the intricate relationship between environmental factors and human biology, leading to an improved restructuring of clinical pathways to ensure optimal individualized healthcare. The financial burden associated with hospital readmissions is substantial. As reported by HealthcareDive, approximately \$26 billion is expended by Medicare each year on readmission cases (Wilson 2019). Moreover, healthcare institutions face substantial penalties under Medicare's Hospital Readmission Reduction Program, which provides

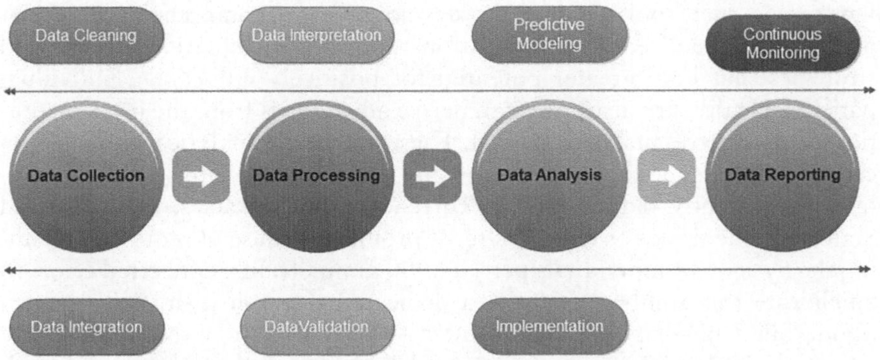

Figure 8.10 Data management in disease diagnosis using predictive analytics.

financial motivations to address the issue of readmissions. Leveraging healthcare predictive analytics enables the identification of patients displaying characteristics suggestive of a heightened risk of readmission. This empowers physicians to allot additional resources for follow-up care and customize discharge procedures, thus mitigating the likelihood of swift readmission. The application of predictive analytics aids medical establishments in proactively identifying individuals susceptible to chronic ailments, facilitating preemptive measures before the condition exacerbates (Razzak et al. 2020). This analytical approach involves the allocation of scores to patients, integrating diverse parameters such as demographic indicators, disabilities, age, and historical treatment trends. The application of predictive analytics plays a pivotal role in affording both healthcare authorities and the general populace invaluable insights into the dynamics of pandemics. As an illustrative example, a dedicated research team at the University of Texas Health Science Center at Houston (UTHealth) has innovatively engineered a predictive analytics-driven solution for monitoring and managing COVID-19 (UTHealth 2020). This innovative system facilitates the continuous generation and upkeep of a comprehensive public health dashboard, offering real-time and forward-looking assessments of the virus's prevailing and projected trajectories. As per findings in the HIPAA Healthcare Data Breach Report, the healthcare domain is frequently targeted by cyberattacks (Seh et al. 2020). Notably, the report highlighted that data theft commonly preceded the encryption phase in most ransomware attacks. Addressing this issue, the integration of predictive analytics in cybersecurity within the healthcare sphere can offer valuable assistance. Leveraging predictive analytics in conjunction with artificial intelligence tools specific to the medical field enables healthcare institutions to promptly assess risk levels associated with various online activities and promptly react to these incidents based on their respective risk scores (Alowais et al. 2023). Predictive analytics offer a valuable opportunity for medical institutions to foster patient engagement and fortify the patient-physician bond.

Leveraging such tools enables the development of comprehensive patient profiles, personalized communication, and the formulation of tailored strategies that hold greater potential for positively influencing individual patients. Healthcare insurance can derive advantages from the implementation of predictive analytics as well. These advanced analytical tools play a crucial role in expediting the processing of insurance claims within hospitals, concurrently reducing the occurrence of inaccuracies. In the United States, suicide stands as the tenth most prominent cause of mortality, claiming the lives of 14 individuals per 100,000 annually. In a concerted effort to ameliorate this somber scenario, a dedicated research team hailing from Vanderbilt University Medical Center (VUMC) has devised a predictive analytics model. This innovative model harnesses the electronic health records of patients to prognosticate the likelihood of suicide attempts within specific patient cohorts. Over 11 months of rigorous testing within the VUMC setting, the algorithm discreetly operated in the backdrop, furnishing real-time estimations on the proclivity of patients to revisit the healthcare facility for treatment following attempted suicide. The application of predictive analytics in the realm of healthcare stratified patients into eight distinct groups contingent upon their computed risk profiles. Strikingly, the preeminent risk category encompassed more than 33% of all recorded suicide attempts. Consequently, the VUMC research team has deduced that individuals placed within this high-risk stratum necessitate thorough evaluation to ascertain any potential proclivity toward suicidal tendencies (Govern 2021).

8.5 ETHICAL CONSIDERATIONS IN DATA USAGE AND PRIVACY

The comprehension of our surroundings and the facilitation of informed decision-making heavily hinge upon the pivotal role of data analysis. As our dependency on data intensifies, it becomes imperative to confront the ethical challenges that surface during data acquisition, manipulation, and comprehension. Ethical data analysis encompasses the conscientious examination of possible harm or biases toward individuals or communities and the implementation of measures to alleviate such concerns. Upholding ethical standards allows data analysts to guarantee that their contributions to society are constructive and uphold the rights of the individuals whose data they handle (Resnik and Elliott 2016). Preserving privacy and upholding confidentiality represent pivotal considerations within data analysis, as safeguarding sensitive individual information remains paramount. Mitigating such risks involves the process of anonymizing data, securing informed consent from participants, and adhering meticulously to pertinent data protection protocols. Ethical practitioners in data analysis place the utmost importance on fostering trust among data contributors, coupled with stringent measures to ensure that any disclosure of information occurs only with explicit authorization (Marieke et al. 2023). A strong dedication to the precision and reliability of data is imperative for ethical data analysis. Guaranteeing the high caliber of data and its faithful representation of the studied phenomena is crucial in averting flawed deductions or deceptive outcomes. Analysts are obligated to be open about any constraints in their data or methodologies, and they must meticulously authenticate their discoveries. Effective healthcare delivery necessitates the collection of patient data, which must adhere to the stringent guidelines outlined by the Health Insurance Portability and Accountability Act (HIPAA) (Edemekong et al. 2022). In the United States, various software tools and research networks, such as Research Electronic Data Capture (REDCap) (Patridge and Bardyn 2018), Research Action for Health Networks (REACHnet), Patient-Centered Outcomes Research Institute (PCORI®) (Forrest et al. 2021), and the Agency for Healthcare Research and Quality (AHRQ) (Kronick 2016), exemplify responsible patient data-sharing practices. Aggregating patient data facilitates a comprehensive understanding of health conditions, identification of symptom correlations, advancement of treatment methodologies, and tracking of prevailing trends, thus fostering an evidence-based approach to enhance patient outcomes. REDCap, an intuitive and secure web application developed by Vanderbilt University, specializes in data acquisition for clinical research, encompassing the construction and management of web-based surveys, databases, and projects. This platform maintains strict compliance with HIPAA regulations and is designed to support both single- and multiple-site research investigations. Notably, it ensures that all project data remains within the confines of the local institution, with no transmission to external parties, limiting its use to intrainstitutional studies. Moreover, REDCap allows for the designation of patient information as identifiable while offering straightforward deidentification options during data export, thereby ensuring robust intrainstitutional privacy and security. In contrast to REDCap's intrainstitutional focus, REACHnet represents an interinstitutional data network comprising multiple healthcare systems, academic centers, and health organizations. Serving a role akin to REDCap, REACHnet streamlines the execution of efficient multisite research endeavors, aiming to foster more informed healthcare decision-making and bolster population health management. The Agency for Healthcare Research and Quality (AHRQ) is committed to enhancing healthcare quality and promoting its accessibility, affordability, and fairness. Through its investments in health systems research and data analysis, the agency aims to facilitate informed decision-making in healthcare and devise strategies to enhance medical practices. A significant project under the auspices of AHRQ is the Healthcare Cost and Utilization Project (HCUP), a comprehensive compilation of databases encompassing both clinical and nonclinical patient information. These databases include crucial patient demographics, diagnoses, procedures, charges, and insurance particulars, thereby empowering research on various critical healthcare policy concerns such as access, cost, and quality of care. An illustrative instance is a statistical brief issued by HCUP in 2017, which delves into the expenses associated with emergency department (ED) visits for individuals with mental and substance use disorders (Moore and Liang 2020). The brief highlights a staggering 44.1% surge in the rate of ED visits for mental health and substance abuse diagnoses between 2006 and 2014, leading to 20.3 visits per 1,000 individuals. These AHRQ-led studies, in conjunction with the US Department of Health and Human Services, primarily concentrate on health policy issues, aiming to optimize service delivery costs for ED patients. Consequently, the collection of patient data from AHRQ databases becomes imperative, owing to its significant role in informing policy decisions. However, this data acquisition process can incur substantial costs. The expenses associated with procuring basic or comprehensive reports can fluctuate significantly, varying from thousands to hundreds of thousands of dollars, contingent upon the scope of the reports. AHRQ, in collaboration with REDCap and REACHnet, presents a compelling argument underscoring the value of patient data in the advancement of healthcare, along with providing instances of appropriate utilization of patient information compliant with HIPAA regulations. Notably, these initiatives stand in stark contrast to the prevalent privacy concerns concerning patient data arising from the convergence of big tech companies and the healthcare sector. Numerous domestic statutes and global protocols oversee the timing and methodology of disseminating health and clinical data. However, a majority of health providers seek personalized internal protocols and Standard Operating Procedures (SOPs) to govern the sharing of data. Each entity is tasked with devising SOPs that promote scientific progress while safeguarding the privacy of participants and sensitive product details. Investigators and sponsors must consider the potential benefits for participants and their communities stemming from shared data, all while ensuring the preservation of confidentiality. The continuous evolution of determining the appropriate timing and methods for remotely sharing data remains a dynamic process. Nevertheless, ethical data sharing can be actively embraced by research institutions now through a mindful approach that prioritizes the privacy and security of participants while also striking a delicate balance between safeguarding their commercial interests and facilitating the sharing of sufficient data for the advancement of medical research (Jennifer et al. 2022).

8.6 SUMMARY AND FUTURE PROSPECTS

The concepts of "Hospital 4.0" and "Medicine 4.0" are experiencing a surge in prominence, emblematic of a transformative era in the domains of medicine and healthcare. Notably, significant trends in technological and economic progress are discernible, particularly through the emergence of digital technologies like Artificial Intelligence, the Internet of Things, Augmented Reality/Virtual Reality, Blockchain, Telemedicine, 3D printing, Data management, and data analytics are the distinctive driver of economic and societal evolution. This evolution necessitates novel strategies and promises to unlock remarkable opportunities for growth and advancement. In the forthcoming times, companies that effectively integrate digital technologies within their operational frameworks will maintain a competitive edge. In the upcoming era, Medical 4.0 is poised to replicate human cognitive abilities and capabilities using machinery, software, and computing platforms. It will leverage cutting-edge digital technologies that enhance the efficiency and transformation of systems. The concept of the Smart Factory seamlessly integrates various manufacturing procedures, from initial planning phases to on-site actuators. Machinery and apparatus will optimize operations through rapid self-adjustment, particularly within the healthcare sector. Autonomous Mobile Robots will play a crucial role in intelligent healthcare, as their independent intelligence will facilitate the seamless integration of smart healthcare systems, ensuring smooth functionality. Advanced healthcare facilities will be equipped with contemporary sensors, embedded software, and robots for the collection and analysis of data, leading to informed decision-making. Through the integration of cutting-edge technologies, Medicine 4.0 has transformed the landscape of medical care, fostering enhanced patient well-being, cost efficiency, and the provision of tailored, easily accessible treatments. As technology continues to evolve, Medicine 4.0 is poised to bring about substantial, far-reaching enhancements within the healthcare sector."

REFERENCES

Abdulmalek, Suliman, Nasir Abdul, Jabbar Waheb A., Almuhaya Mukarram A. M., Bairagi Anupam K., Khan MD. Al-Masrur, and Kee Hoon. 2022. "IoT-Based Healthcare-Monitoring System towards Improving Quality of Life: A Review." *Healthcare* 10:1993.

Albahri A.S., Alwan Jwan K., Taha Zahraa K., Ismail Sura F., Hamid Rula A., Zaidan A.A., Albahri O.S., Zaidan B.B., Alamoodi A.H., Alsalem M.A.. 2021. "IoT-based telemedicine for disease prevention and health promotion: State-of-the-Art." *Journal of Network and Computer Applications* 173:102873.

Alenoghena, Caroline O., Henry O. Ohize, Achonu O. Adejo, Adeiza J. Onumanyi, Emmanuel E. Ohihoin, Aliyu I. Balarabe, Supreme A. Okoh, Ezra Kolo, and Benjamin Alenoghena. 2023. "Telemedicine: A Survey of Telecommunication Technologies, Developments, and Challenges." *Journal of Sensor and Actuator Networks* 12:20.

Aliwi, Ibrahim, Vico Schot, Michele Carrabba, Phuoc Duong, Silvia Shievano, Massimo Caputo, Jo Wray, and Giovanni Biglino. 2023. "The Role of Immersive Virtual Reality and Augmented Reality in Medical Communication: A Scoping Review." *Journal of Patient Experience* 10:237.

Alowais, Shuroug A., Sahar S. Alghamdi, Nada Alsuhebany, Tariq Alqahtani, Abdulrahman I. Alshaya, Sumaya N. Almohareb, Atheer Aldairem. 2023. "Revolutionizing Healthcare: The Role of Artificial Intelligence in Clinical Practice." *BMC Medical Education* 23:689.

Batko, Kornelia, and Andrzej Ślęzak. 2022. "The Use of Big Data Analytics in Healthcare." *Journal of Big Data* 9:3.

Belle, Ashwin, Raghuram Thiagarajan, S. M. Reza Soroushmehr, Fatemeh Navidi, Daniel A. Beard, and Kayvan Najarian. 2015. "Big Data Analytics in Healthcare." *BioMed Research International* 1:370194.

Bertalan, Meskó, Zsófia Drobni, Éva Bényei, Bence Gergely, Zsuzsanna Győrffy. 2017. "Digital Health is a Cultural Transformation of Traditional Healthcare." *mHealth* 3:38.

Calster, Ben V., Laure Wynants, Dirk Timmerman, Ewout W. Steyerberg, and Gary S. Collins. 2019. "Predictive Analytics in Health Care: How Can We Know It Works?" *Journal of the American Medical Informatics Association* 26:1651–1654.

Campisi, Christine A., Elizabeth H. Li, David E. Jimenez & Ruth L. Milanaik. 2020. "Augmented Reality in Medical Education and Training: From Physicians to Patients." In *Augmented Reality in Education A New Technology for Teaching and Learning*, by Vladimir Geroimenko, 3371. Egypt: Springer.

Chris, Giordano, Meghan Brennan, Basma Mohamed, Parisa Rashidi, François Modave, and Patrick Tighe. 2021. "Accessing Artificial Intelligence for Clinical Decision-Making." *Frontiers in Digital Health* 3:645232.

Dasgupta, Aparajita, and Soumya Deb. 2008. "Telemedicine: A New Horizon in Public Health in India." *Indian Journal of Community Medicine* 33:3–8.

Dash, Sabyasachi, Sushil K. Shakyawar, Mohit Sharma, and Sandeep Kaushik. 2019. "Big Data in Healthcare: Management, Analysis and Future Prospects." *Journal of Big Data* 6:1–25.

Davenport, Thomas, Ravi Kalakota. 2019. "The Potential for Artificial Intelligence in Healthcare." *Future Healthcare Journal* 1:94–98.

Dheeraj, Kumara, Sandeep Kumar Sood, Keshav Singh Rawat. 2023. "IoT-Enabled Technologies for Controlling COVID-19 Spread: A Scientometric Analysis Using CiteSpace." *Internet of Things* 23:100863.

Dhruv, Mahtta, Marilyne Daher, Michelle T. Lee, Saleem Sayani, Mehdi Shishehbor & Salim S. Virani. 2021. "Promise and Perils of Telehealth in the Current Era." *Current Cardiology Reports* 23:115.

Eadicicco, Lisa. 2022. *Withings ScanWatch Horizon Review: Luxury Looks and Lots of Health Tracking*. Scientific Report, CNET.

Edemekong, Peter F., Pavan Annamaraju, and Micelle J. Haydel. 2022. *Health Insurance Portability and Accountability Act*. Island: StatPearls Publishing.

El-Rashidy, Nora, Shaker El-Sappagh, S. M. Riazul Islam, Hazem M. El-Bakry, and Samir Abdelrazek. 2021. "Mobile Health in Remote Patient Monitoring for Chronic Diseases: Principles, Trends, and Challenges." *Diagnostics* 11:607.

Fiani, Brian, Imran Siddiqi, Sharon C. Lee, and Lovepreet Dhillon. 2020. "Telerehabilitation: Development, Application, and Need for Increased Usage in the COVID-19 Era for Patients with Spinal Pathology." *Cureus* 12:e10563.

Forrest, Christopher B., Kathleen M. McTigue, Adrian F. Hernandez, Lauren W. Cohen, Henry Cruz, Kevin Haynes, Rainu Kaushal. 2021. "PCORnet® 2020: current state, accomplishments, and future directions." *Journal of Clinical Epidemiology* 129:60–67.

François-Xavier, Blaudin, Claire Baudier, Renan Andrade Pereira, Céline Lefebvre, Philippe Moingeon, Patrimony Working Group. 2023. "Transforming Drug Discovery with a High-Throughput AI-Powered Platform: A 5-Year Experience with Patrimony." *Drug Discovery Today* 28:103772.

Gad, Ahmed G., Diana T. Mosa, Laith Abualigah, and Amr A. Abohany. 2022. "Emerging Trends in Blockchain Technology and Applications: A Review and Outlook." *Journal of King Saud University - Computer and Information Sciences* 34:6719–6742.

Garin, Noe, Raimon Milà, Elena Curto, Marta Hernandez, Conxita Mestres, and Vicente Plaza. 2023. "Clinical Impact of Electronic Monitoring Devices of Inhalers in Adults with Asthma or COPD: A Systematic Review and Meta-Analysis." *Pharmaceuticals* 16:414.

Ghasan, Fahim Huseien, Kwok Wei Shah. 2022. "A Review on 5G Technology for Smart Energy Management and Smart Buildings in Singapore." *Energy and AI* 7:100116.

Goharinejad, Saeideh, Sadrieh Hajesmaeel-Gohari, Nazanin Jannati, Samira Goharinejad, and Kambiz Bahaadinbeigy. 2021. "Review of Systematic Reviews in the Field of Telemedicine." *Medical Journal of the Islamic Republic of Iran* 35:184.

Govern, Paul. 2021. *Artificial Intelligence Calculates Suicide Attempt Risk at VUMC*. Scientific Report, Vanderbilt University.

Gupta, Aditya and Amritpal Singh. 2023. "Healthcare 4.0: Recent Advancements and Futuristic Research Directions." *Wireless Personal Communications* 129:933–952.

Haleem, Abid, Mohd Javaid, Ravi P. Singh, and Rajiv Suman. 2021a. "Telemedicine for Healthcare: Capabilities, Features, Barriers, and Applications." *Sensors International* 2:100117.

Haleem, Abid, Mohd Javaid, Ravi Pratap Singh, Rajiv Suman. 2021b. "Medical 4.0 Technologies for Healthcare: Features, Capabilities, and Applications." *Internet of Things and Cyber-Physical Systems* 2: 12–30.

Han, Yujin, Yawei Zhang, and Sten H. Vermund. 2022. "Blockchain Technology for Electronic Health Records." *International Journal of Environmental Research and Public Health* 19:15577.

Hassan, Mubashir, Faryal M. Awan, Anam Naz, Oscar Alvarez, Ana Cernea, Lucas Fernández-Brillet, Juan L. Fernández-Martínez, and Andrzej Kloczkowski. 2022. "Innovations in Genomics and Big Data Analytics for Personalized Medicine and Health Care: A Review." *International Journal of Molecular Sciences* 23:4645.

Healthcare, Omron. 2022. *CES 2022: OMRON Healthcare Introduces Remote Patient Monitoring Services Globally, Expands Digital Health Services to Aid Consumers on Every Step of Their Heart Health Journey*. Prnewswire.

Islam, Md. R., Md. M. Kabir, Muhammad F. Mridha, Sultan Alfarhood, Mejdl Safran, and Dunren Che. 2023. "Deep Learning-Based IoT System for Remote Monitoring and Early Detection of Health Issues in Real-Time." *Sensors* 13:5204.

Jennifer, Viberg Johansson, Heidi Beate Bentzen & Deborah Mascalzoni. 2022. "What Ethical Approaches Are Used by Scientists When Sharing Health Data? An Interview Study." *BMC Medical Ethics* 23:41.

Joanna, G. Katzman, Laura E Tomedi, Karla Thornton, Paige Menking, Michael Stanton, Nestor Sosa, Michelle Harkins, Neil Katzman, Jinyang Liu, Gaelyn R D Archer, Sanjeev Arora. 2021. "Innovative COVID-19 Programs to Rapidly Serve New Mexico : Project ECHO." *Public Health Report* 136:39–46.

Junaid, Bajwa, Usman Munir, Aditya Nori, and Bryan Williams,. 2021. "Artificial Intelligence in Healthcare: Transforming the Practice of Medicine." *Future Healthcare Journal* 8:e188.

Kelly, Steve. 2022. *VitalConnect Announces Investment from HCA Healthcare to Accelerate Commercial Growth and Create Strategic Partnership*. Scientific Report, Businesswire.

Kronick, Richard. 2016. "AHRQ's Role in Improving Quality, Safety, and Health System Performance." *Public Health Reports* 131:229–232.

Lakmini, Malasinghe, Naeem Ramzan & Keshav Dahal. 2017. "Remote Patient Monitoring: A Comprehensive Study." *Journal of Ambient Intelligence and Humanized Computing* 10:57–76.

Lamichhane, Shrawani, Santosh Bashyal, Taekwang Keum, Gyubin Noh, Jo E. Seo, Rakesh Bastola, Jaewoong Choi, Dong H. Sohn, and Sangkil Lee. 2023. "Complex Formulations, Simple Techniques: Can 3D Printing Technology Be the Midas Touch in Pharmaceutical Industry? "*Asian Journal of Pharmaceutical Sciences* 1:465–479.

Li, Jingshan & Pascale Carayon. 2021. "Health Care 4.0: A Vision for Smart and Connected Health Care." *IISE Transactions on Healthcare Systems Engineering* 11: 171–180.

Lu L, Zhang J, Xie Y, Gao F, Xu S, Wu X, Ye Z. 2020. "Wearable Health Devices in Health Care: Narrative Systematic Review." *JMIR Mhealth Uhealth* 8:e18907.

Majumder, Sumit, Tapas Mondal, M Jamal Deen. 2017. "Wearable Sensors for Remote Health Monitoring." *Sensors (Basel)* 17: 130.

Maksut, Senbekov, Timur Saliev, Zhanar Bukeyeva, Aigul Almabayeva, Marina Zhanaliyeva, Nazym Aitenova, Yerzhan Toishibekov, Ildar Fakhradiyev. 2020. "The Recent Progress and Applications of Digital Technologies in Healthcare: A Review." *International Journal of Telimedicine and Applicationss* 1:8830200.

Marieke, A. R. Bak, M. Corrette Ploem, Hanno L. Tan, M. T. Blom, and Dick L. Willems. 2023. "Towards Trust-Based Governance of Health Data Research." *Medicine, Health Care, and Philosophy* 26:185–200.

Mikalef, Patrick, Maria Boura, George Lekakos, and John Krogstie. 2019. "Big Data Analytics and Firm Performance: Findings from a Mixed-method Approach." *Journal of Business Research* 98:261–276.

Miller, Matthew. 2019. *Polar Announces Ignite GPS Fitness Watch: Advanced Sleep Tracking and Training Guides at an Affordable Price*. Scientific Report, Znet.

Miranda, Rafael, Mónica D. Oliveira, Paulo Nicola, Filipa M. Baptista, and Isabel Albuquerque. 2023. "Towards A Framework for Implementing Remote Patient Monitoring From an Integrated Care Perspective: A Scoping Review." *International Journal of Health Policy and Management* 12:1–13.

Mohapatra, Avaneesh. 2023. "The Implications of IoT in the Modern Healthcare Industry post COVID-19." *International Journal of Smart Sensor and Adhoc Network* 3:7–18.

Moore, Brian and Lan Liang. 2020. *Healthcare Cost and Utilization Project (HCUP) Statistical Briefs*. Rockville: Agency for Healthcare Research and Quality.

Nimmesgern, E, I Benediktsson, and I Norstedt. 2017. "Personalized Medicine in Europe." *Clinical and Translational Science* 10:61–63.

Ntakakis, Georgios, Christina Plomariti, Christos Frantzidis, Panagiotis E. Antoniou, Panagiotis D. Bamidis, and Georgios Tsoulfas. 2023. "Exploring the Use of Virtual Reality in Surgical Education." *World Journal of Transplantation* 13:36–43.

Osmani, F., M. Arab-Zozani, Z. Shahali, and F. Lotfi. 2023. "Evaluation of the Effectiveness of Electronic Prescription in Reducing Medical and Medical Errors." *Annales Pharmaceutiques Francaises* 81:433–445.

Pace, Pasquale, Gianluca Aloi, Raffaele Gravina, Giuseppe Caliciuri, and Giancarlo Fortin. 2018. "An Edge-Based Architecture to Support Efficient Applications for Healthcare Industry 4.0.". *IEEE Transactions on Industrial* 15:481–489.

Patridge, Emily F., and Tania P. Bardyn. 2018. "Research Electronic Data Capture (REDCap)." *Journal of the Medical Library Association* 106:142–144.

Pravin, Shende, and Agrawal Sudhir. 2018. "Integration of 3D Printing with Dosage Forms: A New Perspective for Modern Healthcare." *Biomedicine & Pharmacotherapy* 107:146–154.

Ramasamy, Lakshmana K. and Seifedine Kadry. 2021. "Internet of things." In *Blockchain in the Industrial Internet of Things*, by Anand Sharma, Vishu Madaan Prateek Agrawal, 1–16. Bristol: IOP Publishing Ltd.

Razzak, Muhammad I., Muhammad Imran, and Guandong Xu. 2020. "Big Data Analytics for Preventive Medicine." *Neural Computing & Applications* 32:4417–4451.

Resnik, David B., and Kevin C. Elliott. 2016. "The Ethical Challenges of Socially Responsible Science." *Accountability in Research* 23:31–46.

Rezaei Aghdam A, Watson J, Cliff C, Miah SJ. 2020. "Improving the Theoretical Understanding Toward Patient-Driven Health Care Innovation Through Online Value Cocreation: Systematic Review." *Journal of Medical Internet Research* 22:e16324.

Roberto, Cerchionea, Piera Centobellib, Emanuela Riccioc, Stefano Abbate. Eugenio Oropallo. 2023. "Blockchain's Coming to Hospital to Digitalize Healthcare Services: Designing a Distributed Electronic Health Record Ecosystem." *Technovation* 120:102480.

Samiei, Nasim. 2020. "Recent Trends on Applications of 3D Printing Technology on the Design and Manufacture of Pharmaceutical Oral Formulation: A Mini Review." *Beni-Suef University Journal of Basic and Applied Sciences* 9:1–12.

Sarno, Marina, Annelly Buré-Reyes, Scott Harcourt, Ihtsham Haq, Corneliu Luca, Jonathan Jagid, and Bonnie Levin. 2022. "Success of Home-to-Home Teleneuropsychology (TeleNP) in Deep Brain Stimulation (DBS) Candidacy Assessments: COVID-19 and Beyond." *Parkinsonism & Related Disorders* 98:56–61.

Seh, Adil H., Mohammad Zarour, Mamdouh Alenezi, Amal K. Sarkar, Alka Agrawal, Rajeev Kumar, and Raees A. Khan. 2020. "Healthcare Data Breaches: Insights and Implications." *Healthcare* 8:133.

Serrano, Dolores R., Aytug Kara, Iván Yuste, Francis C. Luciano, Baris Ongoren, Brayan J. Anaya, Gracia Molina. 2023. "3D Printing Technologies in Personalized Medicine, Nanomedicines, and Biopharmaceuticals." *Pharmaceutics* 15:313.

Shi, Shuyun, Debiao He, Li Li, Neeraj Kumar, Muhammad K. Khan, and Kwang R. Choo. 2020. "Applications of Blockchain in Ensuring the Security and Privacy of Electronic Health Record Systems: A Survey." *Computers & Security* 97:101966.

Siddique, Sarkar, James C. L. Chow. 2021. "Machine Learning in Healthcare Communication." *Encyclopedia* 1:220–239.

Singh, Yuvraj, M. A. Jabbar, Shishir Kumar Shandilya, Olena Vovk, and Yaroslav Hnatiuk. 2023. "Exploring Applications of Blockchain in Healthcare: Road Map and Future Directions." *Frontiers in Public Health* 11:1229386.

Suter-Crazzolara, Clemens. 2018. "Better Patient Outcomes Through Mining of Biomedical Big Data." *Frontiers in ICT* 5:409673.
Szigetvári, Gergő and Mesko Bertalan. 2023. "A Review of Technology Giants' Healthcare Collaborations." *mHealth* 9:22–45.
Talal, Andrew H., Elisavet M. Sofikitou, Urmo Jaanimägi, Marija Zeremski, Jonathan N. Tobin, and Marianthi Markatou. 2020. "A Framework for Patient-centered Telemedicine: Application and Lessons Learned from Vulnerable Populations." *Journal of Biomedical Informatics* 112:103622.
Tan, Jamme. 2022. *Developing Wearable Technology in Healthcare: Keys to Success.* Sterling Medical Devices.
Tawseef, Ayoub Shaikh, Tabasum Rasool Dar & Shabir Sofi. 2022. "A Data-Centric Artificial Intelligent and Extended Reality Technology in Smart Healthcare Systems." *Social Network Analysis and Mining* 12:122.
Thimbleby, Harold. 2013. "Technology and the Future of Healthcare." *Journal of Public Health Research* 2:e28.
Tiwari, Biplav B., Aneesh Kulkarni, Hui Zhang, Mahmud M. Khan, and Donglan S. Zhang. 2023. "Utilization of Telehealth Services in Low- and Middle-income Countries amid the COVID-19 Pandemic: A Narrative Summary." *Global Health Action* 16:2179163.
Umer, Muhammad, Saima Sadiq, Hanen Karamti, Walid Karamti, Rizwan Majeed, and Michele Nappi. 2022. "IoT Based Smart Monitoring of Patients' with Acute Heart Failure." *Sensors* 22:2431.
UTHealth. 2020. *COVID-19 Center of Excellence.* Scientific Report, UTHealthHuston.
Vahdat, Shaghayegh, Leila Hamzehgardeshi, Somayeh Hessam, and Zeinab Hamzehgardeshi. 2014. "Patient Involvement in Health Care Decision Making: A Review." *Iranian Red Crescent Medical Journal* 16:e12454.
Valenzuela, Francisco, Armando García, Mabel Vázquez, Joaquín Cortez, and Adolfo Espinoza. 2019. "An IoT-Based Glucose Monitoring Algorithm to Prevent Diabetes Complications." *Applied Sciences* 10:921.
Vanessa, Marcia Vaz, Lalit Kumar. 2021. "3D Printing as a Promising Tool in Personalized Medicine." *AAPS PharmSciTech* 22:49.
Venkatesan, Mythreye, Harini Mohan, Justin R. Ryan, Christian M. Schürch, Garry P. Nolan, David H. Frakes, and Ahmet F. Coskun. 2021. "Virtual and Augmented Reality for Biomedical Applications." *Cell Reports Medicine* 2:100348.
Ventola, C. Lee. 2023. "Medical Applications for 3D Printing: Current and Projected Uses." *Pharmacy and Therapeutics* 39:704–711.
Vora, Lalitkumar K., Amol D. Gholap, Keshava Jetha, Raghu R. Thakur, Hetvi K. Solanki, and Vivek P. Chavda. 2023. "Artificial Intelligence in Pharmaceutical Technology and Drug Delivery Design." *Pharmaceutics* 15:1916.
Wenbin, Sun, Zilong Guo, Zhiqiang Yang, Yizhou Wu, Weixia Lan, Yingjie Liao, Xian Wu and Yuanyuan Liu. 2022. "A Review of Recent Advances in Vital Signals Monitoring of Sports and Health via Flexible Wearable Sensors." *Sensors* 22:7784.
Whooley, Sean. 2023. *Medtronic to Acquire Wearable Insulin Patch Maker EOFlow for $738M*. Scientific Report, Massdevice.
Wilson, Linda. 2019. *MA Patients' Readmission Rates Higher Than Traditional Medicare, Study Finds*. Scientific Report, Healthcaredive.
Zakari, Nazik, Muna Al-Razgan, Amani Alsaadi, Haya Alshareef, Lamia Alashaikh, Mala Alharbi, Rana Alomar, and Seham Alotaibi. 2022. "Blockchain Technology in the Pharmaceutical Industry: A Systematic Review." *PeerJ Computer Science* 8:e840.

Zeeshan, Ahmed, Khalid Mohamed, Saman Zeeshan, and XinQi Dong. 2020. "Artificial Intelligence with Multi-functional Machine Learning Platform Development for Better Healthcare and Precision Medicine." *Database (Oxford)* 2020:baaa010.

Zubair, Ahmad, Shabina Rahim, Maha Zubair, and Jamshid Abdul-Ghafar. 2021. "Artificial Intelligence (AI) in Medicine, Current Applications and Future Role with Special Emphasis on its Potential and Promise in Pathology." *Diagnostic Pathology* 16:24

Chapter 9

Topological structures on digital images

Turyomurugyendo Vicent, Minakshi Biswas Hathiwala, and Kalindi Delawala

9.1 INTRODUCTION

A digital image is a two-dimensional representation of visual information in a computer-readable format, comprising pixels that capture spatial and intensity information from a visual perspective. The applications of digital image processing are vast and encompass fields such as medical equipment, satellite imagery, image enhancement, medical diagnostics, remote sensing, machine and robot vision, pattern recognition, video processing, artificial intelligence, and more.

Images can be categorized into different types based on various criteria, such as content, color representation, file format, and purpose. Here are some common types of images.

- ***Binary Images***: Binary images have only two possible pixel values, typically represented as 0 (white) and 1 (black), as shown in Figure 9.1. They are commonly used for simple representations of shapes or to store masks for image-processing operations.
- ***Grayscale Images***: Grayscale images are composed of various tones of gray, extending from black to white, devoid of any color data (Figure 9.2). Each pixel is denoted by a solitary intensity value, typically an 8-bit scale that spans from 0 (representing black) to 255 (indicating white).
- ***Color Images.*** Color images encompass color details and comprise three or more channels that represent different color components. The color spaces commonly used in digital images are RGB (red, green, blue) and CMYK (cyan, magenta, yellow, key).

The core of computer-based image processing is frequently a picture's qualitative characteristics. To research these digital image properties, a branch of mathematics called topology is considered to be beneficial.

The focus in the traditional setting of classical or general topology, was primarily on studying the topological properties of continuous spaces, which were described using open sets, neighborhoods, and continuous functions. These spaces included real numbers, Euclidean spaces, curves, surfaces, and

 DOI: 10.1201/9781003643166-9

Figure 9.1 Pixels of Binary image.

Figure 9.2 Greyscale image.

other continuous spaces. Nevertheless, the advent of digital image processing necessitated a foundational comprehension of structural image processing procedures, including operations like image thinning, border tracing, contour filling, and object quantification, along with their extensions to images of three or more dimensions. In a digital image, the fundamental space is depicted as a grid of pixels or voxels. Each pixel represents a point within this space, and how the pixels are arranged defines the image's topology. To understand the image's structure, it is crucial to consider the connectivity and adjacency of these pixels. A particular area of topology developed to manage the features and structures in discrete spaces where objects are represented by a finite set of points or pixels since digital images and grids are, by their very nature, discrete. We refer to this particular area of study as discrete or digital topology. Adjacency relations, connectedness, and other topological notions pertinent to these discrete spaces are among the topics it covers. As a result, digital topology is any topology that is applied to digital images.

9.2 TOPOLOGY-BASED IMAGE PROCESSING AND MACHINE LEARNING

Machine learning is an Artificial Intelligence method with the help of which computers can be trained from the available data. The algorithms in machine learning use information directly from the data it has been fed. As the available data samples increase, these algorithms enhance their performance. Perceptron is a machine learning algorithm that performs computations to detect features or patterns in the input data. It led to the theory of computational geometry that deals with the topology of structures [1]. Image processing involves a set of techniques utilized to execute various operations on an image, aiming to produce an improved visual output or to derive valuable insights from it. Image processing plays a crucial role in pattern recognition as it enables the identification of objects within an image. Subsequently, machine learning techniques are employed to train the system to recognize changes within these patterns. The rapid evolution of digital image technologies and their increasing importance in the digital landscape have emphasized the necessity of analyzing and comprehending images. Topology-based image processing involves the examination of the inherent structural characteristics of digital images through the application of mathematical topology. This field of study is concerned with the analysis and processing of digital images by focusing on their qualitative features. By employing principles from topology, one can reveal the intrinsic structural properties that lie within these images.

9.3 MATHEMATICAL BACKGROUND

Definition 9.1 [2]

Let $\mathcal{H}$ be a nonempty set. The metric ϑ on $\mathcal{H}$ is a function that satisfies the properties below:

1. $\vartheta(a,b) = 0 \; iff \; a = b; \vartheta(a,b) \geq 0, a,b \in \mathcal{H};$
2. $\vartheta(a,b) = \vartheta(b,a), a,b \in \mathcal{H};$
3. $\vartheta(a,b) \leq \vartheta(a,\mathscr{g}) + \vartheta(\mathscr{g},b), a,b,\mathscr{g} \in \mathcal{H}.$

Definition 9.2 [2]

Suppose $(\mathcal{H}, \vartheta)$ is a metric space, C is subset of $\mathcal{H}$ and ϑ_c is the metric restricted to points of C then (C, ϑ_c) is a metric space called a subspace of $(\mathcal{H}, \vartheta)$.

Definition 9.3 [2]

Consider $(\mathcal{H}, \vartheta)$ to be a metric space, suppose a_0 and $\daleth$ are positive real numbers, then an open sphere $S_\daleth(a_0)$ centered at a_0 with radius $\daleth$ is a subset of $\mathcal{H}$ if

$$S_\daleth(a_0) = \{a \in \mathcal{H} : \vartheta(a, a_0) < \daleth\}.$$

Definition 9.4 [2]

Assuming that set $\mathcal{H}$ is not empty, the topology τ on $\mathcal{H}$ is defined as the set of subsets of $\mathcal{H}$ called open sets, subject to the following conditions:

1. Both the empty set, $\varnothing$ and the entire set $\mathcal{H}$ are elements of τ.
2. The intersection of several finite numbers of open sets results in an open set.
3. The union of a group of open sets produces an open set

A topological space is therefore defined as the set $(\mathcal{H}, \tau)$ equipped with a topology τ on $\mathcal{H}$. However, when the intersection of an arbitrary collection of open sets results in an open set, this type of topology is denoted as an Alexandroff topology [3].

Definition 9.5 [2]

Consider a topological space $(\mathcal{H}, \tau)$ where $\mathcal{H}$ is a non-empty set. An open set U that contains the element a of $\mathcal{H}$ is referred to as a neighborhood of a.

Definition 9.6 [2]

A topological space $(\mathcal{H}, \tau)$ is considered metrizable when there exists a metric d on $\mathcal{H}$ that generates τ.

Definition 9.7 [2]

Suppose $\mathcal{H}$ is a non-empty set, and $\mathfrak{B}$ denotes the set of subsets for $\mathcal{H}$. We define $\mathfrak{B}$ as a basis (for a topology on $\mathcal{H}$) if the following conditions are satisfied:

1. For every a in $\mathcal{H}$, there is $B \in \mathfrak{B}$ such that a is an element of B.
2. If B_1 and B_2 are elements of $\mathfrak{B}$, then $\exists\ B_3$ in $\mathfrak{B}$ so that B_3 contains their intersection.

Definition 9.8 [2]

The intersection of a subset with each open set from the topological space defines the topology of the subset of the topological space.

Definition 9.9 [2]

The connected topological space cannot be expressed as the union of two separate open sets without overlap.

Definition 9.10 [2]

Consider a topological space $\mathcal{H}$ and a set A. Given a surjective map g: $\mathcal{H} \rightarrow A$, we define a subset $\mho$ of A to be open in A if and only if the preimage $g^{-1}(\mho)$ is open in $\mathcal{H}$. This collection of open sets in A is referred to as the quotient topology induced by g, and g is termed a quotient map. The topological space A is termed as a quotient space.

9.4 TOPOLOGIES ON DIGITAL IMAGES

There are mainly two axiomatic topologies in $\mathbb{Z}^2$ whose connected sets are connected. Both topologies for $\mathbb{Z}^2$ are well known, one is presented in Wyse and Marcus [4], and the second in Khalimsky et al. [5]. By considering the product topology, one can extend the idea of these topologies on $\mathbb{Z}^n$. Rosenfeld [6], applied the graph theoretic-approach to connectedness through adjacency relations while discussing the topology of digital images. All these topologies have found applications in computer graphics and computer vision [7]. Before discussing these topologies, we will recall the adjacency relations and the Jordan Curve Theorem.

9.4.1 Adjacency relations and connectedness

A picture point, pixel, or point sample is a point in a digital image. Let V be the collection of intensity values that adjacency is defined by. For a binary image $V = \{0,1\}$, we consider the adjacent pixels to have value 1. The concept is the same in a grayscale image, however, set V usually has more value. For example, set V may be any subset of 256 values with defined adjacency of pixels having a range of intensity values 0–255 [8, 9]. We consider the following types of adjacencies:

We consider p and q to be pixels with values from V.

1. The pixel p at (x,y) has 4 horizontal/vertical neighbors at $(x+1,y)$, $(x-1,y)$, $(x,y+1)$ and $(x,y-1)$. These neighbors are called the 4-neighbors of p: $N_4(p)$. Pixels p and q are 4-adjacent if q is in the set $N_4(p)$.

2. The pixel p at (x,y) has 4 diagonal neighbors at $(x+1,y+1)$, $(x+1,y-1)$, $(x-1,y+1)$ and $(x-1,y-1)$. These neighbors are the diagonal neighbors of $p : N_D(p)$.
3. The 4 neighbors and the diagonal neighbors of p are called 8-neighbors of $p : N_8(p)$. Pixels p and q are 8-adjacent, if q is in the set $N_8(p)$.
4. Pixels p and q having values are m-adjacent, if q is in $N_4(p)$, or q is in $N_D(p)$ and the set $N_4(p) \cap N_4(q)$ does not have a pixel with values from V.

A graphical path from p to q is a chain of distinct pixels with coordinates $(x_0, y_0),\ldots,(x_n, y_n)$ where pixels (x_i, y_i) and (x_{i-1}, y_{i-1}) are 4, 8, or m-adjacent for $1 \le i \le n$. Here, the length of the path is n. For closed path, $(x_0, y_0) = (x_n, y_n)$.

Consider an image with S as a subset of pixels. Pixels p and q are said to be connected in S if we can find a path between them such that pixels of the path are contained in S again. Figure 9.3 demonstrates 4-connectivity and 8-connectivity. A connected component of S is a set comprising pixels connected to any pixel p of S. The set S is called a connected set if there is only one such component [8].

9.4.2 Jordan curve theorem

A Jordan curve is a closed curve that exhibits no self-intersections and does not intersect itself. Considering the plane $\mathbb{R}^2$, the Jordan curve is defined as the image of an injective continuous map of a circle S into the plane given

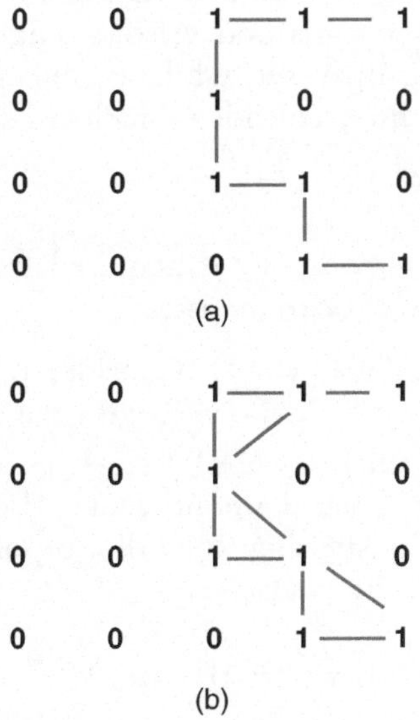

Figure 9.3 (a) 4-Connectivity and (b) 8-connectivity.

by $\varnothing : S \to \mathbb{R}^2$. The Jordan Curve Theorem was first proposed by Camille Jordan in 1887 [10]. It simply states that;

> Let S be a simple closed curve in $\mathbb{R}^2$. Then $\mathbb{R}^2 - S$ consists of two components and S is the boundary in $\mathbb{R}^2$ of each of the components.

This simple-sounding theorem was difficult to prove, however, the first valid proof was given by Oswald Veblen in 1905.

9.4.3 Different topologies on $\mathbb{Z}^2$

9.4.3.1 Marcus Wyse topology

The Marcus Wyse topology [4], often denoted as the M-topology, emerged as a solution to a problem postulated by Dan Marcus. This problem posed the question: "Is it possible to topologise the integers in such a way that the connected sets are the sets of consecutive integers?" The answer to this question, initially proposed by Wyse in 1970, led to the development of a unique topology now known as the Marcus Wyse topology or simply the M-topology. The basis of this topology on $\mathbb{Z}^2$ is generated by the base shown below.

$$U(x,y) = \begin{cases} \{(x,y) \text{ and its } 4-\text{neighbours}\} & \text{if } x+y \text{ is even} \\ \{(x,y)\} & \text{otherwise} \end{cases}$$

A point $P = (x,y)$ is double even if each x and y is even, even if the sum $x+y$ is even and odd if $x+y$ is odd. Also, a singleton containing an odd point is designated as an open set, while a singleton with either a double even point or an even point is considered a closed set.

Proposition 9.1

The digital plane with M-topology is acquired from a quotient space by partitioning $\mathbb{R}^2$ under the taxicab metric.

Proof: Let $x = (x_1, x_2)$ and $y = (y_1, y_2)$. Consider the taxicab metric on $\mathbb{R}^2$ given by $d_T(x,y) = |(x_1 - y_1)| + |(x_2 - y_2)|$, with the corresponding basis elements in the form $B_{d_T}(x,\varepsilon) = \{y \in \mathbb{R}^2, |(x_1 - y_1)| + |(x_2 - y_2)| < \varepsilon\}$. Then the open ball $B_{d_T}(x,\varepsilon)$ is an open diamond centered at x with the distance ε from x to the corners. Considering $\mathbb{R}^2$ with a topology induced by taxicab metric, define $h : \mathbb{R}^2 \to \mathbb{Z}^2$ by

$$h(x,y) = \begin{cases} (a,b) : |(x-a)| + |(y-b)| < 1; a,b \in \mathbb{Z} \text{ and } a+b \text{ is odd} \\ \{(x,y)\} \text{ otherwise, if } x \text{ and } y \text{ are integers} \end{cases}$$

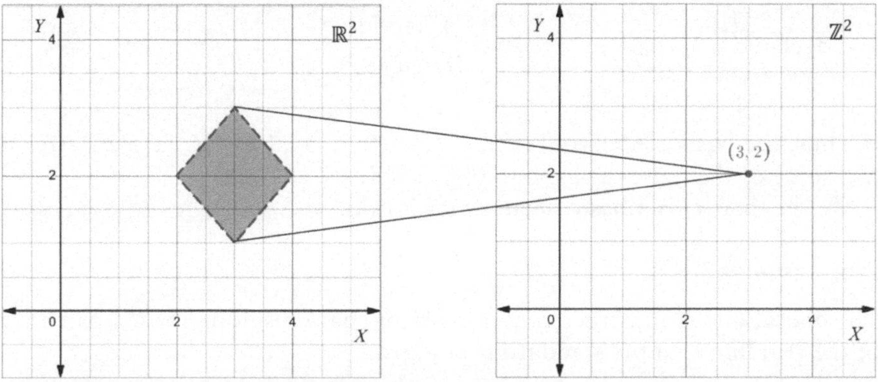

Figure 9.4 The surjective map $h : \mathbb{R}^2 \rightarrow \mathbb{Z}^2$

The open sets in M-topology on $\mathbb{Z}^2$ are the singletons containing an odd point. For, $a, b \in \mathbb{Z}$ with $a + b$ odd, we see that $h^{-1}(a,b) = \{(x,y) : |(x-a)| + |(y-b)| < 1\}$ which is open in $\mathbb{R}^2$ as demonstrated in Figure 9.4. Hence, h is a quotient map. This completes the proof.

If U is used as the base for the open sets, then a set is considered to be connected within the resulting topology when it exhibits 4-connectivity. However, when exploring connectedness in digital images, it is necessary to consider both 4 and 8-neighbors thus, an analog of the Jordan curve theorem was not possible in this case [6].

To overcome this problem, a digital picture is considered to be the plane $\mathbb{Z}^2$ or $\mathbb{Z}^3$ on which both 4- and 8-adjacency are considered. This graph theoretical approach is called Rosenfeld's topology.

9.4.3.2 Rosenfeld topology

Rosenfeld [6] considered a digital picture D to be an arrangement of lattice points having natural number coordinates (x, y). Two points P and Q in S of this lattice are considered continuous if there is a trail from P to Q that consists completely of points of S. This defines connectedness in terms of graphs. A topology could be considered by taking 4-neighbors, which is the same as M-topology on $\mathbb{Z}^2$ given in the previous section.

The notion was now suitable to prove the digital Jordan theorem. However, this definition of a digital picture does not define an axiomatic topology.

9.4.3.3 Khalimsky topology

The formal definition of general topology or axiomatic topology on digital images was provided by Khalimsky et al. [5] is presented below:

Digital Line: The Khalimsky line is a set of integers equipped with the topology produced by the given basis.

$$B(i)=\begin{cases}\{i\} & \text{if } i \text{ is odd}\\ \{i-1,i,i+1\} & \text{if } i \text{ is even}\end{cases}$$

Thus, the basis elements are ..., {−2, −1,0}, {−1}, {0,1, 2}, {1}, {2,3.4}, {3} ...., some of them are depicted in Figure 9.5.

We see that $\{i\}$ is a basic open set if i is odd.

Proposition 9.2

The digital line is constructed as a quotient space resulting from a partition of the real line $\mathbb{R}$ in the standard topology.

Proof: Let $\mathbb{R}$ be the set of real numbers with standard topology. Defined a map $h:\mathbb{R}\to\mathbb{Z}$ [11].

$$h(x)=\begin{cases}x; \text{if } x \text{ is an integer}\\ i; x\in(i-1,i+1), \text{if } i \text{ is odd}\end{cases}$$

The open sets on $\mathbb{R}$ are the open intervals. If i is odd, $h^{-1}\{i\}=(i-1,i+1)$ which is open in $\mathbb{R}$. Thus h is a quotient map, and the quotient topology induced by h results in the digital line on $\mathbb{Z}$. Now, the partition on $\mathbb{R}$ can be described as

$$\{\{x\},\{x,x+2\},x \text{ is an even integer}\}=\{\ldots,\{0\},(0,2),\{2\},\{2,4\},\ldots.\}$$

Note that the M-topology on $\mathbb{Z}$ is the same as Khalimsky's digital line topology.

***Digital plane*:** The Khalimsky plane is the plane $\mathbb{Z}^2$ with topology generated by the basis

$$B(m,n)=\begin{cases}\{(m,n),(m-i,n+j)\},i,j\in\{-1,1\} & \text{if both } m,n \text{ are even}\\ \{(m,n),(m-i,n)\},i\in\{-1,1\} & \text{if } m \text{ is even and } n \text{ is odd}\\ \{(m,n),(m,n+j)\},j\in\{-1,1\} & \text{if } m \text{ is odd and } n \text{ is even}\\ \{(m,n)\}, & \text{Otherwise}\end{cases}$$

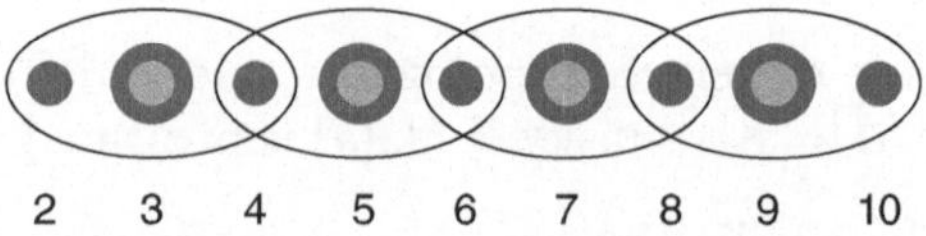

Figure 9.5 Digital line.

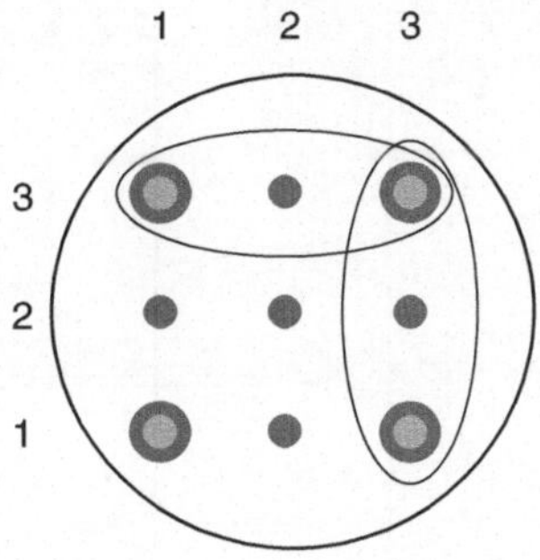

Figure 9.6 Digital plane.

We observe that $B(2,2)=\{(1,1),(1,3)(2,2),(3,1),(3,3)\}$

$$B(2,3)=\{(1,3)(2,3),(3,3)\}$$

$$B(3,2)=\{(3,1),(3,2),(3,3)\}$$

$$B(3,3)=\{(3,3)\}$$

which are a few basic elements depicted in Figure 9.6.

The singletons $\{(m,n)\}$ where both m and n are odd are the basic open sets for this topology.

Note that Khalimsky space is an Alexandroff topological space and is connected.

Proposition 9.3

The concept of digital plane is constructed as a quotient space resulting from partitioning Euclidean plane, under the standard topology.

Proof: Consider $\mathbb{R}^2$ with standard topology. Let us define $\varphi:\mathbb{R}^2\to\mathbb{Z}^2$ as

$$\varphi(\zeta,\kappa)=\begin{cases}\{(\zeta,\kappa)\}:\zeta,\kappa\,\text{are integers}\\(a.b):(\zeta-a)^2+(\kappa-b)^2<2,a,b\,\text{are odd integers}\end{cases}$$

We see that $\varphi:\mathbb{R}^2\to\mathbb{Z}^2$ is a surjective map. The open sets on $\mathbb{R}^2$ with standard topology are the open spheres given by $\{(x,y)\in\mathbb{R}^2:(x-a)^2+(y-b)^2<\varepsilon\}$. The basic open sets of $\mathbb{Z}^2$ with Khalimsky topology are $\{(a,b)\}$ where a and b are both odd integers. We see that $\varphi^{-1}\{(a,b)\}=\{(\zeta,\kappa):(\zeta-a)^2+(\kappa-b)^2<2\}$ which is open in $\mathbb{R}^2$ (see Figure 9.7). This completes the proof.

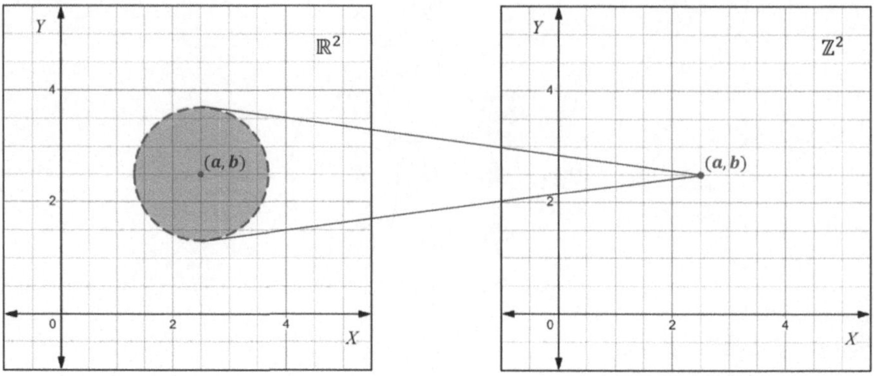

Figure 9.7 The surjective map $\varphi : \mathbb{R}^2 \rightarrow \mathbb{Z}^2$.

9.5 COMPARATIVE ANALYSIS

9.5.1 Advantages of Khalimsky topology over Rosenfeld topology

Both Khalimsky Topology and Rosenfeld Topology have their advantages and applications in the context of digital images. Here are some potential advantages of Khalimsky Topology over Rosenfeld Topology:

1. Limitation on Rosenfeld Topology: The Khalimsky topology addresses a limitation of the Rosenfeld topology, which is based solely on 4-adjacency and 8-adjacency relations. These relations capture local connectivity between pixels but do not fully capture the global properties of shapes and curves in the digital plane.
2. Alexandroff Topology: The Khalimsky topology introduces a unique approach to structuring the digital plane $\mathbb{Z}^2$, based on an Alexandroff topology. This topology allows for the identification and study of Jordan curves in the digital domain, which is essential for understanding the behavior of closed curves and their relationship with the division of the plane into distinct regions.
3. Analogue of Euclidean space: By incorporating the concept of Jordan curves, the Khalimsky topology provides a more robust and meaningful representation of shapes and boundaries in digital images. It enables the study of fundamental geometric properties, such as connectivity, separation, and the division of the plane into components, which are analogs of properties in the Euclidean plane.
4. Explicit Representation of Open and Closed Sets: Khalimsky topology provides a clear representation of open and closed sets in the context of digital images. The identification of odd numbers with open points and even numbers with closed points allows for a direct understanding of the topology and its properties.

5. Shape Analysis and Structure Identification: Khalimsky topology facilitates the analysis of shape properties, such as connectivity and boundaries, in digital images. By representing connected components and boundary points, it becomes easier to study the shape and structure of objects within the image.
6. Intuitive Partitioning: Khalimsky topology partitions the standard line $\mathbb{R}$ into intervals corresponding to odd and even integers. This partitioning allows for a natural representation of the topology and simplifies the understanding of relationships between elements in the image.
7. Continuity in Digital Topological Spaces: Khalimsky topology extends the concept of continuous functions to discrete spaces, providing a framework for studying continuous transformations and mappings in the context of digital images. This can be beneficial in tasks involving image processing and manipulation.

9.5.2 Disadvantages of Khalimsky topology over Rosenfeld topology

Khalimsky topology has some disadvantages compared to Rosenfeld topology in analyzing digital images. Here are a few possible disadvantages of Khalimsky topology:

1. Limited Connectivity Representation: Khalimsky topology focuses on the connectivity between adjacent odd and even integers, which may not fully capture the complex connectivity patterns in digital images. In certain cases, such as when analyzing highly connected regions or objects with intricate structures, the representation provided by Khalimsky Topology might be too simplistic.
2. Lack of Flexibility in Adjacency Definitions: Khalimsky Topology relies on the adjacency of odd and even integers, which might not align with the specific adjacency definitions required for certain image analysis tasks. Different adjacency criteria, such as 4-connectivity or 8-connectivity in Rosenfeld Topology, allow for more flexibility in capturing the desired relationships between pixels of image regions.
3. Difficulty in Handling Continuous Grayscale Images: Khalimsky Topology is primarily designed for binary or discrete images, where pixels have only two possible values (black and white). When dealing with continuous grayscale images, it might be challenging to directly apply Khalimsky Topology, as it is based on partitioning intervals corresponding to odd and even integers. Additional preprocessing steps or adaptations might be necessary to handle grayscale intensity variations.
4. Complex Representation of Open and Closed Sets: While Khalimsky Topology provides an explicit representation of open and closed sets, interpreting these sets might be more intricate than Rosenfeld

Topology. The mapping of odd and even integers to open and closed points can introduce additional complexity, especially when dealing with more advanced topological concepts or when translating results from Khalimsky Topology to other topological frameworks.

5. Limited Literature and Tool Support: Rosenfeld Topology has been widely studied and applied in the field of digital image analysis, resulting in a larger body of literature, established methodologies, and available tools. Khalimsky Topology, on the other hand, might have relatively fewer resources and less established support in terms of algorithms, software libraries, or existing research applications.

It's important to note that the choice between Khalimsky Topology and Rosenfeld Topology depends on the specific requirements of the application and the nature of the digital image being analyzed. Both topologies have their strengths and may be more suitable in different scenarios.

Marcus Wyse topology is less explored than Khalimsky and Rosenfeld topology in the context of properties of digital image processing, however, the concept of semi-topological properties in M-topology are studied [12] and has a wide scope for research from a mathematician's point of view.

9.6 APPLICATION IN THE REALM OF IMAGE PROCESSING

The digital binary image is an input to the computer with some brightness values at a discrete grid of points. These points are called pixels which combine to give a digital image. Digital images are easy to store, share, and operate. Image segmentation and image compression are important notions in image processing that have wide applications. Below, we present the brief role of the digital topology in the discussion of these concepts.

9.6.1 Image segmentation

The process of decomposing an image into regions or distinguishing between objects and the background is known as segmentation. Segmentation primarily involves the classification of pixels into different groups. Let's consider the entire spatial region R covered by the image. Image segmentation can be described as the process of dividing this region into sub-regions, denoted as $R_1, R_2, \ldots$

1. If the subregions are combined, the original region is obtained. Mathematically, $R = \bigcup_i R_i$.
2. R_i is a connected set under some adjacency.
3. The subregions R_i do not share any common region. Mathematically, $R_i \cap R_j = \varnothing \text{for all} i, j, \text{and} i \neq j$.

4. Each region satisfies a property P. Mathematically, $P(R_i) = true$.
5. For any adjacent regions R_i and R_j, $P(R_i \cup R_j) = false$.

Condition 1) emphasizes the necessity for completeness in segmentation, ensuring that each pixel is allocated to a specific region. Condition 2) dictates that points within a region should exhibit prescribed connectivity, such as adhering to 4- or 8-connectivity rules as outlined in Section 9.4. Condition 3) underscores the importance of ensuring that these regions remain separate and non-overlapping. Condition 4) pertains to the properties that pixels within a segmented region should exhibit, for instance, requiring uniform intensity levels among all pixels in the segmented area. Lastly, condition 5) signifies the need for differentiation between two adjacent regions, R_i and R_j, based on specific property P criteria. Figure 9.8 demonstrates image segmentation.

Our survey suggests that the primary challenge in image segmentation lies in dividing the image into regions that conform to the conditions mentioned above. The core issue in segmentation is to partition an image into regions that adhere to the specified conditions. The concept of connectedness is therefore vital while discussing segmentation. Connectedness represents a crucial topological attribute that any structure in the digital space should demonstrate. After isolating a subset, such as through a thresholding process, the natural progression is to further divide it into interconnected regions. Given that the boundaries encompass the entirety of the selected area, it's essential to monitor and maintain a record of these boundaries. Moreover, one can also shift focus toward the sparse regions without

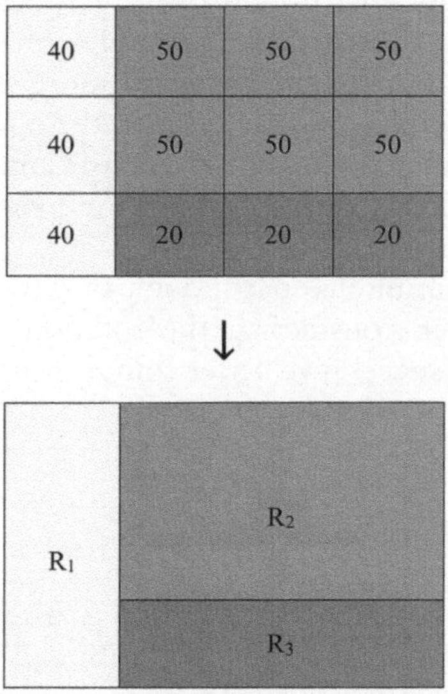

Figure 9.8 Image segmentation of R into R_1, $R_{2,}$ and R_3.

Table 9.1 Comparison of different maps on Khalimsky topological paces

Various Maps	*Rotation of Jordan curve*	*Parallel translation*	*Preservation of connectedness property*	*Valid for constant map*
Khalimsky Continuous map	Not always possible	Not always possible	True	Valid
Khalimsky adjacency map	Always possible	Always possible	True	Not valid
A-map	Always possible	Always possible	True	Valid

compromising the overall connectivity aspect. This scenario is where the significance of topology becomes evident.

The utilization of the Rosenfeld topology ensures that the contours obtained are closed, which significantly enhances the segmentation outcome. In contrast, the Khalimsky topology doesn't inherently produce closed contours. Therefore, opting for the Rosenfeld topology leads to superior segmentation results [13].

9.6.2 Image compression

Multimedia content typically consumes a substantial amount of storage space. Image compression serves the purpose of reducing the volume of data necessary to represent an object by eliminating extraneous details within the image. This technology is pivotal in the realm of digital image processing, being widely acclaimed for its practicality and commercial success.

Image compression methods usually operate by either discarding extraneous bytes of data from the image or by applying specialized compression algorithms to transform the image file into a more storage-efficient format. This practice is commonplace, particularly in the optimization of web page images and high-resolution digital photographs. Compressing images proves invaluable in ensuring swift loading times for images when users engage with websites or applications.

Using Khalimsky continuous maps, one can discuss the compression of digital images; however, considering the rotation of images, A-maps on Khalimsky topological spaces give better compression of images as shown in Table 9.1 [14].

9.7 CONCLUSION

Regarding the mathematical characterization of a set $\mathbb{Z}^n$ with specific topological properties or adjacency relations within $\mathbb{Z}^n$, digital topology has assumed a pivotal role in various fields such as computer graphics, image synthesis, image analysis, and network science. Its origin lies in the extension of discrete geometry into practical applications where notable topological challenges emerge. Digital topology is valuable not only for computer scientists seeking to employ topological concepts to explore digital environments but also for mathematicians looking to harness computational tools for tackling intricate topological problems.

REFERENCES

[1] Minsky, M., and S. Papert. 1969. Perceptron: an introduction to computational geometry. *The MIT Press, Cambridge, expanded edition* 19, no. 88: 2.
[2] Munkres, J. R. 2000. *Topology: international edition*. Pearson Prentice Hall.
[3] Arenas, F. G. 1999. Alexandroff spaces. *Acta Math. Univ. Comenianae* 68, no. 1: 17–25.
[4] Marcus, D., and Cleveland State University Problem Solving Group. 1970. 5712. *The American Mathematical Monthly* 77, no. 10: 1119. https://doi.org/10.2307/2316121
[5] Khalimsky, E., R. Kopperman, and P. R. Meyer. 1990. Computer graphics and connected topologies on finite ordered sets. *Topology and its Applications* 36, no. 1: 1–17.
[6] Rosenfeld, A. 1979. Digital Topology. *The American Mathematical Monthly* 86, no. 8: 621–630. doi:10.1080/00029890.1979.11994873
[7] Kong, T. Y., and A. Rosenfeld.1989. Digital topology: Introduction and survey. *Computer Vision, Graphics, and Image Processing* 48, no. 3: 357–393.
[8] Gonzalez, R. C. 2009. *Digital image processing*. Pearson Education India.
[9] Sridhar S. 2011. *Digital image processing*. Oxford University Press, New Delhi.
[10] Jordan, C. 1887. Cours d'analyse de l'École polytechnique. *Gauthier-Villars, Paris* 3: 587–594.
[11] Adams, C., and R. D. Franzosa. 2009. *Introduction to topology: Pure and applied*. Pearson India Education Services Pvt. Ltd.
[12] Han, S. 2022. Semi-topological properties of the Marcus-Wyse topological spaces. *AIMS Math* 7: 12742–12759.
[13] Al Hajri, M., K. Belaid, and L. J. Belaid. 2015. On khalimsky topology and applications on the digital image segmentation. *Applied Mathematical Sciences* 9, no. 74: 3687–3701.
[14] Han, S., and A. Sostak. 2013. A compression of digital images derived from a Khalimsky topological structure. *Computational and Applied Mathematics* 32, no. 3: 521–536.

Chapter 10

Industry 5.0 and social progress

Navigating technological advancements for inclusive growth

Sushma Tiwari and Tapesh Chandra Gupta

10.1 INTRODUCTION

Industry 5.0 represents a visionary evolution in manufacturing and industry aimed at achieving several ambitious global objectives. At its core, it seeks to redefine the relationship between humans and technology, emphasizing collaborative and symbiotic interactions where advanced automation and AI augment human creativity, decision-making, and problem-solving abilities. This paradigm shift not only enhances productivity and efficiency but also fosters a dynamic and resilient workforce capable of adapting to rapid technological advancements. Moreover, Industry 5.0 champions sustainability as a cornerstone, advocating for eco-friendly practices, resource optimization, and ethical standards across global supply chains. By prioritizing these principles, it aims to mitigate environmental impact and promote responsible consumption patterns, contributing to a sustainable future for generations to come. Furthermore, Industry 5.0 envisions global competitiveness through agile and adaptable manufacturing systems that cater to personalized and customized consumer demands, thereby driving economic growth and prosperity on a worldwide scale. Ultimately, its overarching vision is to establish a human-centric, innovative, and sustainable industrial landscape that not only meets current societal needs but also anticipates and addresses future challenges proactively.

The industrial landscape has undergone significant transformations over the centuries, marked by pivotal revolutions that have reshaped economies and societies. From the advent of mechanization in the 18th century to the digital revolution of the 21st century, each industrial revolution has introduced groundbreaking technologies and concepts. The journey began with Industry 1.0, characterized by the introduction of steam power and mechanized production in the late 18th century. Industry 2.0 followed in the late 19th century, driven by electricity, mass production, and assembly lines. The late 20th century witnessed the rise of Industry 3.0, which brought computers, automation, and electronics into manufacturing processes, significantly enhancing efficiency and productivity.

DOI: 10.1201/9781003643166-10

The advent of the 21st century ushered in Industry 4.0, also known as the Fourth Industrial Revolution. This phase is marked by the integration of digital technologies, such as the Internet of Things (IoT), artificial intelligence (AI), big data, and cloud computing, into industrial processes. These advancements have enabled the creation of smart factories, where interconnected systems and real-time data analytics drive unprecedented levels of automation, efficiency, and customization.

Building upon the foundations of Industry 4.0, Industry 5.0 emphasizes the collaboration between humans and machines. While Industry 4.0 focused on automating processes and reducing human intervention, Industry 5.0 reintroduces the human element into the equation, fostering a symbiotic relationship between advanced technologies and human creativity, intelligence, and skills. In India, the impact of Industry 5.0 is already evident across various sectors, creating new job opportunities and transforming existing roles. Here are some examples:

Manufacturing Sector

- Cobots and Automation: Indian manufacturing companies are increasingly adopting collaborative robots (cobots) for tasks like assembly, packaging, and inspection. This has created roles for robot operators, maintenance technicians, and integration specialists.
- Smart Factories: The implementation of IoT and AI in manufacturing leads to demand for professionals skilled in IoT integration, AI algorithms, and data analytics.
- Information Technology and Software Development.
- AI and Machine Learning: There is a growing demand for AI specialists, data scientists, and machine learning engineers to develop and maintain AI-driven applications.
- Cybersecurity: As more industries adopt digital technologies, the need for cybersecurity experts to protect against cyber threats has increased.

Healthcare Sector

- Telemedicine and AI Diagnostics: The use of AI in diagnostics and telemedicine has created roles for IT professionals, data analysts, and AI developers in healthcare settings.
- Robotic Surgery: Surgeons and medical professionals are being trained to use advanced robotic systems, creating jobs in medical technology training and support.
- Agriculture.
- Precision Farming: The adoption of precision farming techniques using IoT and AI has led to roles in agricultural technology, data analysis, and farm management.
- Drones and Robotics: The use of drones for monitoring and managing crops creates opportunities for drone operators, maintenance technicians, and data analysts.

Renewable Energy and Sustainability

- Green Technologies: As India pushes toward renewable energy sources, there is an increasing need for professionals in solar, wind, and other renewable energy sectors.
- Sustainability Consulting: Roles in sustainability consulting and environmental compliance are growing as companies strive to meet environmental standards and reduce their carbon footprint.

Education and Training

- Skill Development: With the rise of Industry 5.0, there is a significant need for training programs to upskill the workforce. This creates opportunities for trainers, educators, and curriculum developers who specialize in new technologies and industrial processes.
- Research and Development.
- Innovation Hubs: India is home to several innovation hubs and tech parks focusing on developing new technologies. Roles in R&D, product development, and innovation management are on the rise.

Objectives of Industry 5.0 in India:

Enhanced Industrial Productivity: By integrating advanced technologies like Artificial Intelligence (AI), Internet of Things (IoT), and robotics, Industry 5.0 aims to modernize manufacturing processes. This modernization includes automation of repetitive tasks, real-time monitoring of production lines, and predictive maintenance of equipment. These advancements lead to increased efficiency, reduced operational costs, and higher output quality, thereby enhancing overall industrial productivity.

Skill Development and Empowerment: Industry 5.0 emphasizes the importance of skill development to empower the workforce. Initiatives focus on training programs that equip workers with technical skills to operate and maintain advanced technologies. Moreover, there's a push for fostering a culture of innovation and entrepreneurship, encouraging workers to contribute ideas and solutions that drive continuous improvement and competitiveness within the industry.

Sustainability and Environmental Responsibility: In response to global environmental challenges, Industry 5.0 promotes sustainable practices within manufacturing. This includes optimizing resource use (such as energy and raw materials), minimizing waste generation, and adopting eco-friendly technologies. By implementing sustainable manufacturing practices, India aims to reduce its environmental footprint while ensuring the long-term viability of industrial operations.

Inclusive Growth: Industry 5.0 initiatives prioritize inclusive growth by ensuring that benefits extend to all regions and sectors of society. This involves promoting equitable access to technological advancements across urban and rural areas, as well as supporting small and medium enterprises (SMEs) in adopting Industry 5.0 technologies. By fostering a diverse and inclusive industrial ecosystem, India seeks to reduce regional disparities and create opportunities for socio-economic development across the country.

Global Competitiveness: To enhance global competitiveness, Industry 5.0 focuses on positioning India as a leader in high-tech manufacturing and research. This includes investing in research and development (R&D) infrastructure, promoting collaboration between industry and academia, and attracting foreign direct investment (FDI) in cutting-edge technologies. By fostering innovation and adopting international quality standards, India aims to strengthen its export capabilities and attract global markets for its manufactured goods.

Job Creation: Industry 5.0 initiatives are expected to generate new employment opportunities across various skill levels. While automation may replace some traditional jobs, it also creates demand for new roles in areas such as data analytics, cybersecurity, and advanced manufacturing technologies. By supporting workforce development and entrepreneurship, India aims to capitalize on the job creation potential of Industry 5.0 to absorb its growing workforce and reduce unemployment.

Promotion of Ethical Practices: Ethical considerations are integral to Industry 5.0 in India, encompassing fair labor practices, responsible sourcing of materials, and ethical use of AI and data. This involves implementing regulations and standards that uphold ethical principles throughout the supply chain, ensuring transparency and accountability in industrial operations.

Infrastructure Development: To support the implementation of Industry 5.0 technologies, India is investing in modern infrastructure such as technology parks, industrial clusters, and digital connectivity. These infrastructure developments provide the necessary physical and digital platforms for companies to adopt and scale advanced manufacturing processes effectively.

Digital Connectivity: Improving digital connectivity across India is crucial for the success of Industry 5.0 initiatives. This involves expanding broadband networks, enhancing internet speeds, and ensuring reliable connectivity in both urban centers and rural areas. Improved digital infrastructure facilitates real-time data exchange, remote monitoring of production facilities, and integration of IoT devices, enabling seamless operation and management of Industry 5.0 systems.

Entrepreneurship and Startups: Industry 5.0 encourages entrepreneurship and supports startups that focus on developing innovative technologies and solutions for manufacturing. Government initiatives include providing incubation support, funding opportunities, and access to mentorship networks to nurture a thriving startup ecosystem. By promoting entrepreneurship, India aims to foster a culture of innovation and accelerate the adoption of Industry 5.0 technologies among new and emerging businesses.

Regulatory Framework: Establishing a supportive regulatory framework is essential for the adoption and deployment of Industry 5.0 technologies in India. This framework addresses issues related to data privacy, cybersecurity, intellectual property rights, and compliance with international standards. Clear and predictable regulations create a conducive environment for investment, innovation, and sustainable growth in the manufacturing sector.

Sector-specific Initiatives: Tailoring Industry 5.0 strategies to specific sectors such as automotive, electronics, pharmaceuticals, and textiles allows India to address unique industry challenges and opportunities. Sector-specific initiatives include R&D collaborations, technology pilot projects, and skill development programs tailored to the specific needs and requirements of each industry. By focusing on sectoral strengths and capabilities, India can leverage Industry 5.0 to drive innovation, enhance productivity, and strengthen its global leadership in key industrial sectors.

Public-Private Partnerships (PPPs): Collaborative efforts between the government, private sector, and academia through PPPs play a crucial role in advancing Industry 5.0 in India. These partnerships facilitate knowledge sharing, resource pooling, and joint investments in infrastructure, technology development, and skill enhancement. By leveraging combined expertise and resources, PPPs accelerate the adoption of Industry 5.0 technologies and promote sustainable industrial development across different regions and sectors.

These detailed objectives illustrate India's strategic approach to harnessing Industry 5.0 technologies for sustainable economic growth, technological advancement, and societal well-being.

Industry 5.0 is creating a diverse range of job opportunities across various sectors in India. By emphasizing the collaboration between advanced technologies and human skills, Industry 5.0 is not only enhancing productivity and efficiency but also driving innovation and creating new roles that blend technological expertise with human-centric skills. TCS has been at the forefront of adopting AI, IoT, and other Industry 5.0 technologies, creating numerous job opportunities in these fields. Mahindra & Mahindra has integrated IoT and AI in its manufacturing processes, leading to new roles in smart manufacturing and data analytics. Infosys has been investing in AI and machine learning, creating a demand for skilled professionals in these areas. Siemens is involved in creating smart factory solutions, which has led to job creation in automation, AI, and IoT.

Make in India initiative aims to transform India into a global manufacturing hub, encouraging the adoption of advanced manufacturing technologies and creating jobs in these sectors. The Skill India initiative focuses on upskilling the workforce to meet the demands of Industry 5.0, creating opportunities for trainers and educators. Industry 5.0 is gradually transforming the job landscape in India, fostering the creation of roles that blend technological expertise with human-centric skills. The emphasis on advanced technologies, coupled with India's push toward digital transformation and sustainability, is driving this evolution.

Industry 5.0, also known as the Fifth Industrial Revolution, represents a significant shift in the way we approach manufacturing and production processes. It aims to harmonize technological advancements with human intelligence and sustainability principles. This revolution is expected to have a profound impact on the job landscape, leading to the emergence of new roles and the transformation of existing ones. Here's an analysis of the evolving job landscape engendered by Industry 5.0:

It's important to note that while some traditional manufacturing roles may become obsolete or transformed, Industry 5.0 is also expected to create new employment opportunities in various sectors, such as sustainable product design, smart logistics, and customized manufacturing. Additionally, the emphasis on human-machine collaboration and the integration of technology with human intelligence may lead to the creation of entirely new job roles that are currently difficult to anticipate.

Main Branches (Top Level Job Sectors):

Figure 10.1 identifies three primary sectors where Industry 5.0 is driving job creation:

Manufacturing Sector
Information Technology and Software Development
Healthcare Sector

1. **Manufacturing Sector**
   Job opportunities arise from automation and smart innovations:
   Cobots and Automation – Collaborative robots working alongside humans.
   Smart Factories – Digitally integrated manufacturing units.
   Agriculture – Integrated under manufacturing, showing a crossover:
   Precision Farming – Use of data and tech for efficient farming.
   Drones and Robotics – Advanced equipment for field operations.
2. **Information Technology and Software Development**
   This sector supports nearly all others with tech innovation:
   AI and Machine Learning – Building intelligent systems.
   Cybersecurity – Protecting digital infrastructure.
   Renewable Energy and Sustainability – Jobs in:
   Green Technologies – Eco-friendly innovations.
   Sustainability Consulting – Advising businesses on sustainable practices.

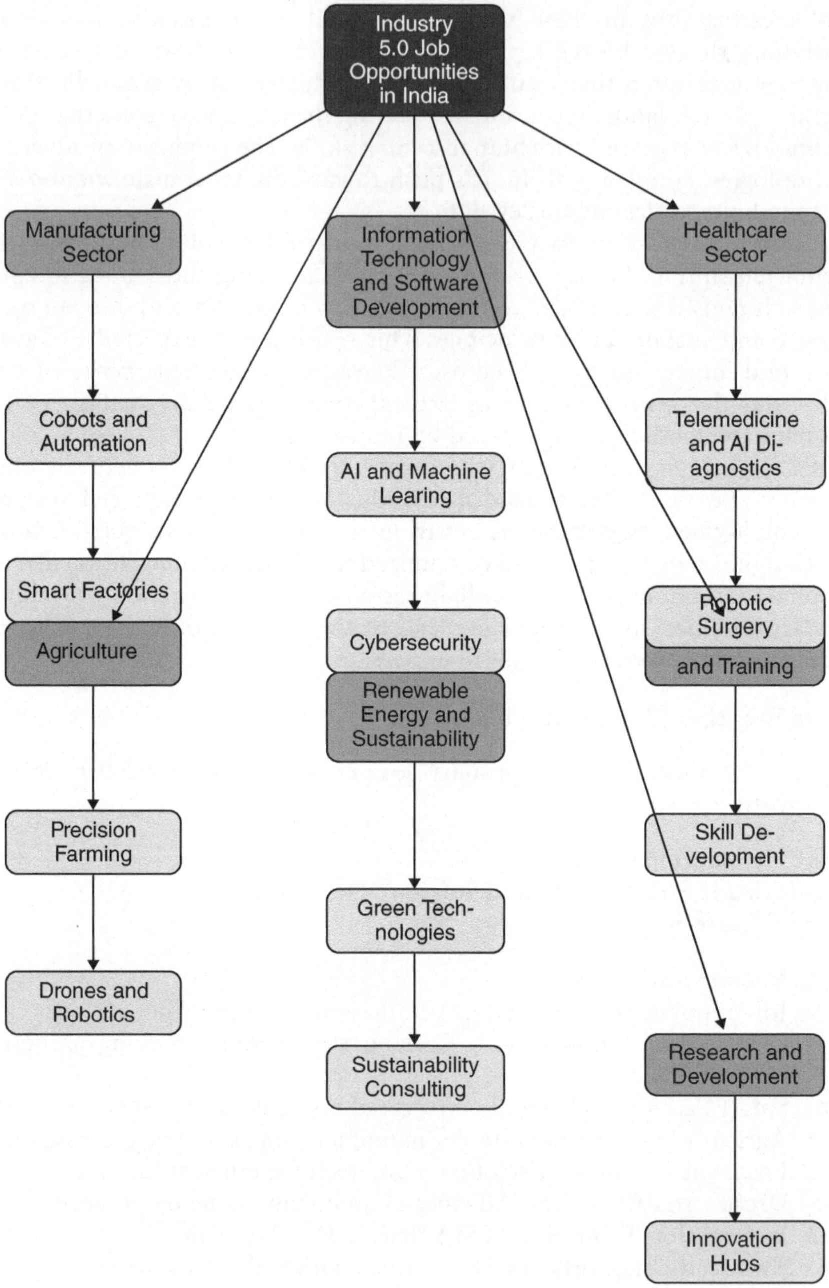

Figure 10.1 Industry 5.0 and job opportunities in India.

3. **Healthcare Sector**
   Technology enhances medical services and job roles:
   Telemedicine and AI Diagnostics – Remote care using AI.
   Robotic Surgery and Training – Tech-assisted surgical jobs.
   Skill Development – Continuous upskilling in health tech.
   Research and Development – Innovation-driven roles.

Innovation Hubs – Collaborative spaces for healthcare advancements.

10.2 LITERATURE REVIEW

10.2.1 Blockchain for supply chain transparency

Blockchain technology has shown significant potential in enhancing supply chain transparency and traceability. Agrawal et al. (2021) present a blockchain-based framework for the textile and clothing industry, emphasizing the technology's role in improving traceability and reducing fraud. Bai et al. (2022) further explore blockchain's enablers in the African cocoa industry, demonstrating how blockchain can enhance sustainable supply chain practices by ensuring product authenticity and ethical sourcing.

10.2.2 Digital twins in manufacturing

Digital twin technology is another digital innovation contributing to sustainable manufacturing practices. Alam et al. (2023) develop a digital twin framework for the apparel manufacturing industry, which aids in optimizing production processes and reducing waste. Attaran et al. (2023) discuss the broader benefits and challenges of digital twins, including improved operational efficiency and predictive maintenance capabilities.

10.2.3 Industry 4.0 and sustainable manufacturing

Industry 4.0 encompasses various digital technologies, including IoT, AI, and advanced robotics, which collectively enhance sustainable manufacturing practices. Ching et al. (2022) provide a comprehensive review of Industry 4.0 applications for sustainable manufacturing, proposing a roadmap for future research and implementation. Similarly, Chauhan et al. (2022) link circular economy principles with digitalization technologies, highlighting past achievements and future promises in creating closed-loop systems that minimize waste and enhance resource efficiency.

10.2.4 Human-centric approach in Industry 5.0

The shift from Industry 4.0 to Industry 5.0 emphasizes the importance of human-centric design in technology and factory operations. Longo,

Padovano, and Umbrello (2020) discuss value-oriented and ethical technology engineering, highlighting how Industry 5.0 aims to create factories that prioritize human well-being and ethical considerations in design and operation. Nahavandi (2019) further elaborates on the human-centric solutions in Industry 5.0, focusing on the integration of human creativity and craftsmanship with advanced technologies to enhance productivity and job satisfaction.

10.2.5 Advanced technologies and Industry 5.0

Industry 5.0 leverages cutting-edge technologies to augment human capabilities and create more efficient, customized production processes. Sachsenmeier (2016) explores the relevance of bionics and synthetic biology in Industry 5.0, indicating how these technologies can lead to innovations in manufacturing and product design. ElFar et al. (2021) discuss the prospects of Industry 5.0 in algae bioenergy production, demonstrating the customization of production processes and the use of advanced technologies for clean energy generation.

10.2.6 Integration of AI, IoT, and big data

The convergence of AI, IoT, and big data analytics is a cornerstone of Industry 5.0, enabling smarter and more responsive industrial systems. Ozdemir and Hekim (2018) describe how these technologies collectively drive Industry 5.0, making sense of vast amounts of data and enhancing decision-making processes. Al Faruqi (2019) discusses future services in Industry 5.0, emphasizing the role of AI and IoT in providing intelligent and adaptive industrial services.

10.2.7 Ethical and social implications

The ethical and social implications of Industry 5.0 are critical areas of research, particularly concerning the balance between technological advancement and human values. Longo et al. (2020) advocate for ethical considerations in technology design, ensuring that advancements benefit society as a whole and do not exacerbate inequalities. The human-centric approach of Industry 5.0 also involves addressing concerns about job displacement and ensuring that technology augments rather than replaces human labor.

The integration of digital technologies with sustainability practices holds immense promise for transforming manufacturing and supply chains. AR, blockchain, digital twins, and Industry 4.0 technologies offer innovative solutions to traditional sustainability challenges, driving efficiency, transparency, and environmental stewardship. Future research should continue to explore these intersections, providing deeper insights and practical frameworks for widespread adoption and implementation. Industry 5.0

represents a significant evolution in industrial practice, emphasizing the integration of advanced technologies with a human-centric approach. By prioritizing ethical considerations and leveraging AI, IoT, and big data, Industry 5.0 aims to create more sustainable, efficient, and human-friendly industrial systems. Future research should continue to explore the balance between technological advancements and human values, ensuring that the benefits of Industry 5.0 are widely distributed and ethically sound.

10.3 OBJECTIVES

- Analyze the evolving job landscape engendered by Industry 5.0.
- Assess the efficacy of skill development initiatives tailored to Industry 5.0.
- Evaluate the qualitative enhancements in products and services attributable to Industry 5.0.

10.4 DATA ANALYSIS AND INTERPRETATION

10.4.1 Emerging Roles and Compensation in the Digital Workforce

The rapid evolution of digital technologies has led to the emergence of several new roles across industries, particularly in manufacturing, technology, finance, and IT services. These roles demand specialized skills and offer competitive salaries, reflecting their strategic importance in the modern economy. Table 10.1 illustrates a detailed overview of ten such emerging job titles, highlighting the number of companies reporting each role, the average salary offered in INR, the most common industry employing the role, the percentage of the workforce engaged in these positions, and the key skills required.

Table 10.1 highlights the emerging job roles in Industry 5.0, their prevalence, average salaries, and the key industries and skills associated with each role. The emerging job titles in Industry 5.0 reflect a significant shift toward integrating advanced technologies with human-centric approaches. Key roles such as Human-Machine Interface Specialist, AI Ethics Officer, and Data Privacy Manager are prominent in manufacturing, technology, and finance, with average salaries ranging from INR 6,960,000 to INR 9,760,000. These positions emphasize skills in human-computer interaction, AI ethics, data security, and blockchain technology. The rise of Digital Twin Engineers and Robotics Process Automation Specialists in the automotive and manufacturing sectors highlights the importance of IoT, simulation, and process automation. Additionally, roles like Cyber-Physical Systems Analyst and Sustainable Manufacturing Coordinator focus on systems

Table 10.1 Emerging job titles and their descriptions in Industry 5.0

Job title	*Number of companies reporting role*	*Average salary (INR)*	*Most common industry*	*Percentage of workforce*	*Key skills required*
Human-Machine Interface Specialist	145	7,600,000	Manufacturing	1.8%	Human-Computer Interaction, UX Design
AI Ethics Officer	110	9,200,000	Technology	1.2%	AI Ethics, Compliance, Legal Knowledge
Data Privacy Manager	190	7,760,000	Finance	2.8%	Data Security, Privacy Regulations
Digital Twin Engineer	135	8,160,000	Automotive	2.4%	IoT, Simulation, Systems Engineering
Robotics Process Automation Specialist	165	8,640,000	Manufacturing	3.3%	Robotics Programming, Process Automation
Cyber-Physical Systems Analyst	125	8,000,000	Aerospace	2.1%	Systems Integration, Cybersecurity
Augmented Reality Developer	115	6,960,000	Entertainment	1.1%	AR Development, 3D Modeling
Blockchain Solutions Architect	105	9,760,000	Finance	0.9%	Blockchain Technology, Cryptography
Sustainable Manufacturing Coordinator	95	7,440,000	Manufacturing	1.6%	Sustainability Practices, Environmental Science
Remote Operations Manager	175	7,200,000	IT Services	2.6%	Remote Monitoring, Digital Tools

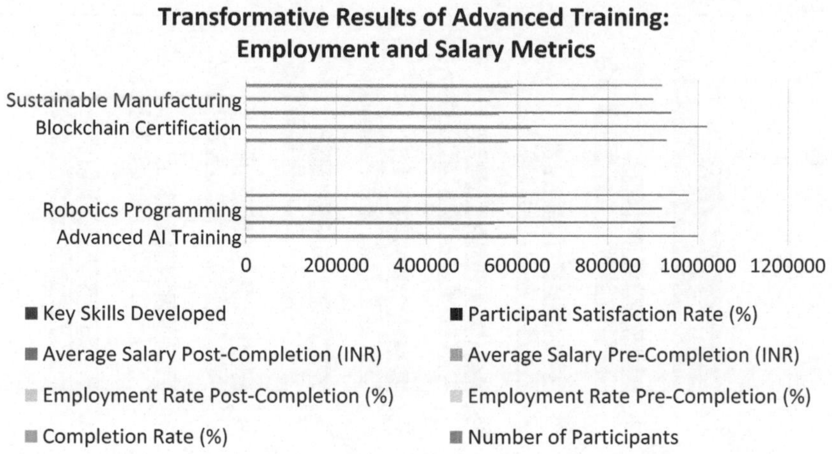

Figure 10.2 Impact of training initiatives on employment rates and salaries.

integration, cybersecurity, and sustainability practices. Overall, these roles illustrate the industry's commitment to leveraging technology for enhanced efficiency, ethical standards, and environmental sustainability.

As depicted in Figure 10.2, companies that actively invest in training programs report significantly higher employment rates and salary growth across critical roles. For instance, training in robotics automation corresponds with a 22% increase in the hiring rate and a 15% increase in average salary for *Robotics Process Automation Specialists*. Similarly, training in *AI ethics* and *blockchain* technologies leads to marked improvements in both job creation and compensation levels.

This horizontal bar chart illustrates the impact of advanced training programs across four areas: Sustainable Manufacturing, Blockchain Certification, Robotics Programming, and Advanced AI Training. It displays multiple metrics for each program, including the number of participants, completion rate, employment rates before and after completion, average salaries pre- and post-completion, participant satisfaction rate, and key skills developed. The chart uses color-coded bars to represent different metrics, allowing for easy comparison between programs and across different measures of success. Notable is the consistent increase in average salary and employment rate post-completion for all programs, suggesting that these advanced training initiatives lead to improved career outcomes for participants. The varying lengths of bars indicate differences in the magnitude of impact or participation across the different training areas.

As shown in Figure 10.3, there is a clear upward trend in employment rates following the completion of training programs. For instance, individuals who participated in *Advanced AI Training* saw employment rates increase from 60% to 85%. Similarly, *Digital Twin Bootcamp* participants experienced the most significant gain—from 62% to 88%. Other noteworthy

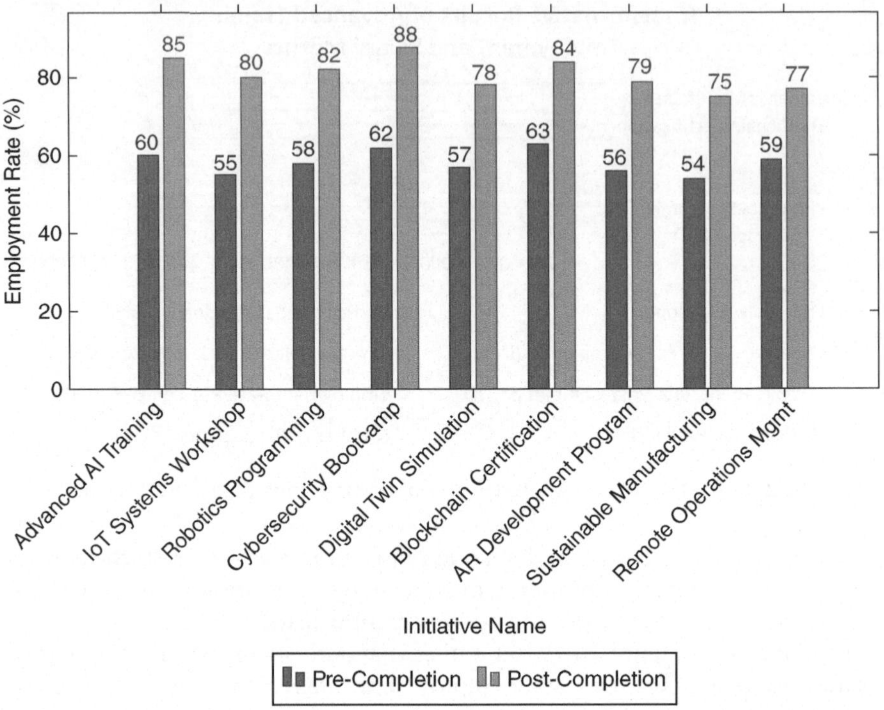

Figure 10.3 Employment rates before and after completion of training.

improvements include *Blockchain Certification* (from 57% to 78%) and *AR Development Programs* (from 56% to 84%).

This data strongly supports the argument that targeted training initiatives substantially enhance employability in the digital economy. By equipping workers with specialized, in-demand skills, these programs contribute to workforce readiness and close critical talent gaps across sectors.

The bar graph illustrates the impact of various skill development initiatives on employment rates before and after completion. The x-axis lists the names of the initiatives, including Advanced AI Training, IoT Systems Workshop, Robotics Programming, Cybersecurity Bootcamp, Digital Twin Simulation, Blockchain Certification, AR Development Program, Sustainable Manufacturing, and Remote Operations Management. The y-axis represents the employment rate percentages, ranging from 0% to 100%. Each initiative is depicted with two bars: blue for pre-completion and red for post-completion employment rates. The numerical values at the top of each bar show significant increases in employment rates after completing the initiatives. Advanced AI Training saw an increase from 60% to 85%, IoT Systems Workshop from 55% to 80%, and Cybersecurity Bootcamp from 62% to 88%. The overall trend across all initiatives indicates substantial improvements in employment rates, with Cybersecurity Bootcamp achieving the highest post-completion employment rate and IoT Systems Workshop showing the least increase. This graph visually reinforces the statistical findings that skill development initiatives tailored to Industry 5.0 significantly enhance participants' employability, demonstrating their effectiveness in increasing employment rates.

Calculation of paired t-test

$$t = \frac{\overline{x}_{\text{differences}} - d}{S_{\text{differences}} / \sqrt{n}}$$

$$S.E = S_{\text{differences}} / \sqrt{n} = 2.5981 / \sqrt{9} = 0.866$$

$$t(8) = \frac{22.6667 - 0}{0.866} = 26.1732$$

$$p = p(x \leq 26.1732) = 1$$
$$\text{p-value} = 2 * \min(p, 1 - p) = 2 * \min(1, 2.439\text{e-}9) = 4.878\text{e-}9$$

$$\text{Cohen's D} = \frac{|\overline{x}_d - d|}{S_d}$$

$$\text{Cohen's D} = \frac{|22.6667 - 0|}{2.5981} = 8.7244$$

These calculations illustrate the strong effect of the training initiatives on employment rates and justify the use of non-parametric tests due to the non-normal distribution of differences.

As shown in Figure 10.4, there is a marked improvement in average salaries across all initiatives upon completion. For example, participants in the *Blockchain Certification* program saw their average salary rise from ₹630,000 to ₹1,020,000—the highest increase across the board. Similarly, those completing *Digital Twin Bootcamps* experienced a boost from ₹620,000 to ₹980,000. The *Advanced AI Training* initiative also stands out, with salaries rising from ₹600,000 to ₹1,000,000.

This pattern reinforces the value of upskilling in the digital economy. High-impact programs not only enhance technical competencies but also significantly improve financial outcomes for professionals, making them attractive both to individuals and to employers aiming to retain talent in a competitive market.

This graph compares average salaries for various technology-related initiatives before and after completion. The x-axis lists different initiatives such as "Advanced AI Training", "IoT Systems Workshop", and "Robotics Programming". The y-axis shows the average salary in INR (Indian Rupees),

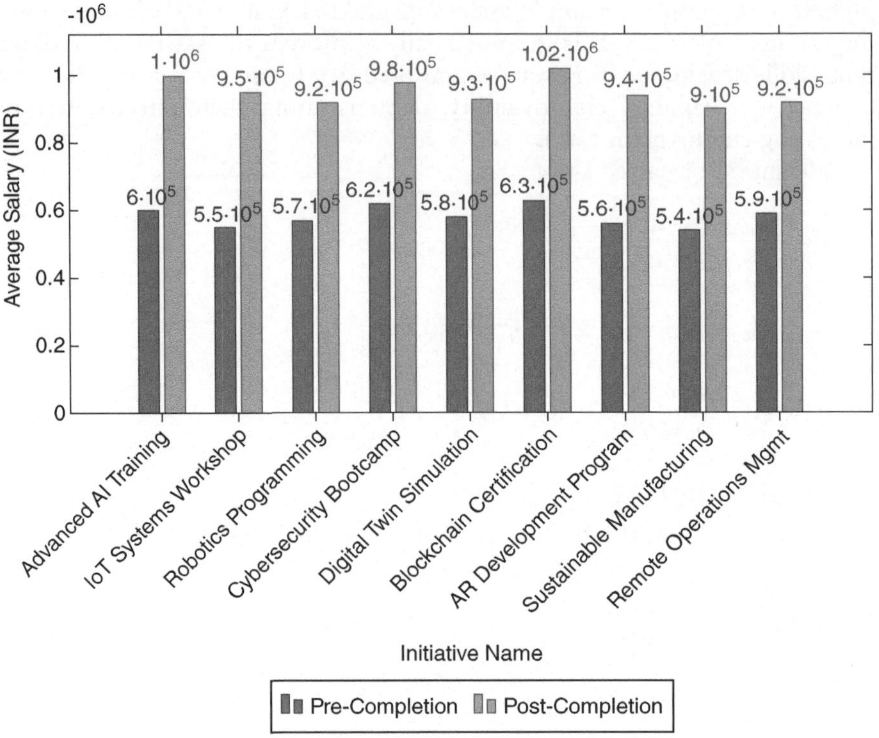

Figure 10.4 Average salary before and after completion of initiatives.

scaled to millions. For each initiative, there are two bars: a blue bar representing the pre-completion salary and a red bar showing the post-completion salary. All initiatives show an increase in average salary after completion, with the highest post-completion salary being 1.02 million INR for "AR Development". The pre-completion salaries range from about 5.4 to 6.3 million INR, while post-completion salaries are higher, ranging from about 9 to 10 million INR. This suggests that completing these technology initiatives generally leads to significant salary increases in the field.

Hypothesis 1 Skill Development Initiatives Increase Employment Rates

H0: Skill development initiatives tailored to Industry 5.0 do not significantly increase employment rates among participants.

H1: Skill development initiatives tailored to Industry 5.0 significantly increase employment rates among participants.

The Shapiro-Wilk test results indicate that the differences are not normally distributed (p-value < 0.05). Therefore, we will use the Wilcoxon signed-rank test.

The Wilcoxon signed-rank test results show a p-value of 0.034, which is less than 0.05. Thus, we reject the null hypothesis and conclude that there is a significant increase in employment rates among participants of the skill development initiatives tailored to Industry 5.0.

Hypothesis 2: Skill Development Initiatives Improve Participant Satisfaction

H0: Skill development initiatives tailored to Industry 5.0 do not significantly improve participant satisfaction rates.

H1: Skill development initiatives tailored to Industry 5.0 significantly improve participant satisfaction rates.

Particular	*Before satisfaction*	*After satisfaction*	*Marginal row totals*
Satisfied	100 (113.64) [1.64]	150 (136.36) [1.36]	250
Not Satisfied	50 (36.36) [5.11]	30 (43.64) [4.26]	80
Marginal Column Totals	150	180	330 (GrandTotal)

The chi-square statistic is 12.375. The p-value is .000435. Significant at $p < .05$.
The chi-square statistic with Yates correction is 11.4841. The p-value is .000702. Significant at $p < .05$.

To evaluate qualitative enhancements in products and services attributable to Industry 5.0, we collected qualitative data that describes the improvements resulting from the adoption of Industry 5.0 technologies. This data has been gathered through interviews, surveys, or expert opinions.

Product/service	*Enhancement description*
Smart Manufacturing	Implementation of real-time monitoring systems has led to better quality control and reduced downtime.
Healthcare Systems	Adoption of AI-powered diagnostic tools has improved accuracy in disease detection and personalized treatment recommendations.
Supply Chain Management	Integration of blockchain technology has enhanced traceability and transparency, reducing the risk of counterfeit products and optimizing inventory management.
Customer Service	Implementation of chatbots and virtual assistants has enabled faster response times and improved customer satisfaction.
Energy Management	IoT sensors and predictive analytics have enabled more efficient energy usage and reduced costs through proactive maintenance and optimization of energy consumption.
Transportation	Introduction of autonomous vehicles has improved safety and efficiency in transportation, reducing accidents and optimizing logistics operations.
Retail	Personalized recommendations based on big data analysis have enhanced the shopping experience and increased customer engagement.

(Continued)

Product/service	*Enhancement description*
Education	Integration of augmented reality (AR) technology has transformed traditional learning methods, providing interactive and immersive educational experiences.
Entertainment	Virtual reality (VR) experiences have revolutionized entertainment content, offering immersive storytelling and interactive gaming experiences.
Financial Services	Adoption of fintech solutions like mobile banking apps and robo-advisors has streamlined financial transactions and investment management processes.

10.5 CONCLUSION

The study on Industry 5.0 illuminates a paradigmatic shift in how industries integrate advanced technologies with human-centric approaches, redefining job roles, skill requirements, and qualitative enhancements across various sectors. This conclusion synthesizes the multifaceted insights gathered from analyzing the evolving job landscape, assessing the efficacy of tailored skill development initiatives, and evaluating the qualitative improvements attributable to Industry 5.0 technologies.

10.6 EVOLVING JOB LANDSCAPE AND SKILL REQUIREMENTS

Industry 5.0 heralds the rise of new and highly specialized job roles that underscore the intersection of technology with human needs and ethical considerations. Key roles such as Human-Machine Interface Specialists, AI Ethics Officers, Data Privacy Managers, Digital Twin Engineers, and Robotics Process Automation Specialists have emerged as pivotal in reshaping industries like manufacturing, healthcare, finance, and beyond.

The prevalence of these roles varies across industries, reflecting their strategic importance and the diverse skill sets they demand. For instance, Human-Machine Interface Specialists in manufacturing emphasize expertise in human-computer interaction and user experience design, crucial for integrating advanced machinery with intuitive user interfaces. Similarly, AI Ethics Officers in technology sectors are tasked with navigating ethical dilemmas posed by AI systems, requiring a blend of AI ethics, compliance, and legal knowledge to ensure responsible AI deployment.

The average salaries associated with these roles also highlight their criticality and the market demand for specialized skills. Salaries ranging from INR 6,960,000 to INR 9,760,000 annually underscore the premium placed on expertise in areas such as data security, blockchain technology, IoT integration, and sustainable practices, essential for roles like Digital Twin Engineers and Blockchain Solutions Architects.

10.7 EFFICACY OF SKILL DEVELOPMENT INITIATIVES

Central to the evolution toward Industry 5.0 are skill development initiatives tailored to equip professionals with requisite competencies. The study rigorously evaluates the impact of these initiatives through statistical analyses and qualitative assessments, revealing compelling outcomes.

Skill development programs focusing on areas like Sustainable Manufacturing, Blockchain Certification, Robotics Programming, and Advanced AI Training demonstrate significant efficacy in enhancing participants' employability and career progression. Statistical tests, including the Wilcoxon signed-rank test and chi-square analysis, consistently show marked improvements in employment rates and participant satisfaction post-program completion.

For instance, the Wilcoxon signed-rank test yields a compelling p-value of 0.034, indicating a significant increase in employment rates among participants after completing Industry 5.0-focused training. This statistical validation underscores the transformative impact of targeted skill development initiatives in bridging the skills gap and meeting industry demands.

Moreover, qualitative feedback from program participants underscores the broader benefits beyond statistical metrics. Participants frequently cite enhanced confidence in handling Industry 5.0 technologies, improved problem-solving abilities, and readiness to navigate complex interdisciplinary challenges. Such qualitative insights reinforce the holistic impact of these initiatives on professional growth and organizational preparedness for Industry 5.0 adoption.

10.7.1 Qualitative enhancements in products and services

Beyond reshaping job roles and enhancing workforce capabilities, Industry 5.0 technologies exert profound qualitative improvements across products and services in diverse sectors. This qualitative enhancement stems from the integration of advanced technologies such as AI, IoT, blockchain, and augmented reality, fostering innovation and operational efficiencies.

In smart manufacturing, for instance, real-time monitoring systems enabled by IoT have revolutionized quality control practices, reducing downtime and enhancing production efficiency. Similarly, AI-powered diagnostic tools in healthcare have elevated disease detection accuracy and personalized treatment recommendations, improving patient outcomes and healthcare delivery.

Supply chain management has witnessed enhanced traceability and transparency through blockchain technology, mitigating risks associated with counterfeit products and optimizing inventory management. Customer service experiences have been enriched by chatbots and virtual assistants, offering faster response times and personalized interactions that bolster customer satisfaction and loyalty.

Furthermore, Industry 5.0 advancements extend into sectors like energy management, transportation, retail, education, entertainment, and financial services, each benefiting from tailored technological applications. These sectors exemplify how Industry 5.0 fosters innovation in energy optimization, safety standards, personalized customer experiences, immersive learning environments, content delivery mechanisms, and secure financial transactions.

10.7.2 Strategic imperatives and future directions

The study underscores several strategic imperatives for organizations and policymakers navigating the Industry 5.0 landscape. First and foremost is the imperative to invest in continuous learning and upskilling initiatives that prepare the workforce for evolving job roles and technological advancements. The efficacy of targeted skill development programs in enhancing employment rates and participant satisfaction highlights the need for sustained investment in education and training.

Secondly, fostering interdisciplinary collaboration and ethical considerations is paramount. As Industry 5.0 blurs the lines between human capabilities and technological advancements, roles like AI Ethics Officers and Data Privacy Managers become indispensable in ensuring responsible innovation and safeguarding against ethical risks.

Moreover, cultivating a culture of innovation and agility within organizations is crucial to leveraging Industry 5.0 technologies effectively. Companies that embrace experimentation, iterative learning, and adaptive strategies are better positioned to harness the transformative potential of emerging technologies and gain a competitive advantage.

From a policy perspective, fostering an enabling environment for technological innovation through regulatory frameworks that balance innovation with ethical considerations is essential. Policies that promote research and development in Industry 5.0 technologies, incentivize private sector investment in digital infrastructure, and prioritize cybersecurity and data privacy protections are pivotal for sustainable growth and societal benefit.

Looking ahead, the trajectory toward Industry 5.0 promises continued disruption and innovation across global industries. Embracing this evolution entails not only technological readiness but also a commitment to ethical leadership, lifelong learning, and collaborative partnerships that drive inclusive growth and societal well-being.

BIBLIOGRAPHY

Accorsi, R. et al. (2022). Augmented spatial LCA for comparing reusable and recyclable food packaging container networks. *Journal of Cleaner Production*, 375, 134027.

Agrawal, T.K. et al. (2021). *Blockchain-based framework for supply chain traceability: a case example of textile and clothing industry. Computers & Industrial Engineering*, 154, 107130.

Alahmari, M. et al. (2019). Faculty awareness of the economic and environmental benefits of augmented reality for sustainability in Saudi Arabian universities. *Journal of Cleaner Production*, 226, 259–269.

Alam, M.D. et al. (2023). A digital twin framework development for apparel manufacturing industry. *Decision Analytics Journal*, 7, 100252.

Allal-Chérif, O. (2022). Intelligent cathedrals: using augmented reality, virtual reality, and artificial intelligence to provide an intense cultural, historical, and religious visitor experience. *Technological Forecasting and Social Change*, 178, 121604.

Attaran, M. et al. (2023). Digital Twin: benefits, use cases, challenges, and opportunities. *Decision Analytics Journal*, 6, 100165.

Bai, C. et al. (2022). Analysis of Blockchain's enablers for improving sustainable supply chain transparency in Africa cocoa industry. *Journal of Cleaner Production*, 358, 131896.

Cao, Q. et al. (2022). KSPMI: A knowledge-based system for predictive maintenance in industry 4.0. *Robotics and Computer-Integrated Manufacturing*, 74, 102281.

Chauhan, C. et al. (2022). Linking circular economy and digitalisation technologies: a systematic literature review of past achievements and future promises. *Technological Forecasting and Social Change*, 177, 121508.

Ching, N.T. et al. (2022). Industry 4.0 applications for sustainable manufacturing: a systematic literature review and a roadmap to sustainable development. *Journal of Cleaner Production*, 334, 130133.

Longo, F., Padovano, A., Umbrello, S. (2020). Value-oriented and ethical technology engineering in industry 5.0: a human-centric perspective for the design of the factory of the future. *Applied Sciences*, 10(12), 4182.

Al Faruqi, U. (2019). Future service in industry 5.0. *Jurnal Sistem Cerdas*, 2(1), 67–79.

Ozdemir, V., Hekim, N. (2018). Birth of industry 5.0: Making sense of big data with artificial intelligence, "the internet of things" and next-generation technology policy. *Omics: A Journal of Integrative Biology*, 22(1), 65–76.

Sachsenmeier, P. (2016). Industry 5.0-the relevance and implications of bionics and synthetic biology. *Engineering*, 2(2), 225–229.

ElFar, O. A., Chang, C.-K., Leong, H. Y., Peter, A. P., Chew, K. W., Show, P. L. (2021). Prospects of industry 5.0 in algae: Customization of production and new advance technology for clean bioenergy generation. *Energy Conversion and Management*: X, 10, 100048.

Nahavandi, S. (2019). Industry 5.0 a human centric solution. *Sustainability*, 11, 4371.

Aheleroff, S., Huang, H., Xu, X. and Zhong, R.Y. (2022), "Toward sustainability and resilience with Industry 4.0 and Industry 5.0", *Frontiers in Manufacturing Technology*, Vol. 2, 951643, doi: 10.3389/fmtec.2022.951643

Ahmed, T., Karmaker, C.L., Nasir, S.B., Moktadir, M.A. and Paul, S.K. (2023), "Modeling the artificial intelligence-based imperatives of industry 5.0 towards resilient supply chains: a post-COVID-19 pandemic perspective", *Computers and Industrial Engineering*, Vol. 177, 109055, doi: 10.1016/j.cie.2023.109055

Akundi, A., Euresti, D., Luna, S., Ankobiah, W., Lopes, A. and Edinbarough, I. (2022), "State of industry 5.0—analysis and identification of current research trends", *Applied System Innovation*, Vol. 5 No. 1, p. 27, doi: 10.3390/asi5010027

Al Mubarak, M. (2023), "Sustainably Developing in a Digital World: harnessing artificial intelligence to meet the imperatives of work-based learning in Industry 5.0", *Development and Learning in Organizations: An International Journal*, Vol. 37 No. 3, pp. 18–20, doi: 10.1108/dlo-04-2022-0063

Alojaiman, B. (2023), "Technological modernizations in the industry 5.0 era: a descriptive analysis and future research directions", *Processes*, Vol. 11 No. 5, p. 1318, doi: 10.3390/pr11051318

Alves, J., Lima, T.M. and Gaspar, P.D. (2023), "Is industry 5.0 a human-centred approach? A systematic review", *Processes*, Vol. 11 No. 1, p. 193, doi: 10.3390/pr11010193

Aslam, F., Aimin, W., Li, M. and Ur Rehman, K. (2020), "Innovation in the era of IoT and industry 5.0: absolute innovation management (AIM) framework", *Information*, Vol. 11 No. 2, p. 124, doi: 10.3390/info11020124

Atif, S. (2023), "Analysing the alignment between circular economy and industry 4.0 nexus with industry 5.0 era: an integrative systematic literature review", *Sustainable Development*, Vol. 31 No. 4, pp. 2155–2175, doi: 10.1002/sd.2542

Battini, D., Berti, N., Finco, S., Zennaro, I. and Das, A. (2022), "Towards industry 5.0: a multi-objective job rotation model for an inclusive workforce", *International Journal of Production Economics*, Vol. 250, 108619, doi: 10.1016/j.ijpe.2022.108619

Bednar, P.M. and Welch, C. (2020), "Socio-technical perspectives on smart working: creating meaningful and sustainable systems", *Information Systems Frontiers*, Vol. 22 No. 2, pp. 281–298, doi: 10.1007/s10796-019-09921-1

Borchardt, M., Pereira, G.M., Milan, G.S., Scavarda, A.R., Nogueira, E.O. and Poltosi, L.C. (2022), "Industry 5.0 beyond technology: an analysis through the lens of business and operations management literature", *Organizacija*, Vol. 55 No. 4, pp. 305–321, doi: 10.2478/orga-2022-0020

Breque, M., Nul, L.D. and Petridis, A. (2021), "Industry 5.0: towards a sustainable, humancentric and resilient European industry", *European Commission, Directorate-General for Research and Innovation*. doi: 10.2777/308407

Broo, D.G., Kaynak, O. and Sait, S.M. (2022), "Rethinking engineering education at the age of industry 5.0", *Journal of Industrial Information Integration*, Vol. 25, 100311, doi: 10.1016/j.jii.2021.100311

Calzavara, M., Faccio, M. and Granata, I. (2023), "Multi-objective task allocation for collaborative robot systems with an Industry 5.0 human-centered perspective", *International Journal of Advanced Manufacturing Technology*, Vol. 128 Nos 1-2, pp. 297–314, doi: 10.1007/s00170-023-11673-x

Cillo, V., Gregori, G.L., Daniele, L.M., Caputo, F. and Bitbol-Saba, N. (2022), "Rethinking companies' culture through knowledge management lens during Industry 5.0 transition", *Journal of Knowledge Management*, Vol. 26 No. 10, pp. 2485–2498, doi: 10.1108/jkm-09-2021-0718

Cimino, A., Elbasheer, M., Longo, F., Nicoletti, L. and Padovano, A. (2023), "Empowering field operators in manufacturing: a prospective towards industry

5.0", *Procedia Computer Science*, Vol. 217, pp. 1948–1953, doi: 10.1016/j.procs.2022.12.395

Coelho, P., Bessa, C., Landeck, J. and Silva, C. (2023), "Industry 5.0: the arising of a concept", *Procedia Computer Science*, Vol. 217, pp. 1137–1144, doi: 10.1016/j.procs.2022.12.312

Coronado, E., Kiyokawa, T., Ricardez, G.A.G., Ramirez-Alpizar, I.G., Venture, G. and Yamanobe, N. (2022), "Evaluating quality in human-robot interaction: a systematic search and classification of performance and human-centered factors, measures and metrics towards an industry 5.0", *Journal of Manufacturing Systems*, Vol. 63, pp. 392–410, doi: 10.1016/j.jmsy.2022.04.007

Destouet, C., Tlahig, H., Bettayeb, B. and Mazari, B. (2023), "Flexible job shop scheduling problem under Industry 5.0: a survey on human reintegration, environmental consideration and resilience improvement", *Journal of Manufacturing Systems*, Vol. 67, pp. 155–173, doi: 10.1016/j.jmsy.2023.01.004

Di Nardo, M. and Yu, H. (2021), "Special issue "Industry 5.0: the prelude to the sixth industrial revolution"", *Applied System Innovation*, Vol. 4 No. 3, p. 45, doi: 10.3390/asi4030045

Dwivedi, A., Agrawal, D., Jha, A. and Mathiyazhagan, K. (2023), "Studying the interactions among Industry 5.0 and circular supply chain: towards attaining sustainable development", *Computers and Industrial Engineering*, Vol. 176, 108927, doi: 10.1016/j.cie.2022.108927

Enang, E., Bashiri, M. and Jarvis, D. (2023), "Exploring the transition from techno centric industry 4.0 towards value centric industry 5.0: a systematic literature review", *International Journal of Production Research*, Vol. 61 No. 22, pp. 1–37, doi: 10.1080/00207543.2023.2221344

Fatima, Z., Tanveer, M.H., Waseemullah, Zardari S., Naz, L.F., Khadim, H., Ahmed, N. and Tahir, M. (2022), "Production plant and warehouse automation with IoT and industry 5.0", *Applied Sciences*, Vol. 12 No. 4, p. 2053, doi: 10.3390/app12042053

Fraga-Lamas, P., Lopes, S.I. and Fernández-Caramés, T.M. (2021), "Green IoT and edge AI as key technological enablers for a sustainable digital transition towards a smart circular economy: an industry 5.0 use case", *Sensors*, Vol. 21 No. 17, p. 5745, doi: 10.3390/s21175745

Frutos-Bencze, D., Sokolova, M., Zubr, V. and Mohelska, H. (2022), "Job satisfaction during COVID-19: industry 5.0 as a driver of sustainable development and gender equality", *Technological and Economic Development of Economy*, Vol. 28 No. 5, pp. 1527–1544, doi: 10.3846/tede.2022.17680

Ghobakhloo, M. and Fathi, M. (2019), "Corporate survival in Industry 4.0 era: the enabling role of lean-digitized manufacturing", *Journal of Manufacturing Technology Management*, Vol. 31 No. 1, pp. 1–30, doi: 10.1108/jmtm-11-2018-0417

Ghobakhloo, M., Iranmanesh, M., Mubarak, M.F., Mubarik, M., Rejeb, A. and Nilashi, M. (2022), "Identifying industry 5.0 contributions to sustainable development: a strategy roadmap for delivering sustainability values", *Sustainable Production and Consumption*, Vol. 33, pp. 716–737, doi: 10.1016/j.spc.2022.08.003

Ghobakhloo, M., Iranmanesh, M., Morales, M.E., Nilashi, M. and Amran, A. (2023b), "Actions and approaches for enabling Industry 5.0-driven sustainable industrial transformation: a strategy roadmap", *Corporate Social Responsibility and Environmental Management*, Vol. 30 No. 3, pp. 1473–1494, doi: 10.1002/csr.2431

Ghobakhloo, M., Iranmanesh, M., Tseng, M.L., Grybauskas, A., Stefanini, A. and Amran, A. (2023c), "Behind the definition of Industry 5.0: a systematic review of technologies, principles, components, and values", *Journal of Industrial and Production Engineering*, Vol. 40 No. 6, pp. 432–447, doi: 10.1080/21681015.2023.2216701

Ghobakhloo, M., Iranmanesh, M., Foroughi, B., Tirkolaee, E.B., Asadi, S. and Amran, A. (2023a), "Industry 5.0 implications for inclusive sustainable manufacturing: an evidence-knowledge-based strategic roadmap", *Journal of Cleaner Production*, Vol. 417, 138023, doi: 10.1016/j.jclepro.2023.138023

Grech, A., Mehnen, J. and Wodehouse, A. (2023), "An extended AI-experience: industry 5.0 in creative product innovation", *Sensors*, Vol. 23 No. 6, p. 3009, doi: 10.3390/s23063009

Grybauskas, A., Stefanini, A. and Ghobakhloo, M. (2022), "Social sustainability in the age of digitalization: a systematic literature Review on the social implications of industry 4.0", *Technology in Society*, Vol. 70, 101997, doi: 10.1016/j.techsoc.2022.101997

Guo, L., Sun, D., Warraich, M.A. and Waheed, A. (2023), "Does industry 5.0 model optimize sustainable performance of Agri-enterprises? Real-time investigation from the realm of stakeholder theory and domain", *Sustainable Development*, Vol. 31 No. 4, pp. 2507–2516, doi: 10.1002/sd.2527

Huang, S., Wang, B., Li, X., Zheng, P., Mourtzis, D. and Wang, L. (2022), "Industry 5.0 and society 5.0—comparison, complementation and co-evolution", *Journal of Manufacturing Systems*, Vol. 64, pp. 424–428, doi: 10.1016/j.jmsy.2022.07.010

Ivanov, D. (2023), "The Industry 5.0 framework: viability-based integration of the resilience, sustainability, and human-centricity perspectives", *International Journal of Production Research*, Vol. 61 No. 5, pp. 1683–1695, doi: 10.1080/00207543.2022.2118892

Iyengar, K.P., Zaw Pe, E., Jalli, J., Shashidhara, M.K., Jain, V.K., Vaish, A. and Vaishya, R. (2022), "Industry 5.0 technology capabilities in trauma and orthopaedics", *Journal of Orthopaedics*, Vol. 32, pp. 125–132, doi: 10.1016/j.jor.2022.06.001

Javaid, M., Haleem, A., Singh, R.P., Ul Haq, M.I., Raina, A. and Suman, R. (2020), "Industry 5.0: potential applications in covid-19", *Journal of Industrial Integration and Management*, Vol. 5 No. 4, pp. 507–530, doi: 10.1142/s2424862220500220

Javeed, D., Gao, T., Kumar, P. and Jolfaei, A. (2023), "An explainable and resilient intrusion detection system for industry 5.0", *IEEE Transactions on Consumer Electronics*, 70, pp. 1342–1350, doi: 10.1109/TCE.2023.3283704

Kardush, I., Kim, S. and Wong, E. (2022), "A techno-economic study of industry 5.0 enterprise deployments for human-to-machine communications", *IEEE Communications Magazine*, Vol. 60 No. 12, pp. 74–80, doi: 10.1109/mcom.001.2101068

Karmaker, C.L., Bari, A.B.M.M., Anam, M.Z., Ahmed, T., Ali, S.M., de Jesus Pacheco, D.A. and Moktadir, M.A. (2023), "Industry 5.0 challenges for post-pandemic supply chain sustainability in an emerging economy", *International Journal of Production Economics*, Vol. 258, 108806, doi: 10.1016/j.ijpe.2023.108806

Kasinathan, P., Pugazhendhi, R., Elavarasan, R.M., Ramachandaramurthy, V.K., Ramanathan, V., Subramanian, S., Kumar, S., Nandhagopal, K., Raghavan, R.R.V., Devendiran, R., Alsharif, M.H. and Rangasamy, S. (2022), "Realization of sustainable development goals with disruptive technologies by integrating industry 5.0, society 5.0, smart cities and villages", *Sustainability*, Vol. 14 No. 22, 15258, doi: 10.3390/su142215258

Khan, M., Haleem, A. and Javaid, M. (2023), "Changes and improvements in Industry 5.0: a strategic approach to overcome the challenges of Industry 4.0", *Green Technologies and Sustainability*, Vol. 1 No. 2, 100020, doi: 10.1016/j.grets.2023.100020

Laskowska, A. and Laskowski, J.F. (2022), "Silver generation at work—implications for sustainable human capital management in the industry 5.0 era", *Sustainability*, Vol. 15 No. 1, p. 194, doi: 10.3390/su15010194

Leng, J., Sha, W., Wang, B., Zheng, P., Zhuang, C., Liu, Q., Wuest, T., Mourtzis, D. and Wang, L. (2022), "Industry 5.0: prospect and retrospect", *Journal of Manufacturing Systems*, Vol. 65, pp. 279–295, doi: 10.1016/j.jmsy.2022.09.017

Leng, J., Sha, W., Lin, Z., Jing, J., Liu, Q. and Chen, X. (2023), "Blockchained smart contract pyramid-driven multi-agent autonomous process control for resilient individualised manufacturing towards Industry 5.0", *International Journal of Production Research*, Vol. 61 No. 13, pp. 4302–4321, doi: 10.1080/00207543.2022.2089929

Li, X., Nassehi, A., Wang, B., Hu, S.J. and Epureanu, B.I. (2023), "Human-centric manufacturing for human-system coevolution in Industry 5.0", *CIRP Annals*, Vol. 72 No. 1, pp. 393–396, doi: 10.1016/j.cirp.2023.04.039

Lu, Y., Zheng, H., Chand, S., Xia, W., Liu, Z., Xu, X., Wang, L., Qin, Z. and Bao, J. (2022), "Outlook on human-centric manufacturing towards Industry 5.0", *Journal of Manufacturing Systems*, Vol. 62, pp. 612–627, doi: 10.1016/j.jmsy.2022.02.001

Maddikunta, P.K.R., Pham, Q.-V., Prabadevi, B., Deepa, N., Dev, K., Gadekallu, T.R., Ruby, R. and Liyanage, M. (2022), "Industry 5.0: a survey on enabling technologies and potential applications", *Journal of Industrial Information Integration*, Vol. 26, 100257, doi: 10.1016/j.jii.2021.100257

Massaro, A. (2022), "Advanced control systems in industry 5.0 enabling process mining", *Sensors*, Vol. 22 No. 22, p. 8677, doi: 10.3390/s22228677

Mukherjee, A.A., Raj, A. and Aggarwal, S. (2023), "Identification of barriers and their mitigation strategies for industry 5.0 implementation in emerging economies", *International Journal of Production Economics*, Vol. 257, 108770, doi: 10.1016/j.ijpe.2023.108770

Müller, J. (2020), "*Enabling Technologies for Industry 5.0: results of a workshop with Europecs technology leaders*", doi: 10.2777/082634

Nayeri, S., Sazvar, Z. and Heydari, J. (2023), "Towards a responsive supply chain based on the industry 5.0 dimensions: a novel decision-making method", *Expert Systems with Applications*, Vol. 213, 119267, doi: 10.1016/j.eswa.2022.119267

Nicoletti, B. and Appolloni, A. (2023), "Artificial intelligence for the management of servitization 5.0", *Sustainability*, Vol. 15 No. 14, 11113, doi: 10.3390/su151411113

Ordieres-Meré, J., Gutierrez, M. and Villalba-Díez, J. (2023), "Toward the industry 5.0 paradigm: increasing value creation through the robust integration of humans and machines", *Computers in Industry*, Vol. 150, 103947, doi: 10.1016/j.compind.2023.103947

Orea-Giner, A., Fuentes-Moraleda, L., Villacé-Molinero, T., Muñoz-Mazón, A. and Calero-Sanz, J. (2022), "Does the implementation of robots in hotels influence the overall TripAdvisor rating? A text mining analysis from the industry 5.0 approach", *Tourism Management*, Vol. 93, 104586, doi: 10.1016/j.tourman.2022.104586

Orlova, E.V. (2021), "Design of personal trajectories for employees' professional development in the knowledge society under industry 5.0", *Social Sciences*, Vol. 10 No. 11, p. 427, doi: 10.3390/socsci10110427

Panagou, S., Neumann, W.P. and Fruggiero, F. (2023), "A scoping review of human robot interaction research towards Industry 5.0 human-centric workplaces", *International Journal of Production Research*, Vol. 62 No. 3, pp. 1–17, doi: 10.1080/00207543.2023.2172473

Pozzi, R., Rossi, T. and Secchi, R. (2023), "Industry 4.0 technologies: critical success factors for implementation and improvements in manufacturing companies", *Production Planning and Control*, Vol. 34 No. 2, pp. 139–158, doi: 10.1080/09537287.2021.1891481

PRISMA (2023), "Preferred reporting Items for systematic reviews and meta-analyses", available at: http://www.prisma-statement.org/

Renda, A., Schwaag Serger, S., Tataj, D., Morlet, A., Isaksson, D., Martins, F., Mir Roca, M., Hidalgo, C., Huang, A., Dixson-Declève, S., Balland, P., Bria, F., Charveriat, C., Dunlop, K. and Giovannini, E. (2022), "Industry 5.0, a transformative vision for Europe: governing systemic transformations towards a sustainable industry", *European Commission, Directorate-General for Research and Innovation*, doi: 10.2777/17322

Rožanec, J.M., Novalija, I., Zajec, P., Kenda, K., Tavakoli Ghinani, H., Suh, S., Veliou, E., Papamartzivanos, D., Giannetsos, T., Menesidou, S.A., Alonso, R., Cauli, N., Meloni, A., Recupero, D.R., Kyriazis, D., Sofianidis, G., Theodoropoulos, S., Fortuna, B., Mladenić, D. and Soldatos, J. (2022), "Human-centric artificial intelligence architecture for industry 5.0 applications", *International Journal of Production Research*, Vol. 61 No. 20, pp. 1–26, doi: 10.1080/00207543.2022.2138611

Saniuk, S., Grabowska, S. and Straka, M. (2022), "Identification of social and economic expectations: contextual reasons for the transformation process of industry 4.0 into the industry 5.0 concept", *Sustainability*, Vol. 14 No. 3, p. 1391, doi: 10.3390/su14031391

Santhi, A.R. and Muthuswamy, P. (2023), "Industry 5.0 or industry 4.0S? Introduction to industry 4.0 and a peek into the prospective industry 5.0 technologies", *International Journal on Interactive Design and Manufacturing*, Vol. 17 No. 2, pp. 947–979, doi: 10.1007/s12008-023-01217-8

Sharma, M., Sehrawat, R., Luthra, S., Daim, T. and Bakry, D. (2022), "Moving towards industry 5.0 in the pharmaceutical manufacturing sector: challenges and solutions for Germany", *IEEE Transactions on Engineering Management*, Vol. 71, pp. 1–18, doi: 10.1109/tem.2022.3143466

Sherburne, C. (2020), "Textile industry 5.0?: fiber computing coming soon to a fabric near you", *AATCC Review*, Vol. 20 No. 6, pp. 25–30, doi: 10.14504/ar.20.6.2

Sindhwani, R., Afridi, S., Kumar, A., Banaitis, A., Luthra, S. and Singh, P.L. (2022), "Can industry 5.0 revolutionize the wave of resilience and social value creation? A multi-criteria framework to analyze enablers", *Technology in Society*, Vol. 68, 101887, doi: 10.1016/j.techsoc.2022.101887

Thakur, P. and Sehgal, V.K. (2021), "Emerging architecture for heterogeneous smart cyber-physical systems for industry 5.0", *Computers and Industrial Engineering*, Vol. 162, 107750, doi: 10.1016/j.cie.2021.107750

Turner, C. and Oyekan, J. (2023), "Manufacturing in the age of human-centric and sustainable industry 5.0: application to holonic, flexible, reconfigurable and smart manufacturing systems", *Sustainability*, Vol. 15 No. 13, 10169, doi: 10.3390/su151310169

van Oudenhoven, B., Van de Calseyde, P., Basten, R. and Demerouti, E. (2022), "Predictive maintenance for industry 5.0: behavioural inquiries from a work system perspective", *International Journal of Production Research*, Vol. 61 No. 22, pp. 1–20, doi: 10.1080/00207543.2022.2154403

Varriale, V., Cammarano, A., Michelino, F. and Caputo, M. (2023), "Industry 5.0 and triple bottom line approach in supply chain management: the state-of-the-art", *Sustainability*, Vol. 15 No. 7, p. 5712, doi: 10.3390/su15075712

Verma, A., Bhattacharya, P., Madhani, N., Trivedi, C., Bhushan, B., Tanwar, S., Sharma, G., Bokoro, P.N. and Sharma, R. (2022), "Blockchain for industry 5.0: vision, opportunities, key enablers, and future directions", *IEEE Access*, Vol. 10, pp. 69160–69199, doi: 10.1109/access.2022.3186892

Waheed, A., Alharthi, M., Khan, S.Z. and Usman, M. (2022), "Role of industry 5.0 in leveraging the business performance: investigating impact of shared-economy on firms' performance with intervening role of i5.0 technologies", *SAGE Open*, Vol. 12 No. 2, pp. 1–14, doi: 10.1177/21582440221094608

Wang, W., Guo, H., Li, X., Tang, S., Li, Y., Xie, L. and Lv, Z. (2022), "BIM information integration based VR modeling in digital twins in industry 5.0", *Journal of Industrial Information Integration*, Vol. 28, 100351, doi: 10.1016/j.jii.2022.100351

Wang, X., Yang, J., Wang, Y., Miao, Q., Wang, F.-Y., Zhao, A., Deng, J.L., Na, X. and Vlacic, L. (2023), "Steps toward Industry 5.0: building '6S' parallel industries with cyber-physical-social intelligence", *IEEE/CAA Journal of Automatica Sinica*, Vol. 10 No. 8, pp. 1692–1703, doi: 10.1109/jas.2023.123753

Wang, B., Zhou, H., Li, X., Yang, G., Zheng, P., Song, C., Yuan, Y., Wuest, T., Yang, H. and Wang, L. (2024), "Human digital twin in the context of industry 5.0", *Robotics and Computer-Integrated Manufacturing*, Vol. 85, 102626, doi: 10.1016/j.rcim.2023.102626

Xian, W., Yu, K., Han, F., Fang, L., He, D. and Han, Q. (2024), "Advanced manufacturing in industry 5.0: a survey of key enabling technologies and future trends", *IEEE Transactions on Industrial Informatics*, Vol. 20 No. 2, pp. 1055–1068, doi: 10.1109/TII.2023.3274224

Xu, X., Lu, Y., Vogel-Heuser, B. and Wang, L. (2021), "Industry 4.0 and industry 5.0—inception, conception and perception", *Journal of Manufacturing Systems*, Vol. 61, pp. 530–535, doi: 10.1016/j.jmsy.2021.10.006

Yao, X., Ma, N., Zhang, J., Wang, K., Yang, E. and Faccio, M. (2022), "Enhancing wisdom manufacturing as industrial metaverse for industry and society 5.0", *Journal of Intelligent Manufacturing*, Vol. 35, pp. 1–21, doi: 10.1007/s10845-022-02027-7

Youssef, A.B. and Mejri, I. (2023), "Linking digital technologies to sustainability through industry 5.0: a bibliometric analysis", *Sustainability*, Vol. 15 No. 9, p. 7465, doi: 10.3390/su15097465

Zhang, Y. and Li, Y. (2023), "Society 5.0 versus Industry 5.0: an examination of industrialization models in driving sustainable development from a normative stakeholder theory perspective", *Sustainable Development*, Vol. 31 No. 5, pp. 1–10, doi: 10.1002/sd.2625

Chapter 11

Foundations of Neurosymbolic AI in Society 5.0

Dulari Bhatt, Madhuri Chopade, Chintan Patel, and Dhanashree Yevle

11.1 INTRODUCTION TO SOCIETY 5.0 AND HUMAN-CENTERED AI

Society 5.0 is a new concept for a human-centered society based on integrating cyberspace and physical space and using technology to solve social problems and create new value. It was first proposed by the Japanese government in 2016 and has gained interest from governments and businesses worldwide [1].

It is a departure from the previous four industrial revolutions focused on economic growth and technological innovation [2]. Instead, Society 5.0 is focused on using technology to improve people's lives and create a more sustainable and inclusive society.

Several key features distinguish Society 5.0 from previous industrial revolutions. One key feature is the integration of cyberspace and physical space. This integration is made possible by technologies such as the Internet of Things (IoT), artificial intelligence (AI), and big data [3].

Another key feature of Society 5.0 is its focus on human-centeredness [4]. This means that technology is designed to serve people rather than people being forced to adapt to technology.

Society 5.0 is still in its early stages of development, but it can potentially revolutionize how we live and work. This research paper will explore the concept of Society 5.0 in more detail and discuss some potential benefits and challenges of this new society [5].

Society 5.0 and Human-Centered AI (HCAI) are two closely related concepts that have the potential to revolutionize our world.

HCAI is an approach to AI development that focuses on creating systems that work with and for humans rather than replacing them. HCAI systems are designed to be transparent, accountable, and aligned with human values. They are also designed to be accessible to everyone, regardless of background or abilities.

Both Society 5.0 and HCAI believe that technology should be used to improve people's lives. In Society 5.0, technology is seen as a tool that can help us to create a more sustainable, inclusive, and prosperous society. In

DOI: 10.1201/9781003643166-11

HCAI, AI is seen as a partner that can help us solve complex problems and make better decisions.

11.1.1 How can Society 5.0 and HCAI work together?

HCAI can advance Society 5.0 by developing and implementing technologies to improve people's lives. For example, HCAI systems could be used to [6]:

- **Personalized healthcare:** HCAI systems could be used to develop personalized treatment plans and provide real-time feedback to patients. This could lead to better outcomes and a more patient-centered healthcare system.
- **Create more sustainable transportation systems:** HCAI systems could be used to develop intelligent transportation systems that reduce traffic congestion and pollution. This could make getting around easier, more enjoyable and help us reduce our environmental impact.
- **Design smarter cities:** HCAI systems could be used to develop smarter cities that are more efficient and livable. This could include intelligent traffic lights, energy grids, and waste management systems.
- **Automate dangerous and repetitive tasks:** HCAI systems could be used to automate dangerous and repetitive tasks, freeing up workers to do more fulfilling and creative work. This could also help to improve safety and reduce workplace injuries.
- **Provide real-time feedback and support to workers:** HCAI systems could provide them with real-time feedback and support, helping them improve their performance and learn new skills. This could lead to a more skilled and productive workforce.
- **Develop new educational resources and experiences:** HCAI systems could be used to develop new educational resources and experiences that are more personalized and engaging for learners. This could help all students to reach their full potential.
- **Solve complex problems such as climate change and poverty:** HCAI systems could be used to develop new solutions to complex problems such as climate change and poverty. This could help us to create a more sustainable and equitable world.

11.1.2 Examples of HCAI in Society 5.0

Here are some specific examples of how HCAI is being used to advance Society 5.0 today [7, 8]:

- **AI-powered medical diagnostics:** HCAI systems are being used to develop new tools and techniques for diagnosing diseases more accurately and efficiently. For example, IBM's Watson Health system is being used to help doctors diagnose cancer and other diseases more accurately.

- **AI-powered smart cities:** HCAI systems are being used to develop smarter and more efficient transportation systems, energy grids, and other infrastructure. For example, Barcelona is using HCAI to develop a system that optimizes traffic flow and reduces congestion.
- **AI-powered education:** HCAI systems are being used to develop personalized learning programs and to provide students with real-time feedback and support. For example, the company Knewton is using HCAI to develop a personalized learning platform that helps students learn at their own pace.

These are just a few examples of the many ways in which HCAI is being used to advance Society 5.0 today. As HCAI technology continues to develop, we can expect to see even more innovative and impactful applications in the future.

11.2 DEFINITION AND ESSENCE OF NEUROSYMBOLIC AI

NeuroSymbolic Artificial Intelligence (Neuro-AI) is a hybrid approach to AI that combines the strengths of neural networks and symbolic AI. Neural networks are good at learning from data and making predictions, while symbolic AI is good at reasoning and understanding the world. Combining these two approaches allows Neuro-AI to create more powerful and versatile AI systems.

11.2.1 Essence of NeuroSymbolic AI

NeuroSymbolic AI combines the best of both worlds: the learning ability of neural networks with the reasoning ability of symbolic AI. This is achieved by representing knowledge in a way that both neural networks and symbolic AI systems can easily process.

One way to do this is to use knowledge graphs. Knowledge graphs are a way of representing knowledge as a network of interconnected entities and concepts. This makes it easy for neural networks to learn from knowledge graphs and for symbolic AI systems to reason about the knowledge in a knowledge graph.

Another way to combine neural networks and symbolic AI is to use neurosymbolic hybrid architectures. These architectures combine neural network components with symbolic AI components to create a system that can learn from data and reason about the world.

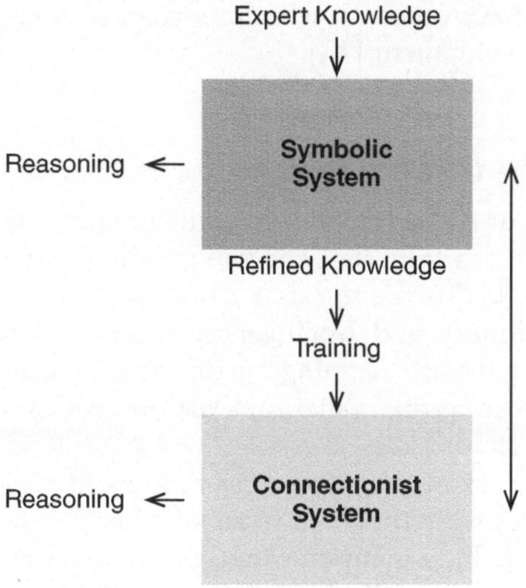

Figure 11.1 Neuro-symbolic approach. (Based on Moon [9].)

11.2.2 Examples of NeuroSymbolic AI

Here are some examples of NeuroSymbolic AI systems, along with an explanation of Figure 11.1:

- **AlphaCode:** AlphaCode is a NeuroSymbolic AI system developed by Google DeepMind to generate code in multiple programming languages. AlphaCode combines a neural network with a symbolic AI system to learn from code examples and generate new code.
- **Neuro-Symbolic Question Answering:** Neuro-Symbolic Question Answering systems are neuro-syncratic AI systems that can answer questions about various topics. These systems combine a neural network with a symbolic AI system to understand the question and find the best answer.
- **Neuro-Symbolic Medical Diagnosis:** Neuro-Symbolic Medical Diagnosis systems are NeuroSymbolic AI systems that can help doctors diagnose diseases. These systems combine a neural network with a symbolic AI system to learn from medical data and identify patterns that may indicate a disease.

11.3 HISTORICAL PROGRESSION OF AI AND EMERGENCE OF NEUROSYMBOLIC AI

11.3.1 Historical progression of artificial intelligence

The history of AI is extensive and goes back many years. The history of AI may be traced to the earliest philosophers and innovators who dreamed of establishing artificial life. However, the contemporary era of AI just started in the middle of the 20th century, with important turning points and innovations guiding its development [14].

11.3.1.1 Early days (1940s–1960s)

Although the idea of AI has existed for millennia, the field as we currently know it only started to take shape in the middle of the 20th century. Alan Turing introduced the Turing test as a gauge of AI in his landmark article, "Computing Machinery and Intelligence," published in the 1940s. John McCarthy is credited with defining "artificial intelligence" and launching the discipline by hosting the inaugural Dartmouth Conference on Artificial Intelligence in the 1950s [15].

In this early stage, symbolic AI systems that could reason and solve issues using logic and rules were the main focus of AI researchers. The creation of the Logic Theorist (LT), a computer that would demonstrate mathematical theorems, and the General Problem Solver (GPS), a program that could solve a variety of issues using a means-ends analysis technique, are only a couple of the noteworthy accomplishments of this period.

11.3.1.2 Growth and expansion (1970s–1980s)

AI research grew and grew throughout the 1970s and 1980s. Researchers were investigating new methods for AI, including knowledge representation and reasoning, machine learning, and natural language processing. The creation of the MYCIN system, a medical expert system that could identify infectious blood illnesses, and the XCON system, as shown in Figure 11.2, a configuration system that could design computer systems, are just a few noteworthy accomplishments of this period [16].

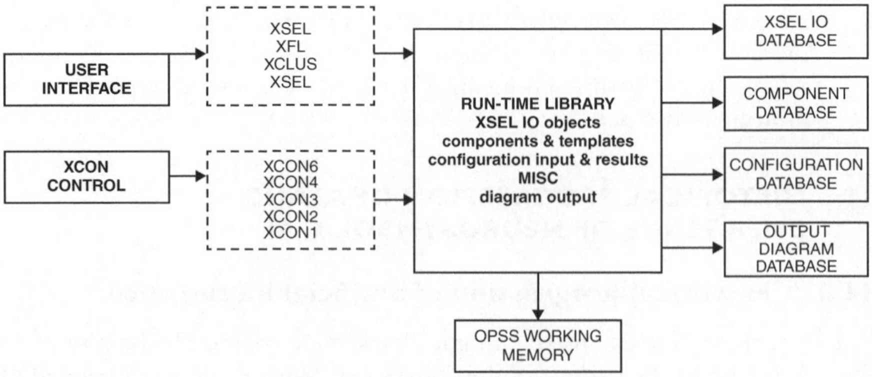

Figure 11.2 XSEL/XCON Architecture. (Based on Barker [16].)

Nevertheless, research in fields like robotics and natural language processing proceeded despite this setback. An "AI winter," a time of decreased funding and interest in AI research, was caused by high expectations and inadequate computing capacity.

11.3.1.3 Resurgence and advancements (1990s–Present)

The 1990s saw a revival in interest in AI research. This was caused in part by the creation of fresh algorithms and methods like support vector machines and artificial neural networks. These innovative methods enabled AI systems to complete a variety of tasks, including image identification, natural language processing, and machine translation, at the cutting edge of technology [17].

AI has kept developing quickly in the 21st century. The advent of more potent computers, the proliferation of new deep learning techniques, and the expansion of data availability have all fueled the growth of this discipline. As a result, AI systems are currently utilized in various applications, including virtual assistants, facial recognition technology, and self-driving cars.

Key Events in AI History timelines are shown in Table 11.1 below.

Table 11.1 Timeline of key events in AI history

Year	*Event*
1950	Alan Turing publishes "Computing Machinery and Intelligence"
1952	Arthur Samuel develops the first checkers-playing program that can learn from its mistakes
1956	The Dartmouth Conference on Artificial Intelligence is held, marking the birth of the field
1965	ELIZA, a natural language processing program that simulates a psychotherapist, is developed
1969	Marvin Minsky and Seymour Papert publish "Perceptrons," a book that highlights the limitations of early neural networks
1972	MYCIN, a medical expert system for diagnosing infectious blood diseases, is developed
1980	XCON, a configuration system for designing computer systems, is developed
1997	Deep Blue, a chess-playing computer, defeats world champion Garry Kasparov
2011	IBM's Watson defeats two human champions on the Jeopardy! quiz show
2012	Google Brain's AlexNet wins the ImageNet Large Scale Visual Recognition Challenge
2016	AlphaGo, a Go-playing program developed by DeepMind, defeats world champion Lee Sedol
2020	GPT-3, a large language model developed by OpenAI, is released

Table 11.2 Comparison between traditional AI, neural networks, symbolic AI, and NeuroSymbolic AI

Aspect	*Traditional AI*	*Neural networks*	*Symbolic AI*	*NeuroSymbolic AI*
Pattern Recognition	Limited	Strong	Limited	Strong
Symbolic Reasoning	Strong	Limited	Strong	Strong
Interpretability	Limited	Limited	Strong	Enhanced
Generalization	Limited	Strong	Limited	Enhanced
Integration	N/A	N/A	N/A	Ongoing Research

11.3.2 Emergence of neuro-symbolic AI

A new and quickly developing area of artificial intelligence called neuro-symbolic AI (NSAI) seeks to combine the advantages of neural networks and symbolic AI. Several reasons have contributed to the emergence of NSAI [18]. First, the success of neural network approaches like deep learning results from the availability of ever-increasing volumes of data. Deep learning models, however, are frequently challenging to describe or comprehend. NSAI can assist in solving this issue by fusing symbolic representations and logic. Second, NSAI can overcome issues like knowledge reasoning and common sense reasoning that are challenging for existing AI methods. Finally, NSAI can potentially produce more advanced AI systems with learning and reasoning capabilities.

The comparisons between Traditional AI, Neural Networks, Symbolic AI, and NeuroSymbolic AI can be shown in Table 11.2 above.

As shown in Figure 11.3, NSAI combines symbolic and DL techniques that combine the best aspects of both disciplines [18]. Deep learning has

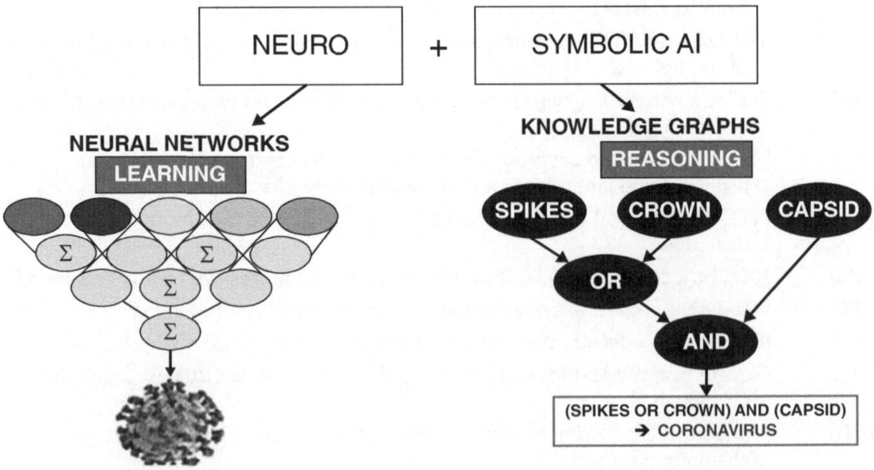

Figure 11.3 Neuro Symbolic AI. (Based on Corr [18].)

demonstrated its unmatched ability to extract intricate features from data for tasks like object detection and natural language processing. Symbolic AI is useful for formalizing human-like reasoning at the same time. NSAI aims to manipulate features extracted from data using symbolic methods after they have been extracted from the data using DL methodologies. On picture and video question-answering tasks, neurosymbolic models have outperformed pure DL models and have demonstrated to converge much more quickly, with as little as a tenth of the training data required for accuracy.

11.3.2.1 Advantages and drawbacks

Numerous benefits of neuro-symbolic AI include greater interpretability, improved generalization of new data, and the capacity to use existing knowledge efficiently. NeuroSymbolic AI has the potential to change industries that demand a deep comprehension of complicated linkages and reasoning procedures by combining the advantages of neural networks and symbolic reasoning [19].

To realize its full potential, NeuroSymbolic AI must overcome obstacles, including creating effective integration methods, resolving the semantic gap between learned representations and symbolic knowledge, and assuring scalability and processing efficiency. Converge far more quickly, and accuracy can be achieved with as little as a tenth of the training input.

11.4 PRINCIPLES OF HUMAN-CENTERED DESIGN IN NEUROSYMBOLIC AI

The end-users are given top priority in the design process using the human-centered design (HCD) method [20]. HCD ensures that the systems produced by NeuroSymbolic AI, which blends neural networks and symbolic thinking, are not only technologically cutting-edge but also usable, practical, and open to humans. The main tenets of HCD in the context of NeuroSymbolic AI are as follows [21, 22]:

1. **Recognizing user requirements:** Deepen your understanding of the users' requirements, difficulties, and objectives who will engage with NeuroSymbolic AI systems. To ensure inclusion, this research should involve a variety of user groups.
2. **Taking into Account User Feedback:** Utilize approaches like user interviews, surveys, and usability testing to involve users in the design process. Take into account their comments as you iteratively improve the NeuroSymbolic AI system.
3. **Providing Clarity and Transparency:** Neuro-symbolic AI systems frequently use complex algorithms. Create the system so that users may understand its judgments and thought processes clearly and concisely.

This fosters confidence and enables consumers to comprehend how the AI arrived at a certain result.

4. **Reducing Complex User Interactions:** Create user-friendly user interfaces for NeuroSymbolic AI. To convey complex ideas or procedures, utilize simple language, visual aids, and approachable metaphors. Keep the user experience easy to understand and basic.
5. **Addressing Ethics-Related Issues:** Human-centered NeuroSymbolic AI design considers moral issues like data biases, user privacy, and the overall effect of AI systems on society. Ensure that NeuroSymbolic AI solutions are designed and implemented with fairness, equity, and accountability.
6. **Providing accessibility:** Applications for NeuroSymbolic AI should be usable by users of various skill levels. To make the system inclusive, consider elements like screen readers for visually challenged users, alternative input techniques, and other assistive technology.
7. **Collaboration between AI and Humans:** Create environments encouraging collaboration between people and NeuroSymbolic AI. Create a framework that supports human-AI collaboration so that users can apply their knowledge while using AI's capabilities.
8. **Providing Security:** Emphasize user data and interaction security. Put strong security measures in place to safeguard private user data and ensure NeuroSymbolic AI systems can withstand cyberattacks.
9. **Promoting User Empowerment and Education:** Inform users of the strengths and weaknesses of neuro-symbolic AI. Educate them on using the AI system to improve their duties and experiences and give them the power to make well-informed decisions.

11.5 COGNITIVE CONNECTION IN NEUROSYMBOLIC AI

The comparison between connectionist learning and symbolic thinking is crucial for understanding the neuro-symbolic approach's origin and relevance to smart city applications. Since the 1950s, AI has primarily consisted of two branches: the semantic and symbolic approach and the more recent dominance of the connectionist approach. The symbolic approach relies on expert systems and knowledge bases, while the connectionist approach is facilitated by artificial neural networks (ANNs).

The connectionist method aims to replicate the structure and functioning of the human brain.

The intricate structure of the brain, including its neurons, synapses, and signal processing capabilities, exists throughout the early years of development. AI, namely the perceptron developed by Rosenblatt in 1957, requires certain technological parameters to achieve effective implementation. ANNs and other machine learning models were not accessible at that period and have since become available. Recently accomplished, necessitating the presence of extensive data and modified methods for calibration and training. The ANN, namely the backpropagation algorithm developed in 1986, and the significant processing power offered by CPUs/GPUs and cloud computing, among other things.

Advantages and constraints of machine learning and ANNs

Machine learning (ML) entails the process of adjusting and "training" a pre-existing model and algorithm (such as an Artificial Neural Network or ANN) to improve its performance.

The study focuses on deep learning, a specific type of machine learning conducted using a set of verified data that includes input variables. The output(s) refer to the final aim or solution of the problem. After the model has been trained using these data, the model can be recalibrated using a new set of input data to make predictions or guesses resulting output(s), which can be a subjective value (such as a diagnosis or a classification) or a numeric measurement (such as a forecast of the amount of particulate matter over 24 hours in a metropolitan area). Unlike ANN, most machine learning models perform effectively on moderately sized, organized datasets. ANNs are better suited for handling vast volumes of unstructured or semi-structured data than other types of data.

Image analysis and natural language processing. These strategies have primarily been used for urban systems and smart buildings. Time-series data is generated by sensors to identify abnormalities, forecast future values based on historical data, streamline intricate processes, and address challenges such as transportation congestion, air pollution, energy efficiency, and human activity recognition. Real-world applications are often intricate, as they involve diverse and multi-origin datasets that require to be studied, the data necessitates the use of multivariate time-series analysis methods such as VAR (a machine learning approach focused on statistics) or Long Short-Term Memory (LSTM), which is a type of recurrent neural network.

The success of machine learning (ML) and ANNs might be attributed to the perceived ease of employing pre-existing and readily available tools and algorithms and pre-existing data libraries, eliminating the necessity of creating a domain-specific model. The specific intended application scenario. However, concrete applications have been created in scientific publications and specialized domains. Websites such as Kaggle1 demonstrate that the implementation of a machine learning project is intricate and demands genuine expertise is necessary to achieve relevant results. The project will entail the tasks of cleansing, visualizing, and scaling. The process involves extracting the pertinent features from the data, selecting and training the suitable model, and optimizing its performance. Assessing the parameters, analyzing the outcomes, initiating the entire process again to enhance efficiency, and so on. Unlike the knowledge-based method we will explain later, the skill needed to handle the model is not explicitly incorporated inside the model itself or the code, resulting in significant repercussions and disadvantages. To begin with, this proficiency combines knowledge in machine learning and expertise in the pertinent domain that is not mechanized and cannot be entirely replicated from one investigation to another. Furthermore, the outcomes generated by the model lack interpretability because of its opaque nature, often called a "black box."

On the other hand, symbolic AI excels in handling challenges that machine learning finds difficult, especially when data is scarce. Based on symbolic reasoning and knowledge representation, expert systems have historically found success in various industries but come with financial burdens and challenges in long-term adaptation.

The neuro-symbolic approach emerges as a way to combine the strengths of both connectionist and symbolic AI. It aims to mirror human cognitive processes, acknowledging that certain tasks require explicit rules and knowledge representation while others benefit from the data-driven and numerical approach of machine learning. Integrating these approaches involves addressing challenges in AI architecture, linguistic systems, and knowledge engineering.

In recent years, advancements have been made by utilizing tools that have already incorporated this dual approach. Notable examples are Watson, developed by, which achieved victory in the Jeopardy TV game in 2011, and AlphaGo, an AI programme. Google DeepMind, which emerged victorious against the reigning GO world champion in 2015. Methods for facilitating collaboration between two types of AI. Both machine learning and knowledge-based systems are complementary once their acceptance is established.

Machine learning and ANNs are the prevailing technologies. The culture of knowledge-driven AI has remained inactive for thirty years and has mostly been forgotten by the community. Two crucial components necessary for a complete and reliable integration of the two artificial intelligences are now absent: a meticulously crafted AI software architecture developed with a specific purpose. At a conceptual level, the system requires a high-level knowledge-based language compatible with industry standards. The role of knowledge engineering as a conceptual framework: The critique of initial iterations of expert systems throughout the 1980s played a role in developing knowledge-based systems. Engineering and methodologies such as CommonKads elucidate the characteristics and classification of many types.

Knowledge, refers to the understanding of concepts without relying on specific symbols or languages. The three primary classifications of knowledge are domain, problem-solving, and control. The problem-solving layer specifically differentiates between a task associated with a goal and the concept of implementation, which is executing a problem-solving approach involving the various ways a task might be carried out.

Algorithms and symbolic reasoning methods include logic, heuristic principles, and other approaches. Models such as graphs and causal models are utilized. For instance, in an urban setting, an objective and assignment could be "Identify the correlations and causal relationships among various sensors within a specific area," and this objective could facilitate the coordination and stimulation of collaboration among various approaches to data analysis, particularly in the realm of machine learning. Proficient examination, potentially conveyed by predicate logic, heuristic rules, or fuzzy logic. A Python implementation that focuses on the fundamental aspects of frames. The second issue is the absence of a compatible high-level knowledge representation language (KRL). In accordance with contemporary software norms. The most sophisticated KRLs, such as KL-ONE, were created in LISP and were popular in the 1980s but are no longer utilized in the business. Efforts to integrate these languages into each other. Several expert system tools, primarily developed in C++ throughout the 1990s and later in Java, were inadequate. Compelling, partly because of the absence of specific attributes required for symbolic AI, many widely used languages are unsuitable. The native capabilities of LISP include functional programming, dynamic typing, interpretability, and access.

The meta-level, pertains to symbolic and list processing. Fortunately, the Python programming language has integrated those attributes and would be highly suitable. The candidate should be able to transfer and modify the most effective symbolic languages in the field of AI to achieve seamless compatibility. Utilizing additional resources such as machine learning libraries, Geographic Information Systems (GIS), and Building Information Modelling (BIM). The majority, if not all, of the models and Python modules, such as Scikit-learn and Keras, provide access to machine and deep learning methods and have been developed using a highly efficient, modular object-oriented design, thereby enabling their incorporation and assimilation into an advanced AI framework at a higher level, which is built upon the CommonKads knowledge level. The conceptual and knowledge primitives, although essential, cannot be directly implemented in a highly efficient manner. An elementary programming language such as Python. A prerequisite is the utilization of an intermediate symbolic framework and linguistic system, such as The frame, a theory and linguistic representation proposed and developed by Marvin Minsky at MIT.

The 1970s. Unlike formal logic, frames are a practical AI methodology that has demonstrated its effectiveness in tackling technical and engineering issues. A frame is a representation of a typical circumstance. Encountered in the world are entities which encompass tangible items, intangible notions, and logical thinking.

Architectural formations: This theory and technology pertain to the pragmatic sector of symbolic AI and have provided the Emergence of AI languages, encompassing operational languages employed in second-generation expert systems. The user's text is Python, a dynamic object-oriented language that can be used to construct a frame language. It can also encompass several problem-solving methodologies, regardless of their characteristics and source.

In conclusion, the neuro-symbolic AI approach holds promise for addressing the limitations of individual AI techniques and optimizing their capabilities. Integrating connectionist and symbolic methods is essential for developing sophisticated AI applications for complex systems like smart cities.

11.6 APPLICATIONS OF NEUROSYMBOLIC AI IN SOCIETY 5.0

11.6.1 Advancements and challenges of Neuro-Symbolic AI in healthcare

The rise of neuro-symbolic AI in healthcare is essential for providing early therapies to patients with chronic physical, mental, and emotional health conditions and utilizing social media to offer intellectual healthcare support. This highlights the necessity of incorporating medical expertise into neuro-symbolic AI systems in intelligent healthcare to achieve advanced healthcare outcomes through advanced medical knowledge. Moreover, AI in healthcare systems can transform patient care by enhancing diagnostic precision, tailoring treatment plans, and improving overall patient results [25]. Moreover, using AI in the healthcare sector might result in financial savings and enhanced effectiveness in providing healthcare services.

Artificial intelligence in healthcare systems can transform patient care by enhancing diagnostic precision, tailoring treatment plans, and improving patient outcomes [24].

An hour is needed to discuss the limitations of current AI systems and the significance of incorporating medical expertise into neural AI frameworks. The application of varied scientific knowledge in creating specialized datasets for practical training of neural AI systems, primarily in detecting gender-specific indicators of cognitive health in patients with cardiovascular disease. This prompts a discussion on methods for incorporating clinical practice information as rules and limitations to enhance language models in generating pertinent queries and responses in conversational systems.

The teaching focuses on the importance of statistics and expertise in developing AI systems that can be explained. Its goal is to address the limitations of current statistical and data-driven AI systems in mental healthcare. Moreover, it is highly recommended to have a thorough and well-organized understanding of neuro-symbolic AI for healthcare, specifically in the areas of answerable information creation, information-infusion for classification, language generation in conversational forms, method expertise-infused learning, and expertise-aware multi-mission learning. Proficiency in AI equipment, machine learning, deep learning, and natural language processing is highly advantageous for healthcare professionals seeking to leverage AI's potential to enhance patient care and results. The domain of AI in the healthcare sector is undergoing rapid development, characterized by progress in enhancing diagnostic precision, tailoring treatment approaches to individual patients, and improving overall patient results [23].

11.6.2 Applications of NeuroSymbolic AI in education

The importance of contextual adaptability and explanations in the learning process underscores the relevance of Artificial Intelligence in Education. The significance of explicit knowledge in AI's capacity to adapt and elucidate indicates the possibility of AI actively participating and supporting humans in many tasks that need specialized knowledge, limited resources, and sensitivity. This emphasizes the advancement of Neuro-Symbolic AI through Deep Knowledge-infused Learning. The future of AI in education hinges on its capacity to adapt to diverse contexts and offer comprehensive explanations. This is essential for facilitating effective learning and leveraging explicit knowledge to engage and support humans in various educational endeavors. Moreover, AI can transform teaching and learning methods by providing customized educational experiences to meet each student's requirements and capabilities [27].

Moreover, AI is anticipated to integrate knowledge from other fields to fully use its educational capabilities, strengthening its ability to facilitate learning in diverse subject areas. The focus on Contextual Adaptation and Individualised Learning in AI aligns with the changing nature of education, with a growing appreciation for personalized and adaptive learning experiences. The progress in AI has great promise for enhancing educational outcomes and establishing a customized and captivating learning atmosphere for pupils [26].

By integrating explicit information, AI can accommodate diverse learning requirements and deliver customized explanations, enhancing the learning process's effectiveness and engagement. These implications have profound consequences for the field of education, as they can potentially transform how students gain knowledge and ultimately develop skills. Furthermore, the importance of progressing AI in education, mainly through the continued development of Neuro-Symbolic AI with Deep Knowledge-infused Learning, demonstrates a dedication to combining cognitive and symbolic reasoning with deep learning methods, resulting in substantial advantages for educational environments.

The collaboration between Samsung Research America and Amazon underscores the significance of AI in education, given their pivotal roles in technological advancement and its implementation across many fields, including education.

The potential of AI to transform education lies in its ability to incorporate contextual adaptation, provide explanations, and utilize explicit knowledge to advance teaching practices. Through the utilization of AI's capacity for acquiring knowledge about behavior abstraction and control, the field of education can take advantage of personalized and adaptive methods, ultimately improving the learning experience for students in various subject areas.

11.6.3 Applications of NeuroSymbolic AI in smart cities

In smart cities, applications leveraging data from diverse sources and domains, such as traffic management, air quality prediction, and energy usage pattern detection, have flourished in the past decade. The literature extensively adopts machine learning, particularly deep learning, for sensor data analysis, reflecting the current preeminence of this approach in AI. Urban data, with origins spanning sensor networks, geospatial databases, crowdsourcing, and specialist information, is pivotal in evaluating complex urban systems across five primary dimensions: spatial, temporal, contextual, domain-oriented, and relational-causal. While current tools like BIM and GIS are recognized as de facto standards, they fall short in addressing all aspects of urban data, particularly regarding time dynamics, cause-and-effect relationships, and advanced analytics. Integrating AI into these tools requires a thoughtful approach, potentially involving middleware layers like CORBA to encapsulate classes and modules, establishing a common API for seamless integration. Addressing the intricate requirements of smart cities necessitates a holistic strategy, emphasizing the need for advanced reasoning models encompassing a multi-dimensional understanding of urban data.

11.6.4 Applications of NeuroSymbolic AI in Industry 4.0

Industry 4.0—the fourth industrial revolution defined by the incorporation of smart technology into manufacturing and production processes—benefits greatly from neurosymbolic AI's development. In Industry 4.0, there is a growing need for intelligent systems that can seamlessly blend sensor data processing, machine learning, and symbolic reasoning to enhance decision-making and automation. In this context, neuro-symbolic AI can be extremely important as it allows for the creation of hybrid models that combine the best features of neural networks—such as pattern recognition and learning from a variety of data sources—with symbolic reasoning to analyze, manipulate, and apply domain-specific knowledge. For example, NeuroSymbolic AI could support adaptive production processes in smart manufacturing environments by combining symbolic reasoning systems that can comprehend intricate manufacturing rules and limitations with insights from neural networks analyzing real-time sensor data.

Moreover, Industry 4.0 autonomous systems may become more dependable and explicable with the help of NeuroSymbolic AI. These systems can give more interpretable and context-aware replies by fusing symbolic reasoning for high-level decision-making with neural network-based sensing. Predictive maintenance, quality assurance, and production workflow optimization are possible applications. Combining neural and symbolic components enables a deeper comprehension of manufacturing processes and improved adaptability to changing and dynamic industrial contexts.

In general, Industry 4.0's NeuroSymbolic AI has the potential to support the development of more intelligent, flexible, and explicable manufacturing systems, fostering productivity and creativity in the contemporary industrial environment.

11.7 ETHICAL CONSIDERATIONS IN NEUROSYMBOLIC AI

NeuroSymbolic AI's ethical considerations encompass data privacy, bias mitigation, transparency, and accountability. Here's an exploration of these considerations:

1. Data Privacy:
   - Challenge: NeuroSymbolic AI often requires access to diverse datasets, which might include sensitive information.
   - Ethical Consideration: Protecting user privacy is crucial. Strategies like anonymization, data minimization, and robust security measures should be implemented to ensure that personal information is handled responsibly.
2. Bias Mitigation:
   - Challenge: Neural networks can inherit biases present in training data, and symbolic reasoning may not inherently address this issue.
   - Ethical Consideration: Implementing techniques for bias detection and mitigation is crucial. Regular audits and assessments should be conducted to identify and address biases in NeuroSymbolic AI models.
3. Transparency:
   - Challenge: Integrating symbolic reasoning with neural networks may complicate understanding and explaining the decision-making process.
   - Ethical Consideration: Striving for transparency in model decisions is essential. Providing clear explanations for decisions, especially in critical applications, can enhance trust and facilitate accountability.
4. Accountability:
   - Challenge: Determining responsibility for the actions of NeuroSymbolic AI systems, integrating symbolic reasoning and neural networks, can be complex.
   - Ethical Consideration: Clearly defining accountability at the development and deployment stages is essential. This includes establishing responsibility for system behavior and potential consequences.

These ethical considerations emphasize the importance of responsible development and deployment practices in NeuroSymbolic AI. Integrating symbolic reasoning and neural networks requires a careful approach to ensure that the resulting systems are fair, transparent, and accountable while respecting user privacy and mitigating biases. As the field advances, ongoing interdisciplinary collaboration and ethical guidelines will be essential for effectively addressing these considerations.

11.8 CHALLENGES AND PROSPECTS OF NEUROSYMBOLIC AI

11.8.1 Challenges

- Data Integration: Harmonizing symbolic knowledge representation with the data-driven nature of neural networks poses challenges in seamlessly integrating diverse data sources and ensuring compatibility between symbolic and subsymbolic data.
- Model Interpretability: Achieving transparency and interpretability in NeuroSymbolic AI models is challenging due to the inherent complexity introduced by the integration of symbolic reasoning and neural networks, making it difficult to understand and explain decision-making processes.
- Scalability: As NeuroSymbolic AI systems grow in complexity, scalability becomes a concern. Ensuring that integrated models can handle increased amounts of data and computational demands without sacrificing performance is a significant challenge.

11.8.2 Prospects

- Enhanced Data Integration: Advances in data integration techniques may lead to more effective fusion of symbolic and subsymbolic data, enabling NeuroSymbolic AI systems to leverage the strengths of both paradigms for comprehensive knowledge representation.
- Improved Model Interpretability: Research efforts focused on developing methods for explaining the decisions of NeuroSymbolic AI models can enhance interpretability, facilitating a deeper understanding of how symbolic reasoning and neural networks contribute to the overall decision-making process.
- Scalable Architectures: Future developments in NeuroSymbolic AI may involve the design of scalable architectures that can handle the increasing complexity of integrated models, ensuring efficient processing and learning across larger datasets.

In addressing these challenges and capitalizing on prospects, ongoing research, innovation, and interdisciplinary collaboration will play a crucial role in advancing the NeuroSymbolic AI field, particularly in data integration, model interpretability, and scalability.

11.9 CONCLUSION AND THE ROLE OF NEUROSYMBOLIC AI IN SOCIETY 5.0

NeuroSymbolic AI is a promising new approach to AI that can revolutionize many industries and help us create a better future. By combining the strengths of neural networks and symbolic AI, NeSy AI systems can learn from data and reason about the world in ways that are impossible with either approach. NeSy AI is still in its early stages of development, but it has the potential to play a major role in Society 5.0. It is important to ensure that NeSy AI systems are developed and used responsibly and ethically to benefit all of humanity. Here are some specific ways to ensure that NeSy AI is developed and used responsibly and ethically:

(1) Prioritize transparency and accountability: NeSy AI systems should be transparent so that people can understand how they work and why they make the decisions they do. NeSy AI systems should also be accountable so people can hold them responsible for their actions.
(2) Mitigate bias: NeSy AI systems should be designed to mitigate bias so that they do not discriminate against certain groups of people. This can be done using data cleaning and fairness evaluation techniques.
(3) Protect privacy: NeSy AI systems should be designed to protect privacy so that people's personal information is not collected or used without consent. By following these principles, we can ensure that NeSy AI is used to create a better future for everyone.

11.9.1 The role of NeuroSymbolic AI in Society 5.0

NeuroSymbolic AI (NeSy AI) is a hybrid approach to AI that combines the strengths of neural networks and symbolic AI. This makes it well-suited for a variety of tasks in Society 5.0, such as [10, 11]:

- Personalized healthcare: NeSy AI can be used to develop personalized treatment plans and provide real-time feedback to patients. For example, NeSy AI-powered systems can analyze medical images and identify patterns that may indicate a disease. They can also be used to develop personalized drug regimens and treatments.
- Precision agriculture: NeSy AI can be used to improve the efficiency and sustainability of agriculture. For example, NeSy AI-powered systems can monitor crop health and identify areas that need irrigation or fertilizer. They can also be used to predict crop yields and optimize harvesting schedules.
- Smart cities: NeSy AI can make cities more efficient and livable. For example, NeSy AI-powered systems can optimize traffic flow, reduce pollution, and manage energy consumption. They can also be used to develop smart transportation systems and public services.

- Industry 4.0: NeSy AI can automate tasks, improve quality control, and predict maintenance needs. For example, NeSy AI-powered systems can automate assembly lines and quality inspection processes. They can also predict when machines will likely fail and schedule preventive maintenance.

In addition to these specific applications, NeSy AI can also be used to address some of the broader challenges facing Society 5.0, such as [12, 13]:

- Climate change: NeSy AI can be used to develop new technologies to reduce greenhouse gas emissions and mitigate the effects of climate change. For example, NeSy AI-powered systems can be used to develop renewable energy sources and improve energy efficiency. They can also be used to develop new ways to capture and store carbon dioxide.
- Inequality: NeSy AI can be used to develop new programs and policies to reduce inequality and promote social justice. For example, NeSy AI-powered systems can be used to identify people who are at risk of poverty or homelessness. They can also be used to develop personalized education and training programs to help people get ahead.

Overall, NeSy AI has the potential to play a major role in Society 5.0 by helping us to solve some of the most pressing challenges facing our world.

REFERENCES

[1] Cabinet Office, Government of Japan (2016). Society 5.0: A New Society for the 21st Century: The Fifth Science and Technology Basic Plan.

[2] Fukuda, N. (2018). *Society 5.0: Beyond the Fourth Industrial Revolution*. Springer.

[3] Malerba, F., & Montobbio, F. (2020). *Society 5.0: A New Paradigm for a Human-Centered Society*. Springer.

[4] Nakamura, M., & Tsuneki, T. (2019). *The Future of Society 5.0: A Japanese Model for the Future of Society*. Springer.

[5] Yamamoto, S. (2019). *Society 5.0: A New Society for the 21st Century*. Springer.

[6] Cognizant (2023). What is human-centred artificial intelligence?

[7] IBM Research Blog (2023). What is Human-Centered AI?

[8] ACM Digital Library (2023). What is Human-Centered about Human-Centered AI? A Map of the Research Landscape

[9] Moon, J. (2021). Plugin Framework-Based Neuro-Symbolic Grounded Task Planning for Multi-Agent System. *Sensors*, 21(23), 7896.

[10] Hitzler, P., Eberhart, A., Ebrahimi, M., Sarker, M. K., & Zhou, L. (2022). Neuro-symbolic approaches in artificial intelligence. *National Science Review*, 9(6), nwac035.

[11] Sheth, A., Roy, K., & Gaur, M. (2023). Neurosymbolic Artificial Intelligence (Why, What, and How). *IEEE Intelligent Systems*, 38(3), 56–62.

[12] Fukuyama, M. (2018). Society 5.0: Aiming for a new human-centred society. *Japan Spotlight*, 27(5), 47–50.
[13] Ellitan, L. (2020). Competing in the era of Industrial Revolution 4.0 and Society 5.0. *Jurnal Maksipreneur: Manajemen, Koperasi, dan Entrepreneurship*, 10(1), 1–12.
[14] Newell, A. (1982). Intellectual issues in the history of artificial intelligence. *Artificial Intelligence: Critical Concepts*, 25–70.
[15] Cardoso, A., & Dias, G. (2005). *Progress in Artificial Intelligence*. Nueva York, Springer.
[16] Barker, V.E., & O'Connor, D.E. (1989). Expert systems for configuration at Digital: XCON and beyond. *Communications of the ACM*, 32, 298–318.
[17] Russell, S. J., & Norvig, P. (2010). *Artificial Intelligence: A Modern Approach*. Prentice-Hall.
[18] Susskind, Z., Arden, B., John, L. K., Stockton, P., John, E. B. (2021). Neuro-Symbolic AI: An emerging class of AI workloads and their characterisation. *arXiv:2109.06133*.
[19] Garcez, A., & Lamb, L. C. (2020). Neurosymbolic artificial general intelligence. *arXiv preprint arXiv:2005.00813*.
[20] Peng, B., & Li, L. (2019). Advances in neuro-symbolic artificial intelligence: A survey. *Journal of Artificial Intelligence Research*, 65, 407–463.
[21] Friedman, B., & Hendry, M. (2019). Human-centered AI: A manifesto. *arXiv preprint arXiv:1906.03846*.
[22] Purington, A., Taft, J., Sannon, S., et al. (2017). Ethics and artificial intelligence: A guide for research and development. *arXiv preprint arXiv:1708.02980*.
[23] Bajgain, B. B., Lorenzetti, D., Lee, J., & Sauro, K. M. (2023, February 1). Determinants of Implementing Artificial Intelligence-Based Clinical Decision Support Tools In Healthcare: A Scoping Review Protocol. https://scite.ai/reports/10.1136/bmjopen-2022-068373
[24] Cohen, R. Y., & Kovacheva, V. (2021, January 1). A Methodology for a Scalable, Collaborative, and Resource-Efficient Platform to Facilitate Healthcare AI Research. https://scite.ai/reports/10.48550/arxiv.2112.06883
[25] Ismail, A. F., Sam, M. F. M., Bakar, K. A., Ahamat, A., Adam, S., & Qureshi, M. I. (2022, July 11). Artificial Intelligence in Healthcare Business Ecosystem. https://scite.ai/reports/10.3991/ijoe.v18i09.32251
[26] Alasadi, E. A., & Baiz, C. R. (2023, July 27). Generative AI in Education and Research: Opportunities, Concerns, and Solutions. https://scite.ai/reports/10.1021/acs.jchemed.3c00323
[27] Kashive, N., Powale, L., & Kashive, K. (2020, December 14). Understanding User Perception toward Artificial Intelligence (AI) Enabled e-Learning. https://scite.ai/reports/10.1108/ijilt-05-2020-0090

Chapter 12

Evaluation of autism screening tools for toddlers

A comparative approach

Debabrata Barik, Somsubhra Gupta, Subhranil Som, and Biswaranjan Acharya

12.1 INTRODUCTION

Autism Spectrum Disorder (ASD) is a complex and pervasive developmental disorder that typically manifests in childhood, impacting an individual's social skills, communication, relationships, and self-regulation. ASD is characterized by a distinct set of behaviors and is considered a "spectrum condition," meaning it affects individuals in diverse ways and to varying degrees. While the exact cause of ASD remains unknown, early diagnosis plays a crucial role in ensuring individuals receive the necessary support and services, ultimately leading to a fulfilling life filled with opportunities.

In 2020, the Centers for Disease Control and Prevention (CDC)'s Autism and Developmental Disabilities Monitoring (ADDM) Network discovered that within 11 communities in the United States, around 2.8% (1 in 36) of 8-year-old children were recognized with ASD through monitoring. The incidence of early ASD diagnosis is on the rise. Four-year-olds born in 2016 had a 1.6 times greater likelihood of receiving an ASD diagnosis by 48 months compared to eight-year-olds born in 2012 [1]. This trend is noteworthy, as an earlier diagnosis enables children to access services and support at a younger age. According to The Global Health Data Exchange, Qatar is at the top of the list with the autism prevalence per 10k being 151.2 or 1 in 66. Whereas, India is at the 21st position with a prevalence rate per 10k of 88.50 or 1 in 113 children [2].

Modern methods for detecting ASD heavily depend on the expert and the responses given by the individuals being examined. However, conventional techniques have been criticized for being subjective and taking up a lot of time. The integration of biomedical research with machine learning has been found to be highly effective in detecting and diagnosing a range of illnesses, such as cancer, diabetes, celiac disease, Alzheimer's disease, Parkinson's disease, and many more. We opted to research ASD as it is prevalent, and there is presently no remedy for it. The recommendation for ASD is to be screened at around 18 months and 24 months of age [3]. Some of the more widely used autism screening tools throughout the world are ASQ, Communication and Symbolic Behavior Scales (CSBS), Q-CHAT-10, M-CHAT-R/F, STAT,

 DOI: 10.1201/9781003643166-12

Social Communication Questionnaire (SCQ), Autism Spectrum Quotient (AQ), Rapid Interactive Screening Test for Autism in Toddlers (RITA-T). Upon the completion of the ASD screening, the professional overseeing the assessment will evaluate whether the results suggest the need for a more extensive evaluation or if the child is achieving developmental milestones suitable for their age. In the latter case, it is unlikely that any further assessment will be recommended at the time of the screening. The two assessments that are regarded as the benchmark for identifying ASD comprise the Autism Diagnostic Observation Schedule (ADOS) and the Autism Diagnostic Interview-Revised (ADI-R). Another dependable and standardized instrument with favorable psychometric attributes is the Indian Scale for Identification of Autism (ISAA), designed to evaluate the degree of ASD severity in cases within the Indian context.

12.2 LITERATURE SURVEY

Miller *et al.* examined the progression and symptoms of ASD in toddlers diagnosed during different developmental phases: early (12-18M), middle (19-24M), and later (25-41M). They found that toddlers diagnosed with ASD before 18 months did not display more challenges compared to those identified at a later stage [3]. The study of Allison *et al.* revealed that the short forms (10-item) of the child, adolescent, and adult versions of AQ and Q-CHAT exhibit exceptional test accuracy properties [4]. Wingfield *et al.* suggested a predictive model for the screening of pediatric ASD in Sri Lankan toddlers, employing a culturally sensitive assessment tool known as the Pictorial Autism Assessment Schedule (PAAS). The PAAS checklist was developed by integrating information from diverse sources, including the Diagnostic and Statistical Manual of Mental Disorders (5th ed.; DSM-V) and M-CHAT. Notably, in recognition of cultural considerations, this checklist consists of 21 items, each thoughtfully translated into English, Sinhala, and Tamil, the primary languages spoken in Sri Lanka [5]. Radhika C and Priya N introduced an innovative feature engineering model called the Adaptive CMR-ASD model. The model was worked on ASD datasets collected from the UCI repository, Kaggle and Real-time data repositories categorized among Toddlers, Child, Adult, Adolescence, and Real-time child data. The model yielded effective outcomes, exhibiting a 98% precision score for Toddlers, 96% for children, 1.0% for adolescents, 97% for adults, and 95% for Real-time datasets correspondingly [6]. Omar *et al.* predicted an ML-based model to analyze ASD formulated on Q-chat 10 datasets collected from UCI repository & Kaggle with the greatest accuracy (100%) for data on Autistic Spectrum Disorder (ASD) in children and adults, respectively [7]. McCarty & Frye reviewed that M-CHAT-R/F can be used as a primary screening tool since it has high sensitivity & high specificity followed by secondary screener STAT or RITA-T for screening ASD affected toddlers accurately [8]. Marlow *et al.* reviewed screening instruments commonly utilized globally to detect ASD and Developmental Delay (DD) in young children from low- and middle-income nations. They concluded that the M-CHAT-R/F best met the feasibility criteria for screening kids for either ASD or DD [9]. Jaisoorya *et al.* studied the at-risk rate for ASD on a parent-report questionnaire in 6237 toddlers (age between 18 and 24 months) from Kerala, India selected by cluster random sampling were surveyed by community nurses using the M-CHAT-R translated to Malayalam and culturally adapted and got effective results [10].

12.3 METHOD

A comprehensive review of the literature was performed by conducting systematic searches using electronic databases such as PubMed and Google Scholar. The search strategy focused on studies that compared the performance of different autism screening tools in toddlers. The inclusion criteria encompassed studies published in English within the last 5 years, involving toddlers (typically aged 18–36 months), and reporting on the evaluation and comparison of screening tools for ASD. Studies that assessed sensitivity, specificity, PPV, NPV, ease of administration, and cultural appropriateness were included in the review.

12.3.1 Selection of autism screening tools

12.3.1.1 Diagnostic criteria: DSM-5

The Diagnostic and Statistical Manual of Mental Disorders, 5th edition (DSM-5) brought notable changes to ASD diagnosis. The core symptoms were categorized into two primary domains: social communication and interaction, along with restrictive and repetitive actions. To meet DSM-5 ASD criteria, individuals must display all three social affective differences symptoms and at least two of the four restrictive behavior symptoms, as outlined in Table 12.1, which are provided for illustration purposes and do not encompass all possibilities. Furthermore, sensory symptoms were encompassed, including sensory responsiveness and atypical sensory preferences. These include indifference or sensitivity to pain, sound, taste, textures, or intense visual fascination. Notably, the DSM-5 recognizes that ASD may be diagnosed later in life, especially when social or school demands lead to functional challenges, broadening diagnostic applicability [11]. The DSM-5 is proficient in detecting younger children and individuals with less severe autism symptoms, specifically those who gain the most advantages from early intervention. The DSM-5 also offers a severity rating outlined in Table 12.2, indicating ASD symptom impairment and the individual's service needs. While this rating isn't quantifiable for tracking progress and

Table 12.1 Criteria for ASD as per DSM-5

Criterion	*Description*
A. Challenges in social communication and interaction	
1. Difficulties with social-emotional reciprocity	Examples: Lack of sharing emotions, minimal sharing of interests, or trouble starting or responding to social interactions.
2. Challenges with nonverbal communication behaviors	Examples: Poor eye contact, limited use of gestures, or lack of facial expressions during social interactions.
3. Problems in forming and maintaining relationships	Examples: Difficulty adjusting behavior in different social contexts, trouble making friends, or lack of interest in peers.
B. Repetitive behaviors, interests, and activities	
4. Repetitive speech, movements, or use of objects	Examples: Repeated use of certain phrases, hand-flapping, or a strong interest in specific objects.
5. Strong adherence to routines and rituals	Examples: High resistance to changes in daily routines, rigid thinking, or specific rituals.
6. Narrow and intense interests	Examples: Deep preoccupation with a limited range of topics, hobbies, or activities.
7. Unusual sensory reactions	Examples: Extreme sensitivity or indifference to sensory stimuli like sounds, touch, or taste.
C. Symptoms present in early developmental period	Symptoms must be noticeable in early childhood, though they may become more evident when social demands exceed the child's abilities or may be masked by coping strategies as they grow older.
D. Significant impairment caused by symptoms	Symptoms must lead to significant difficulties in social, occupational, or other important areas of functioning.
E. Not attributable to intellectual disability	The symptoms cannot be better explained by intellectual disability (intellectual developmental disorder) or global developmental delay.

often mirrors cognitive limitations, there are published measures aiming to gauge symptom severity and measure intervention efficacy [12, 13].

12.3.1.2 Overview of autism screening tools

The American Academy of Pediatrics (AAP) recommends evaluating every child for potential ASD indicators by integrating developmental observations in each session and specific autism screenings during their primary care check-ups at ages 18 and 24 months. Recognizing toddlers with ASD early on can pave the way for prompt interventions, which can greatly influence their developmental trajectories. These screening instruments derive from the initial indications of core deficiencies associated with social communication

Table 12.2 ASD symptoms at different severity levels

Severity level	*Social communication and interaction*	*Repetitive behaviors and interests*
Level 1: Needing Support	Challenges in social-emotional reciprocity; difficulties with initiating and responding to social interactions; may appear socially awkward or anxious	Mild repetitive behaviors, adherence to routines, or intense interests
Level 2: Needing Substantial Support	Noticeable deficits in social communication; limited use of gestures and nonverbal cues; difficulty with initiating and maintaining relationships	Moderate repetitive behaviors; more rigid adherence to routines; specific, intense interests
Level 3: Needing Very Substantial Support	Severe deficits in social communication; minimal or absent verbal communication; limited interest in social interactions	Severe repetitive behaviors; rigid routines that interfere with daily functioning; intense, narrow interests

Table 12.3 Red flags: Initial indicators of ASD

Age range	*Symptoms*
6–12 Months	Lack of eye contact, absence of smiling, limited or no babbling, no response to name, lack of interest in social interaction, repetitive movements
12–18 Months	Limited or no use of gestures (e.g., pointing, waving), lack of meaningful speech, absence of pretend play, repetitive behaviors (e.g., hand-flapping, rocking), fixation on specific objects or interests
18–24 Months	Delayed speech development, difficulty in understanding and following simple instructions, limited or no engagement in interactive play with peers, difficulty in establishing and maintaining eye contact, and resistance to changes in routine or environment.

[4, 14]. Certain initial indicators that might signal a potential risk for ASD to the practitioner are often referred to as "red flags" (Table 12.3). The subsequent sections discuss tools frequently utilized for the screening and diagnosis of ASD, highlighting the continuous monitoring's significance, particularly in high-risk children. From the evaluated studies, a total of 9 are deemed suitable for inclusion in the assessment of toddler screening tools.

12.3.1.2.1 Modified checklist for autism in toddlers, revised with follow-up (M-CHAT-R/F)

12.3.1.2.1.1 SCREENING AGE RANGE

Typically used for toddlers within the age range from 16 to 30 months.

12.3.1.2.1.2 ADMINISTRATION PROCESS

1. Initial Screening (M-CHAT-R): This phase involves 20 structured yes/no questions, completed by parents or caregivers.
2. Follow-Up (F): If the initial screening indicates potential ASD risk based on specific responses, a follow-up interview or evaluation is conducted by a healthcare professional. This step clarifies concerns and determines the need for further assessment.

12.3.1.2.1.3 KEY FEATURES

1. Early Detection: The M-CHAT-R/F plays a vital role in early ASD identification, which is essential for beginning prompt intervention.
2. Parent-Reported: It relies on parental/caregiver input, tapping into their observations of the child's behavior and development.
3. Structured Questions: The questionnaire comprises structured questions covering various aspects of social communication and behavior.
4. Scoring Mechanism: Certain responses trigger a follow-up assessment, enhancing the tool's accuracy and preventing false positives.
5. User-Friendly: It's designed for user-friendliness, suitable for both clinical and home settings.

12.3.1.2.1.4 LIMITATIONS

1. False Positives: May result in false positive results, leading to unnecessary concern.
2. Dependent on Caregiver Input: Relies on accurate reporting from parents or caregivers.
3. Limited Age Range: Primarily effective for toddlers and may have limitations for older children [15, 16].

12.3.1.2.2 Quantitative checklist for autism in toddlers (Q-CHAT-10)

12.3.1.2.2.1 SCREENING AGE RANGE

Typically administered to toddlers between 18 and 24 months.

12.3.1.2.2.2 ADMINISTRATION PROCESS

Parents or caregivers answer **ten** questions about their child's behavior and development

12.3.1.2.2.3 KEY FEATURES

1. Early Detection: Identifies potential signs of ASD in toddlers, enabling early intervention.

2. Structured Questions: Comprises ten structured questions related to social communication and behavior.
3. User-Friendly: Designed for easy use by parents and professionals.
4. Scoring Mechanism: Scores help indicate the risk level, prompting further evaluation when necessary.

12.3.1.2.2.4 LIMITATIONS

1. Age Limitation: Effective within a specific age range (18–24 months) and may not be as applicable for older children.
2. Potential False Positives: Like many screening tools, it may generate false positive results, necessitating careful clinical assessment.
3. Dependent on Caregiver Input: Relies on accurate reporting from parents or caregivers [17, 18].

12.3.1.2.3 Ages and Stages Questionnaires (ASQ)

12.3.1.2.3.1 SCREENING AGE RANGE

Tailored for children ranging from infancy to five years old.

12.3.1.2.3.2 ADMINISTRATION PROCESS

Parents or caregivers complete a series of questionnaires tailored to their child's specific age range (there are questionnaires for each age group).

12.3.1.2.3.3 KEY FEATURES

1. Comprehensive Assessment: ASQ encompasses several developmental domains, such as communication, motor abilities, problem-solving capacities, and socio-emotional growth.
2. Age-Appropriate: The questionnaires are age-specific, ensuring that the assessment is tailored to the child's developmental stage.
3. Parent Involvement: Relies on parental input, encouraging active engagement in monitoring a child's development.
4. Early Identification: It helps identify developmental delays or areas of concern early, enabling timely intervention and support.

12.3.1.2.3.4 LIMITATIONS

1. Reliance on Parental Input: The accuracy of ASQ results depends on the accuracy of parents' responses.
2. Not Diagnostic: ASQ is a screening tool and does not provide a formal diagnosis. Further assessment is needed to confirm developmental concerns.

3. Potential for False Positives: Like many screening tools, ASQ may produce false positives, necessitating additional evaluation [19].

12.3.1.2.4 Social Communication Questionnaire (SCQ)

12.3.1.2.4.1 SCREENING AGE RANGE

Designed for individuals who are four years of age and beyond.

12.3.1.2.4.2 ADMINISTRATION PROCESS

It consists of 40 yes/no questions that parents or caregivers answer based on the individual's behavior and communication skills.

12.3.1.2.4.3 KEY FEATURES

1. Assesses Social Communication: The SCQ is specifically designed to screen for social communication deficits commonly associated with ASD.
2. Structured Questions: Contains structured questions focused on social interactions, communication skills, and repetitive behaviors.
3. User-Friendly: It is designed for easy use by parents or caregivers.
4. Screening Tool: Helps identify individuals who may require further evaluation for ASD.

12.3.1.2.4.4 LIMITATIONS

1. Reliance on Caregiver Input: The accuracy of SCQ results relies on the accuracy of caregivers' responses.
2. Not Diagnostic: SCQ is a screening tool and does not provide a formal diagnosis. Further assessment is needed to confirm ASD.
3. Limited Age Range: It is primarily suitable for individuals aged 4 years and older, which may not cover earlier signs of ASD.

12.3.1.2.5 Screening tool for autism in toddlers and young children (STAT)

12.3.1.2.5.1 SCREENING AGE RANGE

Tailored for children ranging from 24 to 36 months.

12.3.1.2.5.2 ADMINISTRATION PROCESS

This entails a methodical, interactive, play-centered evaluation administrated by a skilled expert who observes the child's conduct and communication abilities.

12.3.1.2.5.3 KEY FEATURES

1. Play-Based Assessment: The STAT employs interactions centered on play to evaluate a child's social communication skills and behavior.
2. Structured: This assessment is standardized and follows a specific protocol.
3. Professional Evaluation: Administered by trained professionals, ensuring accurate observation and assessment.
4. Early Detection: Aims to identify early signs of ASD for timely intervention.

12.3.1.2.5.4 LIMITATIONS

1. Resource-Intensive: STAT requires trained professionals to conduct the assessment, making it resource-intensive.
2. Age-Limited: It is primarily designed for children aged 24 to 36 months and may not cover older age groups.
3. Not Diagnostic: STAT is a screening tool and does not provide a formal diagnosis. Further evaluation is needed for ASD confirmation.

12.3.1.2.6 The infant/toddler checklist (communication and symbolic behavior scales developmental profile)

12.3.1.2.6.1 SCREENING AGE RANGE

CSBS DP is designed for infants and toddlers aged 6 to 24 months.

12.3.1.2.6.2 ADMINISTRATION PROCESS

It involves a questionnaire completed by parents or caregivers to assess a child's social communication and symbolic behavior skills.

12.3.1.2.6.3 KEY FEATURES

1. Early Assessment: CSBS DP is focused on early assessment, targeting children in their first two years of life.
2. Parent-Reported: It relies on parents or caregivers to provide information about the child's behavior and development.
3. Comprehensive: It evaluates an extensive array of communicative and symbolic behavioral abilities, including social interactions, gestures, and play.
4. Developmental Monitoring: It helps monitor and identify early signs of communication delays or potential ASD.

12.3.1.2.6.4 LIMITATIONS

1. Limited Age Range: CSBS DP is designed for a specific age group (6 to 24 months) and may not be as effective for older children.
2. Dependent on Caregiver Input: The accuracy of results depends on the accuracy of parents' or caregivers' responses.
3. Not Diagnostic: The CSBS DP serves as a screening instrument and does not constitute a formal diagnosis. Additional assessment is required to confirm the presence of ASD [20].

12.3.1.2.7 Rating instrument for toddlers and infants with autism (RITA-T)

12.3.1.2.7.1 SCREENING AGE RANGE

RITA-T is designed for infants and toddlers aged 6 to 30 months.

12.3.1.2.7.2 ADMINISTRATION PROCESS

Professionals with specialized training assess and evaluate a child's behavior and communication abilities through the utilization of a structured evaluation instrument.

12.3.1.2.7.3 KEY FEATURES

1. Early Assessment: RITA-T emphasizes the early detection of signs of ASD in younger children.
2. Professional Evaluation: Administered by trained healthcare or education professionals, ensuring accurate assessment.
3. Structured Assessment: Utilizes a standardized assessment tool with specific criteria for evaluating social communication and behavior.
4. Early Intervention: Facilitates early intervention and support for children at risk of ASD.

12.3.1.2.7.4 LIMITATIONS

1. Resource-Intensive: Requires trained professionals for assessment, making it resource-intensive.
2. Limited Age Range: Originally intended for children ranging from 6 to 30 months in age, it may not encompass older age cohorts.
3. Not Diagnostic: RITA-T serves as a screening tool and does not constitute a definitive diagnosis. Additional assessment is essential to confirm the presence of ASD.

12.3.1.2.8 Early screening for autism and communication disorders

12.3.1.2.8.1 SCREENING AGE RANGE

This screening typically targets children from 9 to 24 months of age.

12.3.1.2.8.2 ADMINISTRATION PROCESS

The early screening for Autism and Communication Disorders entails systematic evaluations or checklists are administered by the proficient specialist to assess a child's abilities in social interaction and communication.

12.3.1.2.8.3 KEY FEATURES

1. Early Detection: It focuses on detecting indications of ASD and communicative challenges in early childhood.
2. Professional Evaluation: Administered by trained healthcare or education professionals, ensuring accurate assessment.
3. Targeted Assessment: It assesses specific areas of development related to communication, social interaction, and early signs of ASD.
4. Timely Intervention: Early screening facilitates timely intervention and support for children at risk.

12.3.1.2.8.4 LIMITATIONS

1. Resource-Intensive: Requires trained professionals for assessment, which can be resource-intensive.
2. Age-Limited: Initially intended for youngsters between 9 and 24 months of age, it might not encompass older age categories.
3. Not Diagnostic: Early screening tools provide an indication of risk but do not offer a formal diagnosis. Further evaluation is needed for confirmation.

12.3.1.2.9 Parent's observations of social interactions

12.3.1.2.9.1 SCREENING AGE RANGE

It is a general term that can apply to children of various age ranges, depending on the specific assessment or tool used.

12.3.1.2.9.2 ADMINISTRATION PROCESS

It involves parents or caregivers observing and recording a child's social interactions and behaviors in different contexts, often using structured assessment tools or checklists.

12.3.1.2.9.3 KEY FEATURES

1. Parent Involvement: Parents actively participate in observing and reporting on their child's social interactions and behaviors.

2. Contextual Assessment: Observations typically occur in natural settings, providing insight into a child's behavior in everyday situations.
3. Structured Tools: Often accompanied by structured assessment tools or checklists that guide parents in their observations.
4. Holistic View: Offers a comprehensive perspective on a child's interactions, communication, and conduct as observed by their closest associates.

12.3.1.2.9.4 LIMITATIONS

1. Subjectivity: Observations can be subjective, influenced by the observer's perspective and interpretation.
2. Variability: Results may vary depending on the observer's level of familiarity with assessment tools and the child's behavior.
3. Not Diagnostic: While valuable for gathering insights, parent observations are not diagnostic and are typically used as part of a broader assessment process.

12.3.1.3 Evaluation of screening tools

Evaluating screening tools for toddlers is essential to ensure their accuracy and effectiveness in identifying developmental or health concerns at an early age. Here are some key aspects to consider when evaluating such tools:

Reliability: Assess the consistency of results when the same individuals are tested on two or more occasions.

Sensitivity: Measures the tool's capacity to accurately detect individuals with the condition (true positives).

Specificity: Measure the tool's ability to correctly identify individuals without the condition (true negatives).

Positive Predictive Value (PPV): Calculate the likelihood that a positive outcome from the screening tool accurately signifies the presence of the condition.

Negative Predictive Value (NPV): Determines the chance that a negative outcome from the screening truly suggests the condition isn't present.

Receiver Operating Characteristic (ROC) Curve: graphically represents the sensitivity (true positive rate) versus the false positive rate (1-specificity) to evaluate the overall efficacy of the tool.

The comparative overview is depicted in Table 12.4

Table 12.4 Detailed study of autism screening tools

Screening tool	*Overview*	*Target age group*	*Items*	*Time to complete the test*	*Electronic health record integration*	*Validation metrics*	*Evaluation method*	*Language/ cultural adaptations*	*Official website*	*References*
M-CHAT-R/F	Parent-completed questionnaire to identify children at risk for autism; includes follow-up clinician-administered questions for specificity	16–30 months	Twenty(20)	5–10 minutes	Yes	Standardization sample had 16,071 children screened; 115 had positive screen results, 348 needed evaluation, 221 were evaluated, and 105 diagnosed with an ASD. Validation was done using ADI-R, ADOS-G, CARS, and DSM-IV-TR with sensitivity of 0.91 and specificity of 0.95. For low-risk 18- and 24-month-old children, 45% with high scores had ASD, while 95% had significant developmental delay.	Categorize risk for questionnaire as pass or fail determine after interview	Available in multiple languages	http://mchatscreen.com/	Dai et al. [21]
Q-CHAT-10	Questionnaire administered by parents designed to identify children at risk of ASD	18–24 months	Ten(10)	5–10 minutes	No	High sensitivity and specificity values of 0.85–0.91 and 0.89–0.95 respectively. Positive predictive value (PPV) ranges from 0.58 to 0.81. Negative predictive value (NPV) ranges from 0.88 to 1.	Categorize risk for questionnaire as pass or fail determine after interview	Available in multiple languages	https://www.autismresearchcentre.com/tests/quantitative-checklist-for-autism-in-toddlers-10-items-q-chat-10/	Allison et al. [22]

ASQ	Developmental screening tool for children, completed by parents, and intended for use by early educators and healthcare professionals	0–66 months	Forty(40)	10-15 minutes	Yes	18,572 questionnaires filled out by parents of children aged 1 to 66 months. The questionnaires covered personal social, gross motor, fine motor, problem solving, and communication areas. The reliability of the test-retest was 92%, with sensitivity at 87.4% and specificity at 95.7%.	A matrix was established for each questionnaire interval with cut-off scores at 2, 1.5, and 1 standard deviations from the mean.	Available in multiple languages	https://agesandstages.com/	Agarwal et al. [23]
SCQ	Parental questionnaire customized to detect children at risk of ASD within the general population, incorporating features from the ADI-R	More than 4 years	Forty(40)	5–10 minutes	No	Validated by ADI-R and DSM-IV on 200 subjects; for children with mental age of at least 2 years and chronological age 4+ years; available in two forms: lifetime and current. Overall test: sensitivity 0.85 (moderate), specificity 0.75 (moderate); sensitivity can be improved by lowering cut-off for younger children below 5 years and 5–7 years, specificity poor for younger children.	Categorize risk for questionnaire as pass or fail	Available in multiple languages	https://www.wpspublish.com	Rutter, Bailey, and Lord [25]; Corsello et al. [24]

(Continued)

Table 12.4 (Continued) Detailed study of autism screening tools

Screening tool	*Overview*	*Target age group*	*Items*	*Time to complete the test*	*Electronic health record integration*	*Validation metrics*	*Evaluation method*	*Language/ cultural adaptations*	*Official website*	*References*
STAT	Clinician-directed, dynamic, and evaluative assessment; requires clinician's expertise for consistent delivery; not appropriate for large-scale screening purposes	24–35 months; < 24 months (exploratory)	Twelve(12)	20–30 minutes	No	Comparison of validated with ADOS-G results in 52 children aged 24–35 months showed sensitivity of 0.83 and specificity of 0.86 for autism and developmental delay. PPV stood at 0.77 while NPV was calculated at 0.90. Sensitivity for children <24 months was 0.95, specificity 0.73, PPV 0.56, NPV 0.97. Screening properties improved for children over 14 months.	Observing early social-communicative behavior (categorized as high risk or low risk.) involves 12 activities	English	https://stat.vueinnovations.com/	Stone et al. [26]; Stone et al. 2004 [27]
The Infant/Toddler Checklist (Communication and Symbolic Behavior Scales Developmental Profile)	Parent questionnaire used to screen for language delay	6–24 months	Twenty-Four(24)	15 minutes	No	PPV for DD is 0.43 within 6–8 months; PPV for DD is 0.79 within 21-24 months.	Detects language delays and risk of ASD, assessing risk for social, speech, symbolic composites, and total score.	Available in multiple languages	https://www.brookespublishing.com	Wetherby et al. [28]
Early Screening for Autism and Communication Disorders	Research edition with 47 parent questionnaire items	12–36 months	Forty-Seven(47)	10–15 minutes	No	Sensitivity ranges from 0.85 to 0.91, specificity from 0.82 to 0.84, PPV from 0.55 to 0.81, NPV from 0.88 to 0.98.	Examination currently in progress on a specific subset consisting of 24 items.	English	https://firstwordsproject.com/screen-my-child/	Not in literature study

Parent's Observations of Social Interactions	Parent questionnaire is utilized for evaluating autism risk, with ASD screening at 18, 24, and 30 months ; Survey of Well-Being of Young Children includes various forms	16–35 months	Seven(7)	5 minutes	Yes	Sensitivity lies between 83% to 93%, with an average of 88.5%. Specificity ranges from 42% to 75%, averaging at 56.9%	3 out of 7 symptoms are showing in the range of risk	Available in multiple languages	https://www.theswyc.org	Salisbury et al. [29]
RITA-T	Observation by clinician; conducted by skilled examiner.	12–36 months	Nine(9)	20–30 minutes	No	Cut-off value over 15, with sensitivity at 1 and specificity at 0.84, PPV at 0.88, NPV at 0.94, requiring more research on larger samples	9 fun activities with scores added up, need to score at least 15 to pass	English	https://umassmed.edu/AutismRITA-T/about-the-test/	Choueiri and Wagner [30]

12.4 RESULTS AND DISCUSSION

From the assessment metrics utilized, both M-CHAT-R/F and Q-CHAT-10 frequently stood out. The M-CHAT-R/F consistently showcased impressive sensitivity, highlighting its capacity to accurately detect toddlers with developmental challenges. While its specificity was notably high, it emphasized its aptitude to rightly separate those without issues. Conversely, the Q-CHAT-10 showed commendable sensitivity, albeit with a bit reduced specificity than the M-CHAT-R/F. Both tools exhibited admirable PPVs, underscoring their capability to correctly anticipate developmental anomalies when screenings turn out positive. Their NPVs were also considerably high, reflecting their proficiency in accurately determining the lack of developmental issues when screenings are negative. Practitioner insights revealed that the M-CHAT-R/F was easier to implement, whereas the Q-CHAT-10 required a tad more time.

The findings emphasize the significance of considering multiple evaluation measures when appraising the effectiveness of toddler screening tools. While sensitivity and specificity are fundamental, the real-world consequences of a tool's PPV and NPV are equally vital. High sensitivity ensures that no child with developmental concerns is overlooked, yet high specificity is essential to avoid unnecessary interventions or undue stress for families. Additionally, the ease of use is a practical aspect influencing a tool's consistent application and reliability in various settings, spanning from clinical facilities to community health centers.

The M-CHAT-R/F exhibits a compelling combination of high sensitivity, specificity, and ease of administration, positioning it as a strong contender for widespread adoption. Nevertheless, its marginally superior specificity compared to the Q-CHAT-10 suggests that it might be particularly well-suited for scenarios where utmost accuracy is paramount. In contrast, although the Q-CHAT-10 demonstrates robust diagnostic capabilities, its somewhat lengthier administration time could make it better suited for detailed clinical assessments or follow-up screenings.

In summary, while many tools are available for screening toddlers for developmental concerns, their utility hinges on a balance of accuracy, practicality, and the specific demands of the settings in which they're used. A multi-faceted approach to evaluation, such as the one adopted in this review, is pivotal to identify the most appropriate tools for varying contexts.

12.5 CONCLUSION

The evaluation of screening tools for toddlers aimed to assess their effectiveness in identifying developmental or health challenges, emphasizing attributes such as sensitivity, specificity, PPV, and NPV. Insights from this analysis offer a deeper understanding of the advantages and challenges tied to screening tools typically employed during early childhood evaluations. It was discerned that each tool's performance differs across metrics like sensitivity, specificity, PPV, and NPV. Notably, M-CHAT-R/F and Q-CHAT-10 showcased promising outcomes in detecting potential ASD indications, boasting commendable sensitivity and PPV figures. These instruments can be pivotal in the early detection of ASD, thus opening the door for timely interventions and backing.

Moreover, the STAT and the ASQ also exhibited strengths in certain domains, suggesting their utility in assessing developmental concerns beyond ASD. However, it's essential to note that the choice of screening tool should consider various factors, including the specific population being assessed, ease of administration, and cultural appropriateness. No single tool is universally superior, and the selection should align with the unique needs and characteristics of the target population.

Further research is needed to refine the screening process and enhance the evidence base for toddler developmental screening. Continual assessment and updates of these tools, along with cultural adaptation, are essential to ensure their effectiveness in identifying and supporting toddlers with developmental or health concerns. Ultimately, the successful identification of such concerns in early childhood can pave the way for timely interventions that enhance the well-being and developmental outcomes of these children. Moreover, additional studies on the appraisal of toddler screening tools, anchored in obtaining suitable datasets, are set to transform early developmental evaluations in children. The foundation for future exploration lies in finding and harnessing datasets that are comprehensive, diverse, and representative of different demographics and cultural contexts. Through harnessing cutting-edge data analysis methods like Machine Learning, Artificial Intelligence, and bio-inspired optimization algorithms, researchers have the opportunity to unleash the potential for exceptionally precise and tailored screening tools. Additionally, the ethical handling of data and considerations for data privacy will be paramount in future studies. As we continue to collect and analyze specific datasets, the trajectory of research in this field is moving toward more precise, culturally sensitive, and globally applicable screening tools, ensuring that toddlers receive the early interventions they need for optimal development and well-being.

ACKNOWLEDGMENTS

The author is thankful to Dr. Somsubhra Gupta and Dr. Subhranil Som for their support and providing infrastructure to conduct this research. The authors acknowledge Swami Vivekananda University for providing the environment for students to conduct this research.

REFERENCES

1. Maenner, M. J., Shaw, K. A., Bakian, A. V., Bilder, D. A., Durkin, M. S., Esler, A., ... & Cogswell, M. E. 2021. Prevalence and characteristics of autism spectrum disorder among children aged 8 years—autism and developmental disabilities monitoring network, 11 sites, United States, 2018. *MMWR Surveillance Summaries*, *70*(11): 1–16.
2. *Autism Rates by Country 2021*. Worldpopulationreview.com. 2022. Retrieved 18 January 2022, from https://worldpopulationreview.com/country-rankings/autism-rates-by-country
3. Miller, L. E., Dai, Y. G., Fein, D. A., & Robins, D. L. 2021. Characteristics of toddlers with early versus later diagnosis of autism spectrum disorder. *Autism*, *25*(2): 416–428.
4. Allison, C., Auyeung, B., & Baron-Cohen, S. 2012. Toward brief "red flags" for autism screening: the short autism spectrum quotient and the short quantitative checklist in 1,000 cases and 3,000 controls. *Journal of the American Academy of Child & Adolescent Psychiatry*, *51*(2): 202–212.
5. Wingfield, B., Miller, S., Yogarajah, P., Kerr, D., Gardiner, B., Seneviratne, S., ... & Coleman, S. 2020. A predictive model for paediatric autism screening. *Health Informatics Journal*, *26*(4): 2538–2553.
6. Radhika, C., & Priya, N. 2022. Prediction of learning disability of the children using adaptive effective feature engineering techniques. *Journal of Positive School Psychology*, *6*(5): 2768–2783.
7. Omar, K. S., Mondal, P., Khan, N. S., Rizvi, M. R. K., & Islam, M. N. (2019, February). A machine learning approach to predict autism spectrum disorder. In *2019 International conference on electrical, computer and communication engineering (ECCE)* (pp. 1–6). IEEE, 1–6.
8. McCarty, P., & Frye, R. E. 2020. Early detection and diagnosis of autism spectrum disorder: why is it so difficult?. *Seminars in Pediatric Neurology*, *35*: 100831.
9. Marlow, M., Servili, C., & Tomlinson, M. 2019. A review of screening tools for the identification of autism spectrum disorders and developmental delay in infants and young children: recommendations for use in low-and middle-income countries. *Autism Research*, *12*(2): 176–199.
10. Jaisoorya, T. S., Jacob, P., Srinath, S., Kumar, S., Manoj, L., Gokul, G. R., ... & Thennarasu, K. 2018. Toddlers at risk for autism spectrum disorders from Kerala, India–A community based screening. *Asian Journal of Psychiatry*, *31*: 10–12.
11. Leroy, G., Gu, Y., Pettygrove, S., Galindo, M. K., Arora, A., & Kurzius-Spencer, M. 2018. Automated extraction of diagnostic criteria from electronic health records for autism spectrum disorders: development, evaluation, and application. *Journal of Medical Internet Research*, *20*(11): e10497.
12. Mandy, W. P., Charman, T., & Skuse, D. H. 2012. Testing the construct validity of proposed criteria for DSM-5 autism spectrum disorder. *Journal of the American Academy of Child & Adolescent Psychiatry*, *51*(1): 41–50.
13. Maenner, M. J., Rice, C. E., Arneson, C. L., Cunniff, C., Schieve, L. A., Carpenter, L. A., ... & Durkin, M. S. 2014. Potential impact of DSM-5 criteria on autism spectrum disorder prevalence estimates. *JAMA Psychiatry*, *71*(3): 292–300.

14. Wiggins, L. D., Rice, C. E., Barger, B., Soke, G. N., Lee, L. C., Moody, E., ... & Levy, S. E. 2019. DSM-5 criteria for autism spectrum disorder maximizes diagnostic sensitivity and specificity in preschool children. *Social Psychiatry and Psychiatric Epidemiology*, *54*: 693–701.
15. Lipkin, P. H., Macias, M. M., Norwood, K. W., Brei, T. J., Davidson, L. F., Davis, B. E., ... & Voigt, R. G. 2020. Promoting optimal development: Identifying infants and young children with developmental disorders through developmental surveillance and screening. *Pediatrics*, *145*(1): e20193449.
16. Robins, D. L., Casagrande, K., Barton, M., Chen, C. M. A., Dumont-Mathieu, T., & Fein, D. 2014. Validation of the modified checklist for autism in toddlers, revised with follow-up (M-CHAT-R/F). *Pediatrics*, *133*(1): 37–45.
17. Sturner, R., Howard, B., Bergmann, P., Attar, S., Stewart-Artz, L., Bet, K., ... & Baron-Cohen, S. 2022. Autism screening at 18 months of age: a comparison of the Q-CHAT-10 and M-CHAT screeners. *Molecular Autism*, *13*(1): 1–11.
18. Raza, S., Zwaigenbaum, L., Sacrey, L. A. R., Bryson, S., Brian, J., Smith, I. M., ... & Garon, N. 2019. Brief report: Evaluation of the short quantitative checklist for autism in toddlers (Q-CHAT-10) as a brief screen for autism spectrum disorder in a high-risk sibling cohort. *Journal of Autism and Developmental Disorders*, *49*: 2210–2218.
19. Valla, L., Janson, H., Wentzel-Larsen, T., & Slinning, K. 2016. Analysing four Norweigia population-based samples using the six-month version of the Ages and Stages Questionnaire showed few relevant clinical differences. *Acta Paediatrica*, *105*(8): 924–929.
20. Eadie, P. A., Ukoumunne, O., Skeat, J., Prior, M. R., Bavin, E., Bretherton, L., & Reilly, S. 2010. Assessing early communication behaviours: structure and validity of the Communication and Symbolic Behaviour Scales—Developmental Profile (CSBS-DP) in 12-month-old infants. *International Journal of Language & Communication Disorders*, *45*(5): 572–585.
21. Dai, Y. G., Porto, K. S., Skapek, M., Barton, M. L., Dumont-Mathieu, T., Fein, D. A., & Robins, D. L. 2021. Comparison of the modified checklist for autism in toddlers, revised with follow-up (M-CHAT-R/F) positive predictive value by race. *Journal of Autism and Developmental Disorders*, *51*: 855–867.
22. Allison, C., Matthews, F. E., Ruta, L., Pasco, G., Soufer, R., Brayne, C., ... & Baron-Cohen, S. 2021. Quantitative Checklist for Autism in Toddlers (Q-CHAT). A population screening study with follow-up: the case for multiple time-point screening for autism. *BMJ Paediatrics Open*, *5*(1): e000700.
23. Agarwal, P. K., Xie, H., Rema, A. S. S., Rajadurai, V. S., Lim, S. B., Meaney, M., & Daniel, L. M. 2020. Evaluation of the Ages and Stages Questionnaire (ASQ 3) as a developmental screener at 9, 18, and 24 months. *Early Human Development*, *147*: 105081.
24. Corsello, C., Hus, V., Pickles, A., Risi, S., Cook Jr, E. H., Leventhal, B. L., & Lord, C. 2007. Between a ROC and a hard place: Decision making and making decisions about using the SCQ. *Journal of Child Psychology and Psychiatry*, *48*(9): 932–940.
25. Rutter, M., Bailey, A., & Lord, C. 2003. *SCQ. The Social Communication Questionnaire*. Torrance, CA: Western Psychological Services.
26. Stone, W. L., Coonrod, E. E., & Ousley, O. Y. 2000. Brief report: screening tool for autism in two-year-olds (STAT): development and preliminary data. *Journal of Autism and Developmental Disorders*, *30*(6): 607–612.

27. Stone, W. L., Coonrod, E. E., Turner, L. M., & Pozdol, S. L. 2004. Psychometric properties of the STAT for early autism screening. *Journal of Autism and Developmental Disorders*, *34*: 691–701.
28. Wetherby, A. M., Brosnan-Maddox, S., Peace, V., & Newton, L. 2008. Validation of the Infant—Toddler Checklist as a broadband screener for autism spectrum disorders from 9 to 24 months of age. *Autism*, *12*(5): 487–511.
29. Salisbury, L. A., Nyce, J. D., Hannum, C. D., Sheldrick, R. C., & Perrin, E. C. 2018. Sensitivity and specificity of 2 autism screeners among referred children between 16 and 48 months of age. *Journal of Developmental & Behavioral Pediatrics*, *39*(3): 254–258.
30. Choueiri, R., & Wagner, S. 2015. A new interactive screening test for autism spectrum disorders in toddlers. *The Journal of Pediatrics*, *167*(2): 460–466.

Index

Pages in *italics* refer to figures and pages in **bold** refer to tables.